Essential Concepts

Essential Concepts in MRI

Physics, Instrumentation, Spectroscopy, and Imaging

Yang Xia

Oakland University
Rochester, MI, USA

WILEY Blackwell

This edition first published 2022
© 2022 John Wiley & Sons Ltd

Registered Office(s)
John Wiley & Sons, Inc., 111 River Street, Hoboken, NJ 07030, USA
John Wiley & Sons Ltd, The Atrium, Southern Gate, Chichester, West Sussex, PO19 8SQ, UK

Editorial Office
9600 Garsington Road, Oxford, OX4 2DQ, UK

For details of our global editorial offices, customer services, and more information about Wiley products visit us at www.wiley.com.

Wiley also publishes its books in a variety of electronic formats and by print-on-demand. Some content that appears in standard print versions of this book may not be available in other formats.

Library of Congress Cataloging-in-Publication Data
Names: Xia, Yang, author.
Title: Essential concepts in MRI : physics, instrumentation, spectroscopy, and imaging / Yang Xia, Oakland University, Rochester, MI, USA.
Description: First edition. | Hoboken, NJ : John Wiley & Sons, Inc., 2022. | Includes bibliographical references and index.
Identifiers: LCCN 2021025868 (print) | LCCN 2021025869 (ebook) | ISBN 9781119798217 (paperback) | ISBN 9781119798231 (pdf) | ISBN 9781119798248 (epub)
Subjects: LCSH: Magnetic resonance imaging.
Classification: LCC RC78.7.N83 X53 2022 (print) | LCC RC78.7.N83 (ebook) | DDC 616.07/548--dc23
LC record available at https://lccn.loc.gov/2021025868
LC ebook record available at https://lccn.loc.gov/2021025869

Cover image: © EDUARD MUZHEVSKYI/Getty Images
Cover design by Wiley

Set in 10/12 pt Trade Gothic LT Std by Integra Software Services Pvt. Ltd, Pondicherry, India
Printed and bound by CPI Group (UK) Ltd, Croydon, CR0 4YY

C9781119798217_060622

To my parents, 镇澳 and 钺,
for their unconditional love and unwavering support during the turbulent years
when my sister, Xing 星, and I grew up in Shanghai, China,
and to my two wonderful children, Aimee 怡元 and Derek 怡康,
for their love and friendship – it is so great to have you two in my life.

Contents

Preface

In the fall semester of 1994, I became a new assistant professor of physics at Oakland University, in the specialization of medical physics. After receiving my assignment to teach a graduate-level one-semester course in magnetic resonance imaging (MRI) for the next semester, I sat in my nearly empty office and wondered what and how to teach *my* students. As someone who had been working in MRI research for eight years at that time, I knew the importance of the fundamental theory. As someone who had been a hands-dirty experimentalist, I knew the importance of hardware and software that enabled any experiment. As someone who was specialized in quantitative MRI, I loved this field of research where the final result was an image, hopefully a beautiful and useful one. At the same time, I remembered my occasional regret during my imaging career that I did not know much about spectroscopy. I therefore determined to teach my students a little bit of nuclear magnetic resonance (NMR) spectroscopy.

I started to read the books that were available at the time, to find a potential textbook for my students. I wished I had read some of these books earlier, since there was so much that I simply didn't know! As I went over these books for a possible adaptation for my course, I could not find any *single* book that contained what I had in my mind as the four essential and inseparable components of MRI – theory, instrumentation, spectroscopy, and imaging. There were books that were excellent and extensive in each of the four essential components in MRI. I was, however, unable to find one book that introduces all four components that I had in mind. (Asking my students to buy multiple books for one course was not an option.) I eventually realized, painfully, that I would have to put together the materials myself, if I wanted to teach the course as I had planned in my mind. My starting point was two excellent books that were available at that time: P.T. Callaghan's *Principles of Nuclear Magnetic Resonance Microscopy* (Oxford University Press, 1991) and R.K. Harris's *Nuclear Magnetic Resonance Spectroscopy* (Longman Scientific & Technical, 1989). I had the pleasure to communicate with both authors on their books during my teaching. My lecture notes, evolved and revised substantially during the last 26 years, became the basis for this book.

Since my course is for one 14-week semester, I must pick and choose what I could cover within that given time; I simply do not have time to cover all important concepts in all four components in great detail. I, however, determined to cover all four components of MRI: the theory of physics that explains this fascinating phenomenon, the instrumentation and experimental techniques that facilitate the execution of this fascinating phenomenon, the early adaptation of this physics phenomenon in the practice of NMR spectroscopy, and finally MRI. The requirements and time constraints of the course reflect the compromised (or optimized) choices, which are personal, for the topic selections in this book and the words "*Essential Concepts*" in the title of this book.

This book is grouped into five parts. Part I introduces the essential concepts in magnetic resonance, including the use of the classical description and a brief introduction of the quantum mechanical description. It also includes the description for a number of nuclear interactions that are fundamental to magnetic resonance. Part II covers the essential concepts in experimental magnetic resonance, which are common for both NMR spectroscopy and MRI. Part III describes the essential concepts in NMR spectroscopy, which should also be beneficial for MRI researchers. Part IV introduces the essential concepts in MRI. The final part is concerned with the quantitative and creative nature of MRI research. At the end of the book there are several short appendices, which include some background information on several topics in the book, some sample syllabi for possible ways to teach this course, as well as some homework problems.

I owe a great debt to the late Sir Paul T. Callaghan, who was my graduate advisor at Massey University in Palmerston North, New Zealand during 1986–1992. He taught me the art and science of NMR imaging at microscopy resolution (μMRI).

In my own research journey at Oakland University since 1994, I am very grateful for the beautiful works of my graduate students (Jonathan Moody, Hisham Alhadlaq, Jihyun Lee, Farid Badar, Daniel Mittelstaedt, David Kahn, Syeda Batool, Hannah Mantebea, Amanveer Singh, Austin Tetmeyer, Aaron Blanc), the mutual education of my former postdocs in MRI (ShaoKuan Zheng, Nian Wang, Rohit Mahar, Nagaraja Cholashetthalli), and the stimulating exchange of many visiting and sabbatical scientists to my lab (Paul T. Callaghan, Siegfried Stapf, Hisham A. Alhadlaq, Ekrem Cicek, RanHong Xie, ZhiGuo Zhuang, Zhe Chen). I have also benefited in my MRI research from the collaboration and interactions with many professional colleagues in MRI (Eiichi Fukushima, Kenneth Jeffrey, Gregory Furman, Jia Hua, Yong Lu, Quan Jiang, Jiani Hu, Craig Eccles, Mark Mattingly, Dieter Gross, Thomas Oerther, Volker Lehmann). Thank you.

I am grateful for four five-year R01 grants from the National Institutes of Health (NIH NIAMS) to my research lab at Oakland University, much internal support from the Research Excellence Fund in Biotechnology and the Center for Biomedical Research at Oakland University, the Department of Physics at Oakland University, and an NMR instrument endorsement from R.B. and J.N. Bennett (Oakland University), which initiated and supported my micro-imaging adventure at Oakland University.

My special thanks go to several colleagues who contributed directly to this book: Bradley J. Roth (Oakland University) and Siegfried Stapf (Technische Universität Ilmenau), who generously offered to read and comment on a draft of this book; Dylan Twardy (Oakland University), who worked with me during a previous semester to obtain some NMR spectra that are used in the book and also read the spectroscopy chapters; Roman Dembinski (Oakland University), who read the spectroscopy chapters in this book; and Farid Badar (Oakland University), who provided several image examples used in the book. I also thank the students in my classes over the years (in particular, several students in my most recent class, who had the opportunity to use an early version of the typed notes); all of you have made this book better.

My final thanks go to my sister, Xing, my daughter, Aimee, and son, Derek – you have successfully kept the homebound me during the 2020 pandemic sane and productive. You see, I had dreamed about publishing my lecture notes as a book for some 15 years. I started on this journey several times in the past, and each time I dropped it without completion due to the onset of a few work-/family-related tasks. *Yes, these were excuses, I know!* When this pandemic started in the beginning of 2020, I had to prepare to teach this course online. After I transcribed the mostly handwritten notes onto a home computer, I kept revising it using the lockdown months when I was working from home. So, here it is.

To my readers, I would love to hear from you, for any corrections and suggestions you might have.

Yang Xia 夏阳
Distinguished Professor
Professor of Physics
Fellow of the American Physical Society (APS)
Fellow of the International Society for Magnetic Resonance in Medicine (ISMRM)
Fellow of the American Institute for Medical and Biological Engineering (AIMBE)
Fellow of the Orthopaedic Research Society (ORS)
Department of Physics
Oakland University
Rochester, Michigan, USA
xia@oakland.edu
micromri@gmail.com
The first draft 2020.8.2
The second draft 2021.1.20
The final revision 2021.3.31

1

Introduction

1.1 INTRODUCTION

This book explores the physics phenomenon that provides the foundation for and the engineering architectures that facilitate the widespread applications of nuclear magnetic resonance (NMR) spectroscopy and magnetic resonance imaging (MRI). NMR is the physics phenomenon at the basis of every MRI experiment. The first word, "*nuclear*," refers to the core player of this phenomenon – stable atomic nuclei. The protons in a common water molecule are the most useful nuclei because of their high sensitivity and simplicity. *Please make a note that when we say* **proton** *in NMR and MRI literature and in this book, we mean* **hydrogen atom**, *not the nucleon.* Since these nuclei are stable, there is never any radioactivity in *NMR*. The second word, "*magnetic*," refers to the environment that these nuclei must have – the nuclei need to be immersed in a magnetic field, which can be generated in several ways including the use of a permanent magnet. The third word, "*resonance*," refers to a concept in physics where a system has the tendency to oscillate at the maximum amplitude at a certain frequency f (Figure 1.1). This resonance system can be mechanical (e.g., the pendulum studied by Galileo Galilei in 1602, and the collapse of several suspension bridges in Europe in the 1800s by marching soldiers), acoustic (e.g., many musical instruments), and electromagnetic (e.g., an electronic receiver in your radio and television). To receive the signal from a particular channel or station among the tens or hundreds of channels and stations available, the resonant frequency of a receiver in a radio or television set is adjusted either manually by turning a knob/dial in an analog circuit of a classical (i.e., pre-digital) radio or TV, or by scanning automatically over a range of frequencies in digital receivers. When the right frequency is met, the signal can reach the maximum.

Let us clarify the terminology of the NMR phenomenon, since it has several acronyms as well as sub-fields. NMR is the original and full name of the phenomenon, which now commonly refers to its physical principles. NMR spectroscopy is the spectroscopic application of NMR, which seeks the chemical information in the process; this term is used commonly in basic science and in particular in physics and chemistry. NMR imaging is the imaging application of NMR, which mainly seeks the spatial information in the process; this term is used mainly by the non-medical imaging community. MRI is identical in content to NMR imaging, which is the term that is commonly used in the medical community (and by everyone else who is not in basic science). Microscopic MRI (μMRI) and NMR microscopy are the high-resolution versions of MRI.

Essential Concepts in MRI: Physics, Instrumentation, Spectroscopy, and Imaging, First Edition. Yang Xia.
© 2022 John Wiley & Sons Ltd. Published 2022 by John Wiley & Sons Ltd.

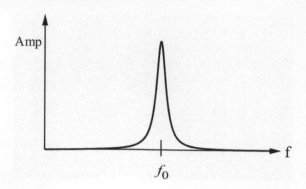

Figure 1.1   The resonance phenomenon, where the signal amplitude reaches a maximum at a particular frequency f_0.

1.2   MAJOR STEPS IN AN NMR OR MRI EXPERIMENT, AND TWO CONVENTIONS IN DIRECTION

The description of NMR and MRI theory would become easier if we first briefly overview what is involved in an NMR experiment. In general, an NMR or MRI experiment consists of three sequential "stages": preparation, excitation, and detection. In the first stage, a sample is placed in an externally applied magnetic field B_0, which allows the nuclear ensemble in the sample (e.g., water molecules in humans or animals or plants or test tubes) to reach the thermal equilibrium state. This preparation stage results in a net macroscopic magnetization in the sample. In the second stage, a perturbation is applied to the sample in order to force the net magnetization away from the thermal equilibrium into a non-equilibrium state. Finally, the response of the net magnetization to this perturbation is recorded via the detector, where the recording is termed as the NMR or MRI signal. Final post-acquisition signal processing generates an NMR spectrum or an MRI image. These three sequential stages in an NMR or MRI experiment are controlled by a list of individual commands, and each occurs at a different time. This list of commands is called a pulse sequence. Chapter 5, Chapter 6, and Chapter 13 will discuss the details of these instrumental and experimental aspects.

A convention in NMR and MRI is that the externally applied magnetic field that is used to establish the net magnetization is *always* named as the B_0 field, which is a vector field and has a direction *always* along the z axis (Figure 1.2), that is, $B_0 = B_0 k$, where k in this expression is the usual unit vector along the z direction in a 3-dimensional (3D) Cartesian coordinate system. The direction of this z axis in Cartesian coordinates, however, can be either in the vertical direction (for vertical-bore superconducting magnets, which are common in research labs, or "open" MRI scanners, which reduce claustrophobia for some patients) or in the horizontal direction (for the electromagnets in research labs, the "vertical donut" magnet MRI, or the horizontal-bore superconducting magnets in common clinical MRI scanners).

(a) Superconducting Magnet (vertical bore)   (b) Magnet in Open MRI

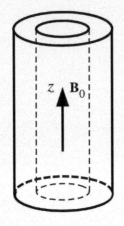

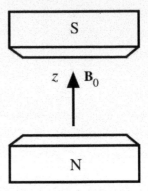

(c) Electromagnet or "Vertical Donut" Magnet   (d) Superconducting Magnet (horizontal bore)

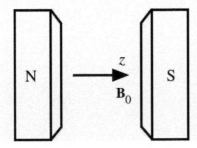

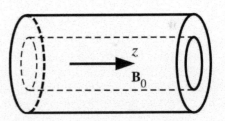

Figure 1.2   The B_0 direction in NMR and MRI. (a) Vertical-bore superconducting magnet, which is common for NMR spectrometers in science and industry laboratories. (b) "Horizontal double-donut" magnet for "open" MRI. (c) Electromagnet or magnet in "vertical double-donut" MRI. (d) Horizontal-bore superconducting magnet, which is common for whole-body imagers for humans or animals.

In addition, this book adapts the convention that the clockwise rotation is positive when one looks into the arrowhead of any axis, shown in Figure 1.3. Among the NMR and MRI literature, this convention for rotation is not consistently adapted (*i.e., some authors use the counterclockwise rotation as the positive rotation*). This inconsistency can lead to either a + or − sign in some equations that describe the motion of the macroscopic magnetization. The notation used in this book is consistent with many books; for example, those by Fukushima and Roeder [1], Callaghan [2], Canet [3], and Haacke *et al.* [4]. We will comment on this issue at several places in Chapter 2.

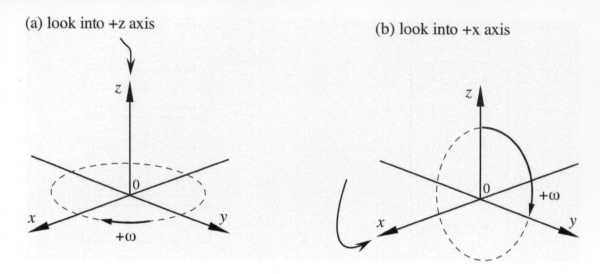

(a) look into +z axis

(b) look into +x axis

Figure 1.3 The positive directions of rotations in a 3D Cartesian coordinate system, (a) when one looks into the +z axis, and (b) when one looks into the +x axis.

1.3 MAJOR MILESTONES IN THE HISTORY OF NMR AND MRI

The physics of NMR started in 1924 when Wolfgang Pauli suggested that hydrogen nuclei might possess a magnetic moment. Pauli made this suggestion based on the observation of optical spectroscopy hyperfine splitting. The first observation of a nuclear magnetic moment was made in 1938 by Isidor I. Rabi, who used molecular-beam magnetic resonance to measure the signs of nuclear magnetic moments in individual atoms and molecules. In 1946, the phenomenon of NMR in liquids and solids was first reported simultaneously by two groups of scientists: Purcell, Torrey, and Pound at Harvard using paraffin as the specimen [5]; and Bloch, Hansen, and Packard at Stanford using water as the specimen [6]. The practical usefulness of NMR was noticed in 1950 by Proctor and Yu [7] and by Dickinson [8], who found that in ammonium nitrate and a variety of fluorine compounds, some kind of chemical effect caused the compounds to have multiple resonant lines. With the publication of the first ethanol spectrum where the three groups of protons in the same ethanol molecules resonated at three different frequencies (Figure 1.4) [9], the power of the NMR technique, being able to measure different chemical environments inside the same molecule (later termed "chemical shift"), initiated the widespread application of NMR in chemistry.

In 1950, Erwin L. Hahn developed a practical way to form a spin echo by using two radio-frequency (rf) pulses [10], which has had a long-lasting influence on NMR experiments, both spectroscopy and imaging. This was significant since once you knew how to use two (or more) pulses to manipulate the spin system, you could truly control the motion of the nuclear spins in the sample to gain insight into the molecular environment. In 1957, Irving Lowe and Richard Norberg demonstrated that the NMR spectrum in the frequency domain is mathematically equivalent to the Fourier transform (FT) of the NMR signal (called the free induction decay, FID) obtained in the time domain [11]. In 1966, Richard R. Ernst and Weston A. Anderson demonstrated the concept of FT NMR [12], which offers several orders of magnitude improvement in the signal-to-noise ratio (SNR) per unit time for a typical proton NMR spectroscopy experiment.

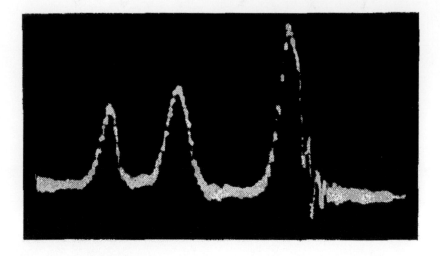

Figure 1.4 The first NMR spectrum of ethanol (CH_3CH_2OH), which demonstrated the huge potential of NMR spectroscopy by identifying three sets of non-equivalent ^1H nuclei in the same molecule. Three separate peaks corresponded to the resonant frequencies of the ^1H nuclei in the OH, CH_2, and CH_3 groups, respectively. Furthermore, the relative areas under the three peaks corresponded to the number of protons in each different chemical environment. Source: Reproduced with permission from Arnold et al. [9].

Coupled with the then-new development of personal computers and the fast Fourier transform (FFT) algorithm in the 1960s, FT NMR permits the practical use of NMR for non-experts. In 1975, Ernst demonstrated a new class of multidimensional NMR spectroscopy, now termed as 2-dimensional (2D) NMR spectroscopy (e.g., COSY, NOESY, etc.), which permits the study of specimens with a complex molecular environment or large macromolecules.

The first application of NMR to study biological samples was done in 1955 by two Swedish researchers, Erik Odeblad and Gunnar Lindström [13]. Using a primitive NMR instrument that Lindström built for his graduate research at the Nobel Institute for Physics (Stockholm, Sweden), Odeblad and Lindström studied the characteristics of NMR signals in a number of biological tissues and speculated that the signal differences between water and biological tissues could be attributed to the absorption and organization of the water molecules to the proteins in the tissue, which was remarkably accurate. A 2016 paper recounts some fascinating facts about this first biological application of NMR [14].

In 1973, Paul C. Lauterbur demonstrated the construction of 2D images using the NMR technique, which opened a completely new direction in the application of NMR (Figure 1.5) [15]. Several key developments in NMR imaging (i.e., MRI), in particular the use of a pulsed gradient approach for the slice selection by Peter Mansfield in 1974, stimulated the building of NMR scanners for humans since the late 1970s. Today, whole-body human NMR imagers, which are called whole-body MRI scanners, are the indisputable diagnostic choice for soft tissue diseases in all hospitals and clinics since MRI is completely non-invasive and totally non-destructive.

While most of the imaging community was geared toward the optimization of NMR scanners for humans, several research groups started to push the resolution of NMR imaging to the other extreme – the microscopic scale. This effort resulted in the 1986 publication of NMR images with structural features smaller than what can be recognized by the human eye (~100 microns) [16, 17]. This high-resolution imaging field has been termed as NMR microscopy (microscopic MRI, μMRI).

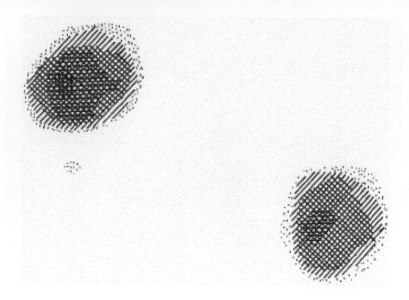

Figure 1.5   The first proton NMR image of two tubes of H$_2$O, which was produced by P.C. Lauterbur by combining four projections taken from different angles from his setup on a Varian A-60 spectrometer, which is currently on display at the State University of New York at Stony Brook. Source: Reproduced with permission from Lauterbur [15].

The latest *"history"* of this fascinating field is still being written as of today in the twenty-first century. NMR and MRI are very active and still evolving, with diverse applications in biology and medicine and various industries. There are many new and exciting developments in recent years, such as the optical pumping in NMR and MRI that improves SNR by more than 1000 times, compressed sensing that can shorten the experimental time tremendously, and exotic pulse sequences that fascinate our imagination. So far, a number of Nobel prizes have been awarded for discoveries related to NMR and MRI, including Rabi (1944) in physics, Bloch and Purcell (1952) in physics, Ernst (1991) in chemistry, Wüthrich (2002) in chemistry, and Lauterbur and Mansfield (2003) in physiology or medicine. By picking up this book, you are learning this fascinating phenomenon and joining this exciting field.

1.4   THE ORGANIZATION FOR A ONE-SEMESTER COURSE

This book is written ultimately for those who are interested in MRI. It contains the four essential components of MRI (Figure 1.6): the theory of physics that is the foundation of this fascinating phenomenon (Part I) [1, 2, 18, 19], the fundamental instrumentation and experimental techniques that facilitate the execution of this phenomenon (Part II) [1, 20], and two main applications – NMR spectroscopy (Part III) [21, 22, 23] and MRI (Part IV and Part V) [2, 4, 24]. Although each of these components could be taught in great detail in one or two semesters, the

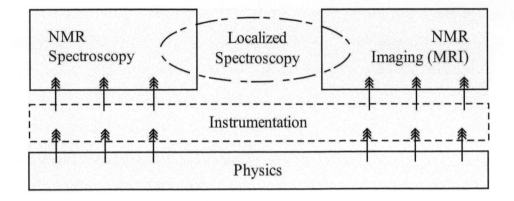

Figure 1.6   Major conceptual components of NMR and MRI.

goal of this book is to cover the essential concepts in all four components in a typical one-semester course, hence the title of the book begins with *Essential*. I trust that you would be well prepared when you need to explore any topic deeper. In addition to the numerous equations, there are about 190 figures in the book that provide the graphical descriptions for the concepts.

For the theory, I first give you the classical description of NMR, since it is easy to understand and visualize and provides a very useful first approximation. (If your goal is to do MRI on water-rich samples, the classical description is mostly sufficient.) I'll then describe NMR in a compact (i.e., abbreviated) quantum mechanical form, so that you will be at least familiar with the basic approach and terminology of the mathematical treatment.

Sandwiched between the fundamental theory and practical applications are the NMR instrumentation and experimental techniques (Part II), which facilitate the execution of this phenomenon. For these techniques, I discuss the basic unit of the NMR system. (The additional hardware in MRI is discussed in Chapter 13.) These get-your-hands-dirty discussions on hardware and experimental techniques will let you see behind the equations and behind the black box, to understand how the experiments are carried out and what are the practical issues in spectroscopy and imaging. Although the *hardware* knowledge will be described in terms of NMR and MRI, it should be useful in other modern technologies involving electronics, computer applications, signal acquisition, and imaging.

The description of NMR spectroscopy aims to supply you with basic knowledge of the topic, which is more than what you can find from any of the MRI books. I truly believe that for any MRI researchers and technical personnel, the knowledge of NMR spectroscopy is critically important. The last two parts (IV and V) cover modern practice in MRI, with an emphasis on quantitative imaging, which is at the center of modern MRI research and diagnostics.

This book can be adapted for a one-semester course in several different formats. For the students who major in science (physics, chemistry, material science, engineering), a course should include all four components of MRI (theory, instrumentation and experiment, spectroscopy, imaging). For this format, one can teach either at the undergraduate senior level or the graduate level. If the students are mainly interested in imaging, a course can be tailored toward MRI, with just a brief introduction to NMR spectroscopy. One can teach this version of the course to students in medical school. Appendix 4 has several sample syllabi for teaching.

References

1. Fukushima E, Roeder SBW. *Experimental Pulse NMR: A Nuts and Bolts Approach*. Reading, MA: Addison-Wesley; 1981.
2. Callaghan PT. *Principles of Nuclear Magnetic Resonance Microscopy*. Oxford: Oxford University Press; 1991.
3. Canet D. *Nuclear Magnetic Resonance – Concepts and Methods*. Chichester: John Wiley & Sons; 1996.
4. Haacke EM, Brown RW, Thompson MR, Venkatesan R. *Magnetic Resonance Imaging: Physical Principles and Sequence Design*. New York: Wiley-Liss; 1999.
5. Purcell EM, Torrey HC, Pound RV. Resonance Absorption by Nuclear Magnetic Moments in a Solid. *Phys Rev*. 1946;69:37–8.
6. Bloch F, Hansen WW, Packard ME. Nuclear Induction. *Phys Rev*. 1946;69:127.
7. Proctor WG, Yu FC. The Dependence of a Nuclear Magnetic Resonance Frequency upon Chemical Compound. *Phys Rev*. 1950;77:717.
8. Dickinson WC. Dependence of the F19 Nuclear Resonance Position on Chemical Compound. *Phys Rev*. 1950;77:736.
9. Arnold JT, Dharmatti SS, Packard ME. Chemical Effects on Nuclear Induction Signals from Organic Compounds. *J Chem Phys*. 1951;19:507.
10. Hahn EL. Spin Echoes. *Phys Rev*. 1950;80:580–94.
11. Lowe IJ, Norberg RE. Free-Induction Decays in Solids. *Phys Rev*. 1957;107:46.
12. Ernst RR, Anderson WA. Application of Fourier Transform Spectroscopy to Magnetic Resonance. *Rev Sci Instrum*. 1966;37(1):93–102.
13. Odeblad E, Lindström G. Some Preliminary Observations on the Proton Magnetic Resonance in Biologic Samples. *Acta Radiol*. 1955;43(6):469–76.
14. Xia Y, Stilbs P. The First Study of Cartilage by Magnetic Resonance: A Historical Account. *Cartilage*. 2016;7(4):293–7.
15. Lauterbur PC. Imaging Formation by Induced Local Interactions: Examples Employing Nuclear Magnetic Resonance. *Nature*. 1973;242:190–1.
16. Johnson GA, Thompson MB, Gewalt SL, Hayes CE. Nuclear Magnetic Resonance Imaging at Microscopic Resolution. *J Magn Reson*. 1986;68:129–37.
17. Eccles CD, Callaghan PT. High-Resolution Imaging: The NMR Microscope. *J Magn Reson*. 1986;68:393–8.
18. Slichter CP. *Principles of Magnetic Resonance*. 3rd ed. Berlin: Springer-Verlag; 1992.
19. Hennel JW, Klinowski J. *Fundamentals of Nuclear Magnetic Resonance*. Essex: Longman Scientific & Technical; 1993.
20. Chen C-N, Hoult D. *Biomedical Magnetic Resonance Technology*. Bristol: Adam Hilger; 1989. 2436–8.
21. Harris RK. *Nuclear Magnetic Resonance Spectroscopy – A Physicochemical View*. Essex: Longman Scientific & Technical; 1983.
22. Bovey FA. *Nuclear Magnetic Resonance Spectroscopy*. 2nd ed. San Diego, CA: Academic Press; 1988.
23. Friebolin H. *Basic One- and Two-Dimensional NMR Spectroscopy*. 2nd ed. New York: VCH; 1993.
24. Morris PG. *NMR Imaging in Biology and Medicine*. Oxford: Oxford University Press; 1986.

Part I

Essential Concepts in NMR

2

Classical Description of Magnetic Resonance

2.1 FUNDAMENTAL ASSUMPTIONS

The states of atomic nuclei are quantum mechanical by nature. This means that the properties that we observe for a single nucleus belong to one in a discrete set of possibilities (i.e., quantum states). A deep understanding of NMR phenomenon therefore requires the assistance of quantum mechanics. In practical NMR and MRI experiments, however, we deal with an extremely large number of nuclei in any specimen (either a human or a tissue block or a drop of liquid). For example, if I give you a glass container that has 18.015 grams of liquid water, do you know how many water molecules are in the container? Well, we know precisely how many are in it: 6.022×10^{23} water molecules (Avogadro's number, since one mole of water is 18.015 grams)! Consider taking just one gram of water from the container, which has a volume of one *milliliter*. If this one milliliter of water is further divided into 100 droplets, each tiny droplet still contains about 3.343×10^{20} water molecules, or about 6.686×10^{20} protons (i.e., *hydrogen atoms*) since each water molecule has two hydrogen atoms. It is an enormous number.

It is fortunate that these protons in a small water droplet act *largely* independently, so that at the macroscopic level, the collection of these protons appears continuous. (If these protons were to act completely independently, NMR would not have much practical use at all. If, on the other hand, these protons were to couple or interact tightly with each other, NMR also would not have much practical usage since we simply do not know how to solve the complex interactions among the enormous number of particles in any practical system.)

The simplest and most common nuclei used in NMR and MRI are hydrogen, or protons, a component of the water molecules in the liquid state. Since a proton is a spin-1/2 particle (another quantum mechanical concept) and often very mobile, we could ignore internuclear dipole interactions and scalar coupling between the protons. Hence all states of the nuclear ensemble may be characterized by a vector quantity that is referred to as the nuclear magnetization (***M***). The adoption of this vector quantity permits a classical description of magnetic resonance phenomena.

A quantum mechanical description is needed when nuclei experience mutual interactions or have a spin > 1/2 (even if they are independent, due to the presence of the quadrupole interaction). A quantum description is especially important in high-resolution NMR spectroscopy and some advance MRI techniques where the understanding of nuclear interactions is essential.

Essential Concepts in MRI: Physics, Instrumentation, Spectroscopy, and Imaging, First Edition. Yang Xia.
© 2022 John Wiley & Sons Ltd. Published 2022 by John Wiley & Sons Ltd.

For the rest of this chapter, the physics of NMR will be described using classical mechanics. Since the classical mechanical description of NMR needs to use a few concepts in quantum mechanics, this type of classical mechanical approach can also be termed as a semi-classical description of NMR.

2.2 NUCLEAR MAGNETIC MOMENT

The theory of electricity and magnetism shows that any motion of a charged body has an associated magnetic field. For example, an electric current is due to the motion of electrons along a conductor on a macroscopic scale. If you bend this conducting wire into a loop, you have just made a coil (Figure 2.1). The coil with an electric current traveling in the wire has an associated magnetic moment, which is the product of the electric current and the area of the coil. When the current-carrying coil is placed in an external magnetic field, the coil will experience a mechanical torque.

This phenomenon can also be extended to the atomic scale: when electrons or nuclei possess angular momentum, there is an associated magnetic moment. Since on the atomic scale, angular momentum is quantized, that is, it can only take certain discrete values (one of the fundamental postulates in modern physics), the magnetic moment is also quantized. (Note that here the angular momentum is a vector quantity since we are using classical mechanics to describe the concept. Later in a quantum mechanical description [Chapter 3], the angular momentum keeps the same symbol *I* but becomes an operator.)

The angular momentum, labeled as *I*, is called the spin angular momentum or simply spin, which should be considered as a fundamental property of the nucleus. One could imagine the nucleus as a finite-sized ball spinning on its axis. (Such a spinning ball picture, however, remains valid only in classical mechanics and should be not taken too literally.) The spin *I* has the following properties:

- *I* may have any half-integer or integer value such as 0, 1/2, 1, 3/2, etc. This value is known as the spin quantum number (*I*).
- *I* will have a fixed value for a given nucleus (due to the even/odd mass and charge number of the nucleus). For example, $I = 0$ for ^{12}C and ^{16}O; $I = 1/2$ for ^{1}H (proton), ^{13}C, ^{19}F, ^{31}P; $I = 1$ for ^{2}H (deuteron) and ^{14}N; and $I = 3/2$ for ^{23}Na.

If $I = 0$, then the nucleus has no spin and cannot be observed by NMR (e.g., there is no NMR for ^{12}C, even though each ^{12}C nucleus contains six protons). If $I > 0$, the nucleus will have an associated magnetic dipole moment $\boldsymbol{\mu}$, given by

$$\boldsymbol{\mu} = \gamma \hbar \boldsymbol{I}, \tag{2.1}$$

where γ is called the gyromagnetic ratio, a characteristic constant for each nuclear species ($\gamma = 2.675 \times 10^8$ rad s^{-1} T^{-1} or 42.576 MHz T^{-1} for protons), and $\hbar$ is the Planck's constant ($6.62607015 \times 10^{-34}$ J s) divided by 2π. (To convert γ between rad s^{-1} T^{-1} to MHz T^{-1}, consider rad/s as the angular velocity equal to 2π times the linear frequency.) The unit of $\boldsymbol{\mu}$ is Joule per Tesla (J T^{-1}). Since γ, $\hbar$, and *I* are all known constants, the magnitude of $\boldsymbol{\mu}$ can be determined accurately.

Note that Eq. (2.1) could be also written as $\gamma = \boldsymbol{\mu}/(\hbar \boldsymbol{I})$, which illustrates the fact that γ is the magnetic dipole moment divided by the angular momentum; hence, γ should be more properly named the *magnetogyric ratio*, not the *gyromagnetic ratio* that implies the inverse of the two

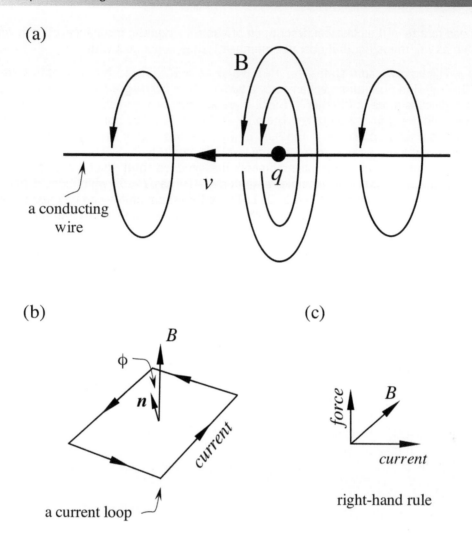

Figure 2.1 (a) Moving charges at velocity v along a conducting wire form an electric current, which has an associated magnetic field **B** by the right-hand rule; moving charges carry momentum. (b) An electric current loop has an area and a current. The magnetic moment of the loop μ equals to the product of the area and current (which is valid for any shaped loop). n is the normal vector of the current loop; ϕ is the angle between n and **B**. The torque that causes the rotation of the loop in the magnetic field B is $\tau = \mu \times B = (\text{area} \times \text{current}) \, B \sin\phi$. (c) The right-hand rule for the direction of the force on a current-carrying wire in the magnetic field.

quantities. The term *magnetogyric ratio* has indeed been used in some books and papers and also recommended by the 2001 International Union of Pure and Applied Chemistry (IUPAC) nomenclature [1]. In modern literature, however, γ is commonly known as the *gyromagnetic ratio*. Both *magnetogyric ratio* and *gyromagnetic ratio* refer to the same value.

The total number of the allowed spin states for a given nucleus is a discrete value, $2I + 1$, which ranges from $-I, -I + 1, -I + 2, \ldots$, to I. (It is another quantum mechanical concept that we need to cite in the classical description of nuclear magnetic resonance.) These values can be written as m_I, the azimuthal quantum number. Hence, a nucleus with

- $I = 1/2$ has two spin states, $m_I = -1/2$ and $m_I = 1/2$. These two spin states are commonly illustrated in quantum mechanics by a two-level energy diagram as in Figure 2.2a, which we will discuss more in Chapter 3. Since there are only two spin states, we may use an arrow to describe the spin-1/2 particles. The $m_I = -1/2$ state is called *spin down* (↓) while the $m_I = 1/2$ state is called *spin up* (↑). The spin-down state has higher energy than the spin-up state, hence is the upper level in the energy level diagram (Figure 2.2a).
- $I > 1/2$ has nuclear quadrupole moments that produce splitting of the resonant lines or line-broadening effects. For example, a deuteron (^{2}H) has $I = 1$, which would have three spin states, corresponding to $m_I = -1, 0, 1$. These three spin states can be used to label a three-level energy diagram as in Figure 2.2b. A quantum mechanical description must be used to understand the behavior of any spin with an angular momentum larger than 1/2.

Note that I is defined in this book as a dimensionless angular momentum; hence, a reduced Planck's constant $\hbar$ has been explicitly included in Eq. (2.1). This use of a dimensionless angular momentum can be found in many books, including those by Callaghan [2] and Hennel and Klinowski [3]. In contrast, I can also be defined as an angular momentum with $\hbar$ in its definition [4]. With this definition, Eq. (2.1) would be written as $\boldsymbol{\mu} = \gamma \boldsymbol{I}$.

Table 2.1 lists some fundamental properties of several common nuclei (more lengthy tables can be found in many books and papers [1, 5]). On paper, the bigger γ yields better sensitivity (the column of relative sensitivity, which is defined for the equal numbers of nuclei at constant field.). In practice, one must consider the normal or natural concentration of a nucleus in the specimen (i.e., its availability). For example, the relative sensitivities of ^{1}H and ^{19}F are similar (1 vs. 0.834); however, a human body is about 60% water (~75% in a newborn, 60–65% in men, and 55–60% in women), while the amount of ^{19}F in human or other biological systems is many orders of magnitude smaller by comparison. This explains the challenges and difficulty in detecting the signals other than protons in biological experiments by NMR and MRI.

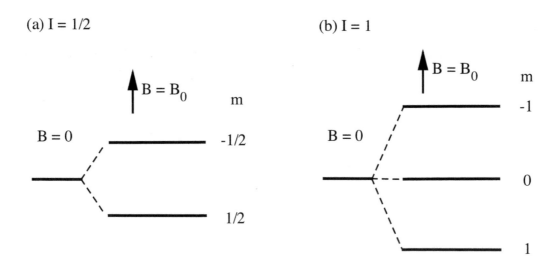

(a) I = 1/2   (b) I = 1

Figure 2.2 The application of an external magnetic field $\boldsymbol{B}_0$ causes (a) a spin-1/2 system to have two discrete energy levels, and (b) a spin-1 system to have three energy levels. Each energy level can be labeled by the individual spin state.

Table 2.1 Properties of common nuclei.

Isotope	Abundance (%)	Spin	γ (10^8 rad s^{-1} T^{-1})	Relative sensitivity	Resonance frequency f_0 at 1T (MHz)
^{1}H	99.9844	1/2	2.6752	1.00	42.577
^{2}H	0.0156	1	0.4107	0.00964	6.536
^{13}C	1.108	1/2	0.6726	0.0159	10.705
^{19}F	100	1/2	2.5167	0.834	40.055
^{31}P	100	1/2	1.0829	0.0664	17.235

2.3 THE TIME EVOLUTION OF NUCLEAR MAGNETIC MOMENT

Now look at the situation when one places a single nucleus in an externally applied magnetic field $\boldsymbol{B}_0$. Because of the dipole moment, the nucleus (which is represented by a magnetic moment $\boldsymbol{\mu}$) will interact with the external field $\boldsymbol{B}_0$, from which the energy of the nucleus in the field will be given by a dot product

$$E = -\boldsymbol{\mu} \cdot \boldsymbol{B}_0. \tag{2.2}$$

Since the nucleus also has an angular momentum, it experiences a torque, which can be expressed as a cross product, $\boldsymbol{\mu} \times \boldsymbol{B}_0$. Since this torque is equal to the time derivative of the angular momentum, by quoting Newton's second law, we have

$$\frac{d(\hbar \boldsymbol{I})}{dt} = \boldsymbol{\mu} \times \boldsymbol{B}_0. \tag{2.3}$$

By multiplying the above equation on both sides with γ, we have

$$\frac{d\boldsymbol{\mu}}{dt} = \gamma \boldsymbol{\mu} \times \boldsymbol{B}_0, \tag{2.4}$$

which is shown in Figure 2.3a. In classical mechanics, this equation describes a precessional motion [6] (Figure 2.3b) of $\boldsymbol{\mu}$ around $\boldsymbol{B}_0$ at an angular frequency ω_0, known as the Larmor frequency,[1]

$$\omega_0 = \gamma \boldsymbol{B}_0, \tag{2.5}$$

where ω is expressed as an angular frequency in rad/s, which can be converted to the temporal/linear frequency f in Hz by noting $\omega = 2\pi f$. Note that when γ is positive (which it is for proton [the nucleus of ^{1}H] and many other nuclei), the rotation will be clockwise, as shown in Figure 2.3b. γ can also be negative (e.g., ^{3}He, ^{15}N), where the rotation becomes counterclockwise.

Although the equation for this Larmor precession seems simple, it is the fundamental equation of the NMR phenomenon. The equation states that (a) the frequency of the nuclear precession is proportional to the externally applied magnetic field $\boldsymbol{B}_0$, and (b) the proportionality constant is γ. Equation (2.4) and Eq. (2.5) are useful and accurate for describing the nuclear precession in the presence of an external field $\boldsymbol{B}_0$.

---

[1] Larmor precession is named after Irish physicist Joseph Larmor, who in 1897 first described the circular motion of the magnetic moment of an orbiting charged object about an external magnetic field.

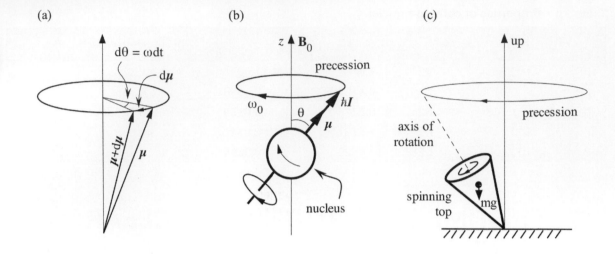

Figure 2.3 (a) Precessional motion in classical mechanics. (b) The Larmor precession of a single nucleus with the aid of classical mechanics. (c) A spinning top can have a precessional motion.

On a macroscopic scale, one can find several analogs for the precessional motion of the nuclear spin. For example, a spinning top (Figure 2.3c) can have a precessional motion when the axis of rotation does not pass through the top's center of gravity (i.e., the top is not standing up perfectly vertical), which yields a torque and an angular momentum for the top that induces it into a precessional motion. In a spinning top, the gravitational force mg that points vertically downward plays the same role of the magnetic field B_0 in NMR. Since a top is a macroscopic object, the tipping angle of a spinning top can vary from 0° to a large angle continuously, while the tipping angle of a nuclear precessional motion is fixed and cannot be varied. A second common example is the precessional motion of our planet Earth in the solar system, which is tipped at a constant angle of 23.4° from an "axis" in space, with a period of the precessional motion of 26,000 years. The torque on Earth is exerted by the sun and the moon.

2.4 MACROSCOPIC MAGNETIZATION

Any practical sample, no matter how small, contains an enormous number of nuclei (remember Avogadro's constant). The macroscopic (or bulk) magnetization M is a spatial density of magnetic moments and can be written as

$$M = \sum_{i=1}^{N} \mu_i,$$ (2.6)

where N is the number of spins in the volume. Since M is a vector, we can therefore, in general, write M in the component form as

$$M = M_x i + M_y j + M_z k,$$ (2.7)

where i, j, and k are the unit vectors along x, y, and z axes, respectively, in the usual 3D Cartesian coordinate system. Equation (2.7) can also be grouped into the transverse and longitudinal forms, as

$$M = M_\perp + M_\parallel,$$ (2.8)

where $M_\perp = M_x i + M_y j$, and $M_\parallel = M_z k$.

In the absence of any external magnetic field, the ensemble average of the nuclear spins in any practical sample should cause the value of magnetization to be averaged to zero, due to the random directions of the magnetic dipoles of the nuclei (since most nuclei act independently).

In the presence of an external field B_0, however, the magnetization M of a sample with non-zero spin will be non-zero (after the specimen has been immersed in B_0 for a sufficient time for the spin system to reach a thermal equilibrium with the environment). This non-zero spin is formed according to Boltzmann's distribution, which describes the probability distribution function of a particle in an energy state E, as $\exp(-E/(k_B T))$, where k_B is the Boltzmann constant (1.380649×10^{-23} m^2 kg s^{-2} K^{-1}) and T is the absolute temperature of the system. Since the spin-up nucleus is at a lower energy state than a spin-down nucleus, the number of nuclei along the direction of the external magnetic field will be more than the nuclei against the direction of the field (Figure 2.4b). Therefore, a non-zero net magnetization, which can be written in magnitude as M_0, occurs due to the population difference between the spin states at the two spin levels.

For spin-1/2 particles such as protons, the net magnetization can be visualized as an ordinary vector, aligned in the direction of B_0, as shown in Figure 2.4c. Since any practical sample has an enormous number of nuclear spins, it is easy to see that at thermal equilibrium the net magnetization has no transverse component (i.e., $M_x = M_y = 0$); the only component of the net magnetization is along the direction of B_0 (i.e., $M_z = M_0$).

Note that in the classical description, the individual magnetic moment μ undergoes precessional motion in an external magnetic field B_0 (as in Figure 2.4a), which is commonly represented graphically by vectors on the surface of a cone (as in Figure 2.4b). The ensemble average of μ in any practical sample is M, which is represented by a stationary vector that aligns with the direction of B_0 (as in Figure 2.4c). At equilibrium, M itself does not precess graphically as μ around B_0; M should be represented by a vector in parallel with B_0, never on the surface of any cone.

(a) A single nucleus   (b) Nuclear ensemble   (c) Macroscopic magnetization

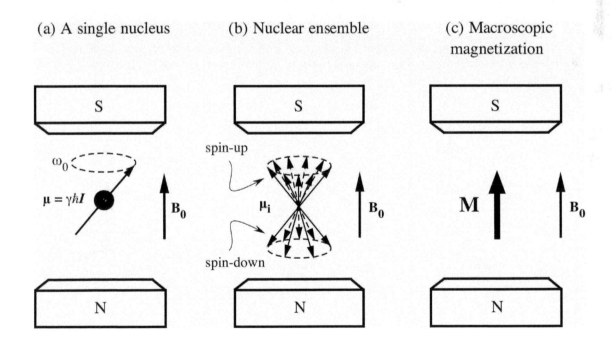

Figure 2.4 (a) A single nucleus in an external magnetic field B_0. (b) A nuclear ensemble that is the collection of a large number of nuclei. (c) The vector average of the nuclear ensemble is represented by a macroscopic magnetization M.

2.5   ROTATING REFERENCE FRAME

The usefulness of M is not its equilibrium state but its time evolution after M is tipped away by an external perturbation (another radio-frequency field) from its thermal equilibrium along the z axis. The evolution of M produces the NMR signal, which reveals the environment of the molecules. In a classical description, the time evolution of the macroscopic magnetization in the presence of a magnetic field can follow the same approach that we used before in deriving Eq. (2.4). By equating the torque to the rate of change of the angular momentum of the macroscopic magnetization M, we have

$$\frac{dM}{dt} = \gamma M \times B. \tag{2.9}$$

Equation (2.9) is a vector equation, which states that the rate of the change of the magnetization has a direction that is at right angles to both the magnetization vector and the magnetic field vector (by the right-hand rule in the cross product of vector analysis, Appendix A1.1). When B is parallel with the z axis as $B_0 k$, the above equation has the same solution as given before in Eq. (2.5), which corresponds to a precessional motion about k at the rate ω_0.

A time evolution of M can be introduced by the application of a small linearly polarized radio-frequency (rf) field $B_1(t)$ that is oscillating in the transverse plane. This B_1 field is actually a superposition of two counter-rotating fields in the transverse plane (Figure 2.5):

$$\begin{aligned} B_1(t) &= i2B_1\cos(\omega't) \\ &= (iB_1\cos(\omega't) - jB_1\sin(\omega't)) + (iB_1\cos(\omega't) + jB_1\sin(\omega't)). \end{aligned} \tag{2.10}$$

One of the two $B_1(t)$ components (which rotates in the same direction as ω_0) will have strong influence on the nuclei despite its small magnitude compared with B_0, while the other component, which rotates in the opposite direction, will have negligible effect on the nuclei (provided $|B_1| \ll B_0$).

It will be very useful to introduce a rotating frame of reference, as shown in Figure 2.6, where the xyz frame is the laboratory frame of reference (stationary) and the $x'y'z'$ frame is the rotating

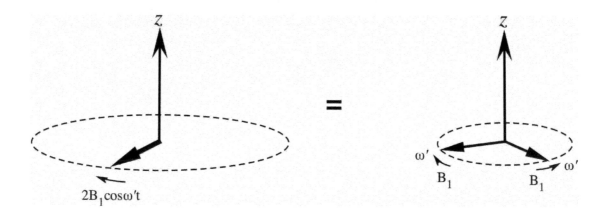

Figure 2.5   Two counter-rotating fields (right) can form a B_1 field precessing in the transverse plane of the laboratory reference frame (left).

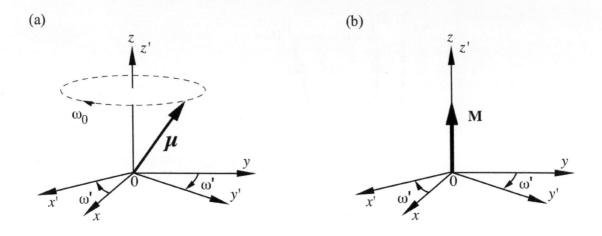

Figure 2.6 (a) A magnetic moment $\boldsymbol{\mu}$ precessing in the stationary (xyz) and rotating ($x'y'z'$) frames. (b) A macroscopic magnetization vector $\boldsymbol{M}$ in the two frames.

frame of reference at an angular frequency ω' in the laboratory frame, in which z and z' are parallel. In addition, one of the $\boldsymbol{B}_1(t)$ components appears stationary in the rotating frame; let's set that stationary component of $\boldsymbol{B}_1(t)$ along the x' axis.

When a magnetic moment is precessing at a rate of ω_0 in the xyz frame, a stationary observer in the $x'y'z'$ frame should see the magnetic moment precessing at a reduced rate $\omega_0 - \omega'$. Since $\omega = \gamma B$, a reduced precession rate implies that the magnetic moment is experiencing a reduced $\boldsymbol{B}_0$, as

$$\omega_0 - \omega' = \gamma(B_0 - \omega'/\gamma). \tag{2.11}$$

In the presence of both $\boldsymbol{B}_0$ and $\boldsymbol{B}_1(t)$ fields, the total vector field $\boldsymbol{B}_{\text{total}}$ is the sum of all fields and the effective field $\boldsymbol{B}_{\text{eff}}$ varies with the frequency of $\boldsymbol{B}_1(t)$, as

$$\boldsymbol{B}_{\text{total}} = \boldsymbol{B}_1(t) + B_0\boldsymbol{k} \tag{2.12}$$

and

$$\boldsymbol{B}_{\text{eff}} = \boldsymbol{B}_1(t) + (B_0 - \omega'/\gamma)\boldsymbol{k}, \tag{2.13}$$

where Eq. (2.12) is for the stationary xyz coordinates and Eq. (2.13) is for the rotating $x'y'z'$ coordinates. The directions of these vector fields are shown schematically in Figure 2.7.

In a typical NMR system, $|\boldsymbol{B}_1|$ is several orders of magnitude smaller than B_0, which means that $\boldsymbol{B}_{\text{total}}$ is almost along the z axis. However, the magnitude and the direction of $\boldsymbol{B}_{\text{eff}}$ will drastically depend upon the frequency of the rotating frame ω'. When ω' approaches the value of ω_0, the effective magnetic field $\boldsymbol{B}_{\text{eff}}$ tips more towards the transverse plane. When $\omega' = \omega_0$ (i.e., the frequency of $\boldsymbol{B}_1(t)$, ω', equals the Larmor frequency ω_0), the second term in Eq. (2.13) becomes zero. Hence $\boldsymbol{B}_1(t)$ becomes the only field in $\boldsymbol{B}_{\text{eff}}$ to interact with the nuclei. $\boldsymbol{M}$ will therefore respond to the effect of $\boldsymbol{B}_1(t)$, which is set along the x' axis (Figure 2.7). This condition, $\omega' = \omega_0$, is termed as the resonance condition.

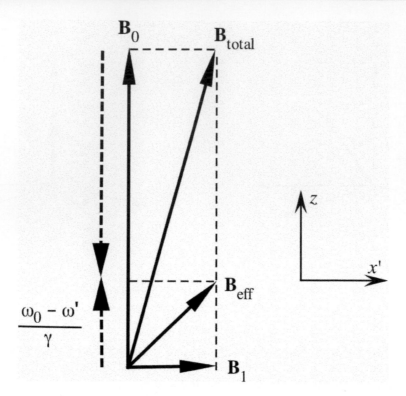

Figure 2.7 In the rotating frame that has a frequency ω', the external magnetic field B_0 appears to be reduced in magnitude, where the reduction is a function of ω'. This reduction results in the tipping of B_{eff} towards the x' axis, which is in parallel with B_1.

Under this resonance condition, the magnetization will be tipped away from its equilibrium orientation towards the transverse plane under the influence of $\boldsymbol{B}_1(t)$, as $\omega_1 = \gamma B_1$ (Figure 2.8). Since the effective component of $\boldsymbol{B}_1(t)$ has been set along the $+x'$ axis, the magnetization will rotate about B_1 in the $y'z'$ plane of the rotating frame. Following the notation for rotation as shown in Figure 1.3b, the direction of the magnetization towards the transverse plane under the influence of a B_1 field set along the $+x'$ axis follows the circular trajectory in Figure 2.8b, towards the $+y'$ axis.

Note that this rotation of the magnetization towards the $+y'$ axis under a B_1 field set at the $+x'$ axis is determined by the convention of positive rotation that is set earlier in Figure 1.3. As we mentioned in Chapter 1.2, different textbooks and academic papers contain inconsistencies in the notation for rotation of $\boldsymbol{M}$, to either the $-y'$ axis or $+y'$ axis upon a B_1 field set along the $+x'$ axis, depending upon which direction is labeled as the positive rotation. Although this discrepancy seems problematic for the graphical illustration of vectoral motion of magnetization at the first appearance, it does not matter as long as one chooses one notation and keeps it for the entire analysis of spin evolution.

Since B_1 is several orders of magnitude smaller than B_0, ω_1 is typically tens to hundreds of kilohertz (instead of a typical ω_0 at tens to hundreds of megahertz). The time to tip the magnetization from the thermal equilibrium towards the transverse plane is typically several to tens of microseconds (μs), depending upon the magnitude of the $\boldsymbol{B}_1$ field. A particular B_1 that is able to

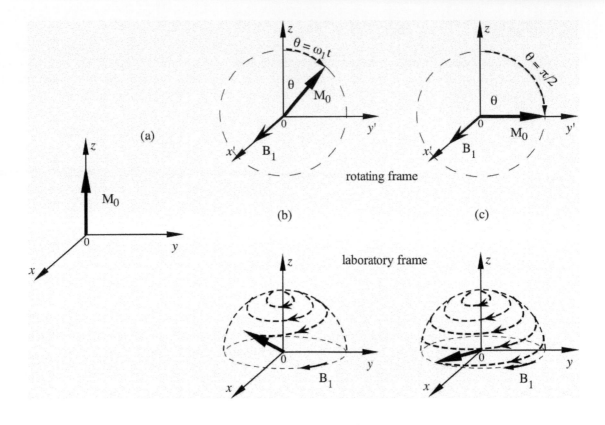

Figure 2.8 Motion of the magnetization in the rotating frame and the laboratory frame, under the influence of a B_1 field set along the x' axis. (a) M is at the thermal equilibrium, along the z axis. (b) M tips towards the transverse plane when B_1 is turned ON. (c) M reaches the transverse plane (which is to say that M has been rotated by 90°).

tip the magnetization from its z axis to the transverse plane (to the $+y'$ axis, as shown in Figure 2.8c in the rotating frame $x'y'z'$) is termed in practice as a 90° B_1 field (or more commonly a 90° B_1 pulse, or a $\pi/2$ pulse). The term *pulse* in the last sentence refers to the short duration of the B_1 field; see Section 2.10 for the definition of rf pulses and Chapter 7 for more description on spin manipulation under various pulse sequences. The motions of the magnetization in the laboratory frame (xyz) are also shown in Figure 2.8, which graphically is a spiraling vector away from its equilibrium position over the envelope of a dome. When the magnetization M reaches the transverse plane, the longitudinal component (M_z) of M becomes zero.

In general, the equation of motion for M in the presence of $B_1(t)$ is given by

$$\frac{d\boldsymbol{M}}{dt} = \gamma \boldsymbol{M} \times \boldsymbol{B}_{\text{eff}} \tag{2.14a}$$

or

$$\frac{d\boldsymbol{M}}{dt} = \gamma \boldsymbol{M} \times ((B_0 - \omega'/\gamma)\boldsymbol{k} + B_1 \boldsymbol{i}) \tag{2.14b}$$

2.6   SPIN RELAXATION PROCESSES

After M has been tipped to the transverse plane, if we switch off the $B_1(t)$ field and sit there watching, what happens to the spin system? As soon as B_1 is switched off, two processes will happen to M, which would eventually lead to the return of M to thermal equilibrium (i.e., $M_z = M_0$, and the zero transverse components of M).

The processes that return the magnetization M to the thermal equilibrium are termed as relaxation, which may be described by two time-constants in the following equations:

$$\frac{dM_z}{dt} = -\frac{M_z - M_0}{T_1}$$
(2.15a)

and

$$\frac{dM_\perp}{dt} = -\frac{M_\perp}{T_2},$$
(2.15b)

where $M_\perp$ refers to the transverse component, defined by Eq. (2.8).

T_1 is known as the spin-lattice (or longitudinal) relaxation time because the relaxation process involves an energy exchange between the spin system and its surrounding thermal reservoir, known as the "lattice." The term "longitudinal" comes from the fact that this relaxation process restores the disturbed magnetization to its thermal equilibrium, being along the longitudinal direction k. T_1 in simple liquids is usually in the range of several seconds.

When the B_1 field rotates the magnetization entirely to the transverse plane (i.e., $M_z = 0$), the magnetization is said to be rotated by 90° (i.e., $\pi/2$). The solution to Eq. (2.15a) becomes

$$M_z(t) = M_0(1 - \exp(-t / T_1)).$$
(2.16a)

If the B_1 field is sufficiently powerful or its duration is sufficiently long, the magnetization can be inverted (i.e., $M_z = -M_0$). Such a B_1 field is said to rotate the magnetization by 180° (a π pulse). Under this condition, the solution to Eq. (2.15a) becomes

$$M_z(t) = M_0(1 - 2\exp(-t / T_1)).$$
(2.16b)

Figure 2.9 shows schematically the motion of the longitudinal magnetization with these two different initial conditions, where $t = 0$ marks the moment that the B_1 field is turned off.

The second relaxation process is characterized by a time constant T_2, which is called the spin-spin (or transverse) relaxation time since it describes the decay in phase coherence between the individual spins in the transverse plane. This decay in phase coherence describes the process in which the spins come to a thermal equilibrium among themselves in the transverse plane, which results in signal loss since NMR and MRI measure the net transverse magnetization. T_2 in biological tissues is usually in the range of tens or hundreds of milliseconds.

The solution to Eq. (2.15b) becomes

$$M_\perp(t) = M_\perp(0)\exp(-t / T_2).$$
(2.17)

The motion of magnetization in spin-spin relaxation is shown schematically in Figure 2.10, where again $t = 0$ marks the moment that the B_1 field is turned off.

(a) The motion of M after a 90° B_1 field

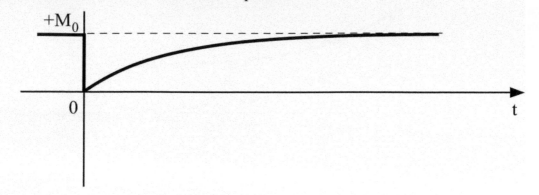

(b) The motion of M after a 180° B_1 field

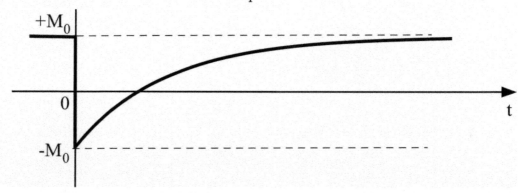

Figure 2.9 The motion of the longitudinal magnetization after it has been tipped by 90° (a) and 180° (b). The B_1 fields that are capable of tipping **M** by 90° and 180° are called a 90° B_1 field/pulse and a 180° B_1 field/pulse, respectively.

The decay of a time-domain signal in the transverse plane leads to a spectral broadening in the frequency domain (cf. Section 2.8 and Appendix A1.2 for Fourier transform). The line broadening due to the relaxation processes is named homogeneous and is inherently irreversible. In practice, the signal decays faster than the *intrinsic* rate due to the T_2 relaxation. Other contributions to the signal decay include, for example, the inhomogeneity of the magnetic field B_0 or low-frequency molecular motions in the specimens (or even a field gradient; cf. Chapter 11). The line broadening due to non-uniformity of the magnetic field is named inhomogeneous and can be eliminated by using an appropriate rf pulse sequence (provided the molecules do not move during the measurement time). T_2 is the time constant that describes the homogeneous broadening, while the term $T_2{}^*$ is used when the decay process contains both T_2 and other (in principle) removable factors. $T_2{}^*$ is always shorter than T_2 (cf. Chapter 7.3 for T_2 and $T_2{}^*$).

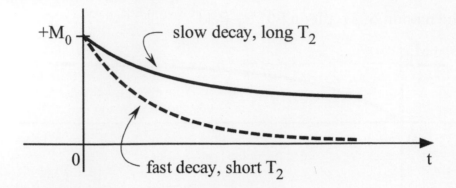

Figure 2.10   The motion of the magnitude of the transverse magnetization after a 90° B_1 field/pulse, where a slow decay leads to a long T_2 value. Note that if Figure 2.9 and Figure 2.10 are plotted together on one graph (i.e., share the same scale in the horizontal axis t), the transverse magnetization would decay to zero much faster than the return of the longitudinal magnetization to its thermal equilibrium (i.e., the maximum), since T_2 is commonly much shorter than T_1.

2.7   BLOCH EQUATION

When it can be assumed that the change of the magnetization following excitation is *independently* caused by external magnetic fields and relaxation processes, the equation of motion of **M** can be written by combining Eq. (2.14) and Eq. (2.15), in the laboratory frame, as

$$\frac{d\boldsymbol{M}}{dt} = \gamma(\boldsymbol{M} \times \boldsymbol{B}) + (-\frac{M_x\boldsymbol{i} + M_y\boldsymbol{j}}{T_2} - \frac{(M_z - M_0)\boldsymbol{k}}{T_1}). \tag{2.18}$$

This is the well-known *Bloch equation* [7]. The first term is due to precessional motion and the second term is due to relaxation. While a precise evaluation of the spin system requires a quantum mechanical treatment, the Bloch equation provides a classical, phenomenological description for liquids and liquid-like systems where the Hamiltonian is of a simple magnetic (vector) form, for example, protons of water molecules in non-viscous liquids and many biological tissues.

Now we are ready to solve the Bloch equation under various conditions. First rewrite the vector equation into the component form, as

$$\frac{dM_x}{dt} = \gamma(M_y B_0 + M_z B_1 \sin(\omega t)) - M_x / T_2, \tag{2.19a}$$

$$\frac{dM_y}{dt} = \gamma(M_z B_1 \cos(\omega t) - M_x B_0) - M_y / T_2, \tag{2.19b}$$

$$\frac{dM_z}{dt} = \gamma(-M_x B_1 \sin(\omega t) - M_y B_1 \cos(\omega t)) - (M_z - M_0)/T_1. \tag{2.19c}$$

The above equations have the usual setup for the magnetic fields as

$$\boldsymbol{B}_0 = B_0 \boldsymbol{k} \tag{2.20a}$$

$$\boldsymbol{B}_1(t) = B_1 \cos(\omega t)\boldsymbol{i} - B_1 \sin(\omega t)\boldsymbol{j}. \tag{2.20b}$$

Thermal equilibrium ensures the initial condition of $\boldsymbol{M}$ in Eq. (2.19) as

$$\boldsymbol{M}(t=0) = M_0\boldsymbol{k}. \tag{2.21}$$

Note that as soon as $\boldsymbol{M}$ is tipped away from its thermal equilibrium state, relaxation processes start. In most analyses, however, we consider only one event at a time – that is, when we use the $\boldsymbol{B}_1$ field to tip the magnetization, we do not consider the relaxation of the magnetization during the tipping process.

In order to better examine the solution of the Bloch equation (more precisely, to examine the spectral shapes of the waveform solutions), we will describe the magnetization in a rotating frame with an angular velocity ω about the z axis. In this $x'y'z'$ frame, we have the component u in the direction of x' and v in the direction of y'. We can use the common rotation matrix in linear algebra to rewrite the transform matrix as

$$u = M_x\cos(\omega t) + M_y\sin(\omega t), \tag{2.22a}$$

$$v = -M_x\sin(\omega t) + M_y\cos(\omega t). \tag{2.22b}$$

Note that Eq. (A1.23) is used in this clockwise rotation, which is consistent with the convention specified in Figure 1.3. (For a counterclockwise rotation, keep both terms of v positive and use a minus sign for the second term of u; see Appendix A1.1).

By writing the Bloch equation in this rotating frame and by setting the time derivatives in the equation equal to zero, we can solve the Bloch equation in the rotating frame. The solutions are

$$u = M_0\frac{\gamma B_1 T_2^2(\omega_0 - \omega)}{1 + T_2^2(\omega_0 - \omega)^2 + T_1 T_2\gamma^2 B_1^2}, \tag{2.23a}$$

$$v = M_0\frac{\gamma B_1 T_2}{1 + T_2^2(\omega_0 - \omega)^2 + T_1 T_2\gamma^2 B_1^2}, \tag{2.23b}$$

$$M_z = M_0\frac{1 + T_2^2(\omega_0 - \omega)^2}{1 + T_2^2(\omega_0 - \omega)^2 + T_1 T_2\gamma^2 B_1^2}, \tag{2.23c}$$

2.8 FOURIER TRANSFORM AND SPECTRAL LINE SHAPES

Before we proceed to analyze the characteristics of the magnetization motion expressed in Eq. (2.23), let us pause for a moment to briefly mention two concepts that are essential to modern NMR, Fourier transform (FT) and spectral line shapes.

2.8.1 Fourier Transform

Fourier transform (defined mathematically in Appendix A1.2) utilizes the properties of sine and cosine functions and allows their periods to approach infinity. An operation by FT mathematically decomposes a generic function (often a function of time) into a complex-valued function of frequency, where each frequency can have a certain amplitude. When discussing the FT, one commonly uses the term *domain* to refer to the "space" where two functions are associated by

FT, such as the time-domain function and its frequency domain counterpart. Figure 2.11 shows a sinusoidal oscillation in time and a delta function in frequency as a pair of Fourier functions. Both functions carry the same amount of information, where the frequency (f_0) of the delta function is determined by the period (T_0) of the oscillation, $f_0 = 1/T_0$. One can perform the FT for a multidimensional function (2D, 3D, ...); one can also carry out an inverse FT that mathematically restores the original time-domain function from its frequency-domain representation. Table 2.2 lists a few functions and their FT products.

Due to the use of sine and cosine functions in FT, there are various symmetries in the two functions associated by FT. For example, an even/odd time-domain function will keep its even/odd property in the frequency-domain function, except when the time-domain function is odd, the frequency-domain function, which is also odd, will exchange the real/imaginary channels [8]. This and other symmetry relationships are very useful in practical applications such as NMR

(a) A time-domain function

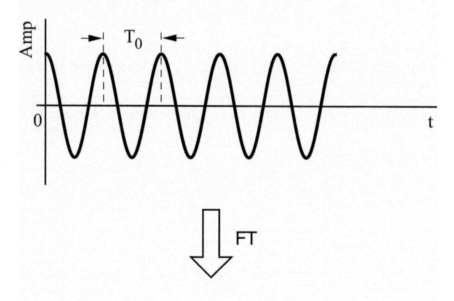

(b) A frequency-domain function

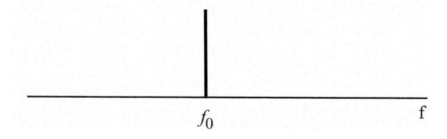

Figure 2.11 Two equivalent functions associated by a Fourier transformation: (a) a sinusoidal oscillation and (b) a delta function. Note that the oscillation frequency in the time-domain function is f_0 in the frequency domain.

Table 2.2 Some functions and their FT representations.

Function in time	FT of the function in frequency
A sine or cosine function [e.g., sin(t)]	A delta function at f
A constant [a DC offset with an amplitude]	A spike at the origin
A square/rectangular pulse	A sinc function [i.e., sin(θ)/θ]
A Lorentzian	An exponential
A Gaussian	A Gaussian

spectroscopy. In addition, modern science and technology use digital computers, where the original analytical function of time needs to be sampled into a digital, that is, discrete form (see Appendix A1.2). To find out how a computer program calculates the FT, one should also consult a highly useful book titled *Numerical Recipes* [9]. (There are several versions of this book, where each version has the *recipes* written in a particular programming language, such as Fortran, C, Pascal.)

While the Fourier transform may be regarded as a purely mathematical operation, it plays an important role in many branches of modern science and technology. For example, a waveform (optical, electrical, or acoustical) and its spectrum are in fact an FT pair, which are appreciated equally as physically picturable and measurable entities. In electronics, the signal $V(t)$ is a single-valued real function of time t while the spectrum $S(f)$ is the frequency version of $V(t)$. Specific filters can be designed so that the output of the amplifier only contains a certain range of frequencies, which are used extensively in electrical power-line and hi-fi audio electronics. In the current context, the NMR signal, known as the free induction decay (FID) in the time domain, and an NMR spectrum in a frequency representation are an FT pair.

2.8.2 Spectral Line Shapes – Lorentzian and Gaussian

Spectral line shapes in NMR describe features of the energy exchange in an atomic system. As shown in Eq. (2.2), a nuclear transition is associated with a specific amount of energy, which would imply an extremely sharp spectral line in NMR. However, the spectral line as measured in NMR is not sharp but broadened considerably. The factors that broaden the spectral line include some fundamental physics principles as well as instrumentation factors.

The fundamental physics principles that leads to the line broadening in high-resolution NMR spectroscopy of solutions is the relaxation process, which causes the NMR signal to decay (hence the naming of the NMR signal as *the free induction decay*). This decay in the NMR signal of liquids is approximately exponential, so the spectral line shape in high-resolution NMR spectroscopy is Lorentzian. The longer the lifetime of the excited states, the narrower the spectral line width (cf. Chapter 3.7). The line shape in spectra of crystallized solids could be a Gaussian or a mixture of both Gaussian and Lorentzian, due to additional nuclear interactions such as dipolar interaction (see Chapter 4).

Table 2.3 summarizes the features of Lorentzian and Gaussian line shapes. Each can be characterized by three parameters, the peak position (x_0), the peak amplitude, and the full width at the half maximum amplitude (FWHM). These two functions can also be normalized; they have the integrals of $f_1(x)$ and $g_1(x)$ equal to 1. When Lorentzian and Gaussian are plotted with the same FWHM (Figure 2.12), a Gaussian curve is a bit wider than a Lorentzian above the half maximum amplitude but drops more rapidly towards the tails/wings below the half maximum amplitude.

Table 2.3 Some features of Lorentzian and Gaussian (A, B, a are constants).

	Lorentzian	Gaussian
Line-shape equation	$f(x) = \dfrac{A}{1 + B^2(x - x_0)^2}$	$g(x) = \exp\left[\dfrac{-(x - x_0)^2}{a^2}\right]$
Line width (FWHM)	$2/B$	$2a\sqrt{\ln(2)} = 1.66511a$
Normalized expression	$f_1(x) = (B/\pi A)f(x)$	$g_1(x) = \dfrac{1}{\sqrt{xa^2}}\,g(x)$
Fourier transform	Exponential	Gaussian

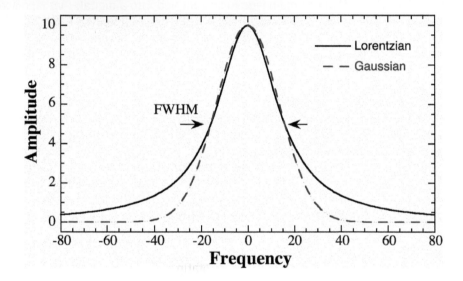

Figure 2.12 Comparison between a Lorentzian and a Gaussian with the same FWHM.

2.9 CW NMR

The earliest NMR experiments ran in a continuous-wave (CW) mode, where the spectrometer is tuned to observe the component of $\boldsymbol{M}$, which is 90° out of phase to the rotating field $\boldsymbol{B}_1$, the so-called absorption mode signal. (Earlier in Section 2.7, we set both $B_1(t)$ and u in the direction of x', and v in the direction of y' in the rotating frame.) During an experiment, the magnetic field $\boldsymbol{B}_0$ is swept slowly through the resonance frequency. As each chemically identical spin group comes into resonance, it undergoes nuclear induction and a voltage is induced in the pick-up coil (cf. the three peaks of ethanol in Figure 1.4). This approach is called the CW method, where the signal of the specimen is recorded continuously on an oscilloscope. Provided that this field sweep is done sufficiently slow, the absorption mode signal at each frequency corresponds to the steady state value of v when $\boldsymbol{M}$ has come to rest in the rotating coordinate system. Hence it is also called the slow passage experiment. Since neither the resonance frequency nor the number of the equivalent groups in a specimen is known, doing an NMR experiment using the CW method could take a long time.

By examining the Bloch equation in the rotating frame [Eq. (2.23)], the following observations can be made:

1. When we are far from the resonance (i.e., $|\omega_0 - \omega|$ is large), we have $u = v = 0$ and $M_z = M_0$.
The non-zero values of u and v appear only in a small interval around ω_0, that is, when there is a resonance.

2. Where $T_1 T_2 (\gamma B_1)^2 \ll 1$ (i.e., the rf power applied is sufficiently low so that the saturation does not occur), v can be simplified as

$$v = M_0 \frac{\gamma B_1 T_2}{1 + T_2^2 (\omega_0 - \omega)^2}. \tag{2.24}$$

By comparing Eq. (2.24) with the line-shape functions in Table 2.3, we see that v is a Lorentzian centered at ω_0 with a line width at half maximum of $1/(\pi T_2)$. Hence, *in principle*, the FWHM of the resonant peak can be used to determine the T_2 relaxation time.

3. When $T_1 T_2 (\gamma B_1)^2$ is not sufficiently smaller than 1, we can have these situations:
   i) when $T_1 T_2 (\gamma B_1)^2 < 1$, the spins are below saturation, and the signal $\propto \gamma B_1 T_2$
   ii) when $T_1 T_2 (\gamma B_1)^2 = 1$, spins are saturated, where both signal and SNR reach the maximum
   iii) when $T_1 T_2 (\gamma B_1)^2 > 1$, the Signal starts to drop.

4. Only the transverse component of the precessing $\boldsymbol{M}$ induces an observable signal in the receiver coil. The transverse component can be written as

$$M_\perp = u + iv \qquad \text{(in rotating frame)} \tag{2.25a}$$

$$\text{or} \quad M_\perp = (u + iv) \exp(i\omega t) \quad \text{(in laboratory frame)}, \tag{2.25b}$$

where the complex i indicates a 90° phase shift (i $= \sqrt{-1}$); v (which is in the direction of y') is called the absorption spectrum, where the signal is proportional to the power absorbed from the EM field; and u (which is in the direction of x') is called the dispersion spectrum (which is a common term in optics). Note that the signal is proportional to B_1 not B_1^2 [10].

2.10   RADIO-FREQUENCY PULSES IN NMR

A much more efficient method in modern NMR experiments is to apply a short but powerful B_1 pulse, which has a duration of several to tens of μs. This rf pulse will be able to cover all possible resonance frequencies simultaneously in the specimen. The reason that a short pulse can excite all possible resonant groups in the specimen is because the frequency range of a 10 μs pulse is about 10^5 Hz (since $f = 1/T$), which is sufficient to cover the range of all resonant peaks due to the differences in their chemical shifts (e.g., the three peaks in the ethanol spectrum in Figure 1.4). Since the Larmor frequency in common NMR magnets is in tens or hundreds of megahertz (e.g., at $B_0 = 1$ Tesla, $f = \gamma B_0/2\pi = 42.6$ MHz), this short B_1 pulse is commonly called a radio-frequency (rf) pulse. It is customary to label a B_1 pulse with the amplitude and duration to tip $\boldsymbol{M}$ by ϕ degrees as a ϕ rf pulse. More precisely, a $90°|_x$ pulse implies the $B_1(t)$ is stationary at the x' axis in the rotating frame and is capable of tipping the magnetization by 90°. By setting the central carrier frequency close to the nominal Larmor frequency ω_0, one single pulse can excite all possible resonance frequencies.

There are two descriptive terms for an rf pulse, whether the pulse is *soft* or *hard* (Figure 2.13), and whether the pulse has a *constant* or *modulated* amplitude. A *soft* rf pulse refers to its narrow band in frequency, which would have a long duration in time (since $f = 1/T$); in contrast, a hard

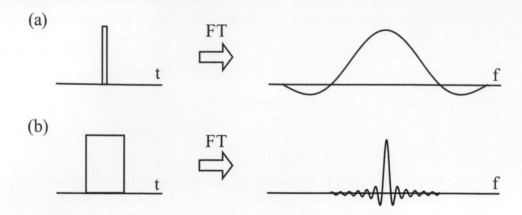

Figure 2.13 Fourier transform of (a) a hard rf pulse that is short in time duration and (b) a soft rf pulse that is long in time duration.

rf pulse would have a narrow time duration and hence a wide band in frequency, which is commonly used in NMR spectroscopy to excite all possible resonance frequencies. An rf pulse with a constant amplitude in time is commonly used in NMR spectroscopy, while an rf pulse with a modulated amplitude in time would result in a particular waveform in frequency, which is commonly used in MRI.

For a square or rectangular pulse where the B_1 field has a constant amplitude during the pulse duration, the tipping angle is given by the area of the pulse, as

$$\phi = \omega_1 t = \gamma B_1 t_p, \tag{2.26}$$

where t_p is the duration of the pulse and B_1 is its magnitude. When $B_1(t)$ does not have a constant amplitude during t_p, the amount of rotation ϕ is then given by the time integral of the amplitude of the rf field, as

$$\phi = \gamma \int_0^{t_p} B_1(t)\, dt. \tag{2.27}$$

For a pulse whose frequency response is not uniform, the central portion of the frequency response (i.e., frequency range close to ω_0) should have a constant amplitude so that all resonant groups centered around ω_0 can experience similar tipping angles.

2.11   FT NMR

In contrast to the situation in CW NMR (Section 2.9) where each particular type of nucleus in a chemical is brought into the resonance individually and sequentially, the use of an rf pulse excites simultaneously all NMR-active nuclei in the chemical. This method of NMR experiments is called FT NMR, which is much more efficient than CW NMR and offers the opportunity of SNR improvement by co-addition of many spectral responses.

In an FT NMR experiment, a resonant rf pulse results in a non-zero transverse component of the magnetization **M**, which precesses in the transverse plane. This precessional motion of **M** can be detected by means of a receiver coil (often called an rf coil, cf. Chapte 5.2), which can sense the transverse *electromotive force* (emf) and produces an FID signal.

Immediately following a $90°|_x$ pulse (i.e., after $B_1(t)$ has been turned *off*), $\textbf{M}$ is given by

$$M_x(0) = M_z(0) = 0 \tag{2.28a}$$

and $\qquad M_y(0) = M_0. \tag{2.28b}$

Subsequently, assuming a uniform field $\textbf{B}_0$, the evolution of magnetization can be obtained by solving the Bloch equation, as

$$M_x(t) = M_0 \exp(-t\,/\,T_2)\,\sin(\omega_0 t). \tag{2.29a}$$

$$M_y(t) = M_0 \exp(-t\,/\,T_2)\,\cos(\omega_0 t). \tag{2.29b}$$

and $\qquad M_z(t) = M_0(1 - \exp(-t\,/\,T_1)), \tag{2.29c}$

Since there is a 90° phase difference between $M_x(t)$ and $M_y(t)$, we can combine the two transverse components (the FID) expressed in Eqs. (2.29a) and (2.29b) into a complex term, as

$$M_\perp(t) = M_0 \exp(i\omega_0 t)\exp(-t\,/\,T_2). \tag{2.30}$$

The complex nature of the NMR signal means that we can measure not only the amplitude of the signal but also its phase, which is one of the remarkable advantages of NMR and MRI. In comparison, several other types of spectroscopy and imaging (e.g., Fourier transform infrared spectroscopy and imaging, computer tomography) can only measure the amplitude of the signal.

The time evolution of the $M_x(t)$ and $M_y(t)$ components are illustrated in Figure 2.14a and Figure 2.14b, where at $t = 0$ (the end of the 90° rf pulse), $M_x(t = 0) = 0$, and $M_y(t = 0) = M_0$. The FT of the FID signal is a Lorentzian in the frequency domain with a line width of $(\pi T_2)^{-1}$, shown in Figure 2.14c and Figure 2.14d, which are commonly termed as the absorption signal and the dispersion signal, respectively.

The time evolution of the $M_z(t)$ component expressed by Eq. (2.29c) is illustrated in Figure 2.14e. Note that since $T_1 > T_2$ in most liquid-containing specimens, it takes much longer for $M_z(t)$ to return to its thermal equilibrium than for $M_y(t)$ and $M_x(t)$ to decay to zero; that is, the time axes in the schematics in Figure 2.14 between (a) and (b) are scaled but between (a) and (e) or (b) and (e) are not scaled.

When a $90°|_{y'}$ pulse is used in the excitation, the solutions of the Bloch equation take the form

$$M_x(t) = M_0 \exp(-t\,/\,T_2)\cos(\omega_0 t), \tag{2.31a}$$

$$M_y(t) = M_0 \exp(-t\,/\,T_2)\sin(\omega_0 t), \tag{2.31b}$$

and $\qquad M_z(t) = M_0(1 - \exp(-t\,/\,T_1)), \tag{2.31c}$

which only switches the oscillation terms between $M_x(t)$ and $M_y(t)$, or in other words the phase of the signal.

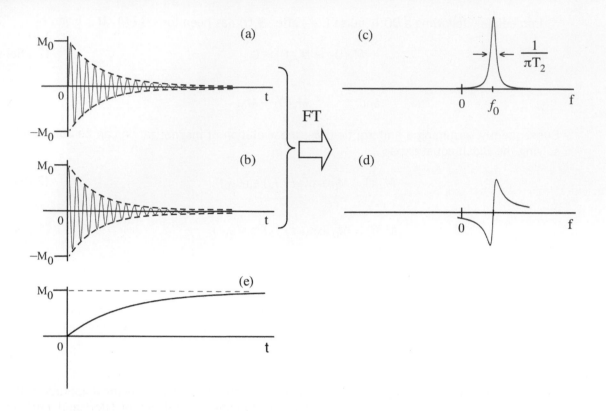

Figure 2.14 (a) and (b) The time-domain NMR signal in the transverse plane (the FID) is complex and contains real and imaginary components. By the way of Fourier transform, the time-domain NMR signal can be represented by the equivalent signals in the frequency domain, as the absorption and dispersion components, shown in (c) and (d). The peak shift f_0 in the frequency domain corresponds to the oscillation of the FID in the time domain. The recovery of the longitudinal magnetization is shown in (e). As noted in Figure 2.10, since T_2 is commonly much shorter than T_1, the transverse magnetization in (a) and (b) should decay to zero long before the longitudinal magnetization (e) returns to its thermal equilibrium [i.e., the time scales in (a) and (b) are comparable but are different from the time scale in (e)].

2.12   SIGNAL DETECTION IN NMR

The frequency ω_0 in Eq. (2.29) is usually too high for the voltage signal to be observed directly after amplification (a good linear amplifier at radio frequency is also more expensive). An electronic process named *heterodyning* is commonly used for signal detection in NMR. This process employs a number of phase-sensitive detectors to reduce the carrier frequency but retain the individual amplitude and phase information. This approach is identical to how we listen to a radio program – we do not really listen to our favorite broadcast program at hundreds of megahertz frequency (radio frequency); we listen to the audio frequency modulation of the radio broadcasting.

When $\Delta\omega$ is used as the offset of the heterodyning signal, the NMR signal in the time domain, previously expressed in Eq. (2.30) in ω_0, becomes

$$S(t) = S_0 \exp(i\phi)\exp(i\Delta\omega t)\exp(-t / T_2), \qquad (2.32)$$

where S_0 is proportional to M_0, and ϕ is a spectrometer parameter called the receiver phase.

With the use of Fourier transformation, we can derive the signal in the frequency domain in both real (Re) and imaginary (Im) parts [2], as

$$\text{Re}\left\{\mathcal{F}\left[S\left(t\right)\right]\right\} = \cos\phi \frac{T_2}{1 + T_2^2\left(\omega - \Delta\omega\right)^2} + \sin\phi \frac{\left(\omega - \Delta\omega\right)T_2^2}{1 + T_2^2\left(\omega - \Delta\omega\right)^2}, \tag{2.33a}$$

$$\text{Im}\left\{\mathcal{F}\left[S\left(t\right)\right]\right\} = \sin\phi \frac{T_2}{1 + T_2^2\left(\omega - \Delta\omega\right)^2} - \cos\phi \frac{\left(\omega - \Delta\omega\right)T_2^2}{1 + T_2^2\left(\omega - \Delta\omega\right)^2}. \tag{2.33b}$$

When $\phi = 0$, the above equations become

$$S_{\text{absorption}} = \text{Re}\left\{\mathcal{F}\left[S\left(t\right)\right]\right\} = \frac{T_2}{1 + T_2^2\left(\omega - \Delta\omega\right)^2}, \tag{2.34a}$$

$$S_{\text{dispersion}} = \text{Im}\left\{\mathcal{F}\left[S\left(t\right)\right]\right\} = -\frac{\left(\omega - \Delta\omega\right)T_2^2}{1 + T_2^2\left(\omega - \Delta\omega\right)^2}. \tag{2.34b}$$

These two equations can be plotted as in Figure 2.14c–d, where the real part is called the absorption signal, while the imaginary part is called the dispersion signal. By comparing Eq. (2.34) with Eq. (2.23) in the CW NMR experiment, we find that the results are of the same form as the absorption and dispersion components. However, the results in Eq. (2.34) are not centered at the Larmor frequency ω_0, but at the difference $\Delta\omega$, the offset frequency.

When $\phi \neq 0$, which is common in practice and means that $\boldsymbol{M}$ is not along any axis in the transverse plane, the real and imaginary parts of the signal contain a mixture of absorption and dispersion components. We call the spectrum "out of phase." We can correct this phase by multiplying the signal by $\exp(-i\phi)$; that is, we apply a 2D rotation matrix to the signal, as we did in Eq. (2.22). This process is termed to "phase" the spectrum in NMR experiments (cf. Chapter 6.10), which illustrates that in actual NMR experiments, the phase of the signal detector can be adjusted continuously.

2.13   PHASES OF THE NMR SIGNAL

In classical physics, a variety of phenomena (e.g., the analysis of Hooke's law in mechanics) can be characterized as simple harmonic motion, which is often visualized with the aid of a rotating disc. Plotting the trajectory of a fixed point on the rotating disc (or in the current context, the trajectory of the tip of a vector $\boldsymbol{M}$ in the 2D xy plane) as a function of time leads to a sinusoidal wave (Figure 2.15a and b). This type of circular motion can be described by a few fundamental equations,

$$f = 1/T, \tag{2.35a}$$

$$\omega = 2\pi f = 2\pi/T, \tag{2.35b}$$

$$m = M\cos\theta = M\cos(\omega t), \tag{2.35c}$$

where m is the amplitude of the motion, M is the maximum amplitude (i.e., the length of the vector), θ is the rotation angle in radian (rad), ω is the angular frequency in rad/s, T is the period of the wave (i.e., the time to complete one cycle) in seconds (s), and f is the frequency of the wave in hertz (Hz).

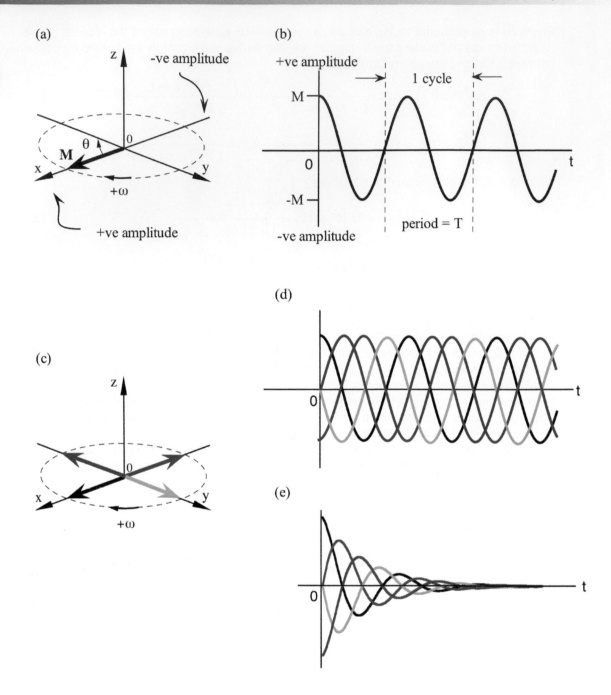

Figure 2.15 (a) A vector **M** rotates in the *xy* plane at a constant angular frequency ω and with a constant amplitude *M*, where the tip of the vector draws a sinusoidal wave in a plot of the amplitude vs. time, as in (b). (c) Four different initial orientations of the rotating vector give rise to four different phases of the wave, as shown in (d). The constant-amplitude oscillation implies no attenuation of the vector length. (e) When the oscillation is attenuated by an exponential decay, the motion of the rotating vector generates a damped oscillatory signal in the transverse plane (the FID). The FID signal will have the same decay envelope, regardless of which of the four initial orientations of the rotating vector shown in (c).

35

If no friction slows down the rotation of the disc and nothing changes the length of the vector, the disc will rotate at a constant frequency indefinitely and the wave will oscillate between +M and −M continuously as the function of time without any attenuation to its amplitude, as shown in Figure 2.15b. If the vector $\boldsymbol{M}$ in Figure 2.15a does not start the motion from being parallel with the x axis but with another axis, the oscillation will have exactly the same frequency and period, only with an extra phase shift (Figure 2.15c and d).

Similar visualization has in fact been used in the illustration of the motion of the magnetization in the laboratory and rotating frames (cf. Figure 1.3, Figure 2.14), except the FID signal has an attenuation term, which modulates the oscillation amplitude in the time domain. So instead of the constant amplitude sinusoidal waves as in Figure 2.15d, the amplitude of the FID oscillation will decay with time, as in Figure 2.15e, which generates the NMR signal. The only difference among the four NMR signals in Figure 2.15e is the four initial orientations of the magnetization vectors in the transverse plane (Figure 2.15c), that is, the initial phases of the magnetization. Even when $\boldsymbol{M}$ does not start from being parallel with any axis in the transverse plane, the real and imaginary parts of the FID both start with some initial values but still oscillate and decay in the same manner as in Figure 2.15e.

With this in mind, we can make some further comments on the laboratory and rotating frames. In the laboratory (stationary) frame, the z axis is firmly defined by the direction of $\boldsymbol{B}_0$ (as shown in Figure 1.2), while the x and y axes are commonly defined by the typical 3D Cartesian coordinates in the 2D transverse plane, which is perpendicular to the z axis (Figure 1.3). In the rotating frame, the direction of the x' axis is quite arbitrary, defined at the instant when the $\boldsymbol{B}_1(t)$ field is switched on. For a single $\boldsymbol{B}_1(t)$ field, whether $\boldsymbol{B}_1$ is along the x' or y' or $-x'$ or $-y'$ axis does not matter at all, since the variation of the $\boldsymbol{B}_1$ direction only results in the exchange of the absorption and dispersion terms in Eq. (2.34). Even when the $\boldsymbol{B}_1$ direction is at an arbitrary angle between the x' and y' axes in the transverse plane, it results in only slightly complicated quadrature signals, where the mixed terms can be and are always phase-adjusted in the phasing process after the signal acquisition [with the use of a simple 2D rotation, as Eq. (A1.23)]. The phase of a $\boldsymbol{B}_1$ field/pulse becomes critical only when this pulse is among a train of $\boldsymbol{B}_1$ pulses, that is, the relative phases among the $\boldsymbol{B}_1$ pulses in a pulse train do matter.

References

1. Harris RK, Becker ED, Cabral De Menezes SM, Goodfellow R, Granger P. NMR Nomenclature. Nuclear Spin Properties and Conventions for Chemical Shifts (IUPAC Recommendations 2001). *Pure Appl Chem.* 2001;73(11):1795–818.
2. Callaghan PT. *Principles of Nuclear Magnetic Resonance Microscopy.* Oxford: Oxford University Press; 1991.
3. Hennel JW, Klinowski J. *Fundamentals of Nuclear Magnetic Resonance.* Essex: Longman Scientific & Technical; 1993.
4. Harris RK. *Nuclear Magnetic Resonance Spectroscopy – A Physicochemical View.* Essex: Longman Scientific & Technical; 1983.
5. Bovey FA. *Nuclear Magnetic Resonance Spectroscopy.* 2nd ed. San Diego, CA: Academic Press; 1988.
6. Meadows M. Precession and Sir Joseph Larmor. *Concepts in Magnetic Resonance.* 1999;11(4):239–41.
7. Bloch F. Nuclear Induction. *Phys Rev.* 1946;70(7–8):460–74.
8. Bracewell R. *The Fourier Transform and Its Applications.* New York: McGraw-Hill Book Company; 1965.
9. Press WH, Flannery BP, Teukolsky SA, Vetterling WT. *Numerical Recipes.* Cambridge: Cambridge University Press; 1989.
10. Hoult DI, Richards RE. The Signal-to-Noise Ratio of the Nuclear Magnetic Resonance Experiment. *J Magn Reson.* 1976;24:71–85.

3

Quantum Mechanical Description of Magnetic Resonance

Although the visualization of a vector **M** moving under the direction of a **B**$_1$ pulse is useful for the understanding of the simplest cases in NMR and MRI, a deep understanding of magnetic resonance [1–6] requires the aid of quantum mechanics, where the essential information of the nuclear systems is contained in the complex wave functions (labeled with Greek letters Ψ, Φ, φ, ψ). These wave functions can be described by the use of a vectoral term called a *ket* and written as |φ>. For each |φ>, one further defines a conjugated vector of a different nature, called a *bra* and written as <φ|. (The terms *bra* and *ket* come from truncations of the word *bracket*.)

In modern physics, the energies (E) and wave functions (ψ) for a molecular or atomic system can be investigated by the use of the Schrödinger equation ($\mathcal{H}\,\psi = E\psi$), where the operator $\mathcal{H}$ is called the Hamiltonian and commonly contains the differential operator ∇^2. (A spin term is usually neglected for the computation of atomic and molecular orbitals because its influence, in terms of energy shift, is negligibly small in the absence of a magnetic field.) A similar quantum mechanical equation can describe the nuclear spins where the Hamiltonian contains the spin angular momentum operator. In NMR, the stable states of quantum mechanics systems are the eigenfunctions of $\mathcal{H}$. Hence, to calculate NMR spectra we must find the eigenvalues of $\mathcal{H}$.

In Chapter 2, the classical description of NMR, spin angular momentum is visualized as a spinning sphere that carries a charge (Figure 2.3b). In a quantum mechanical description of NMR, spin angular momentum is a quantum mechanical quantity without a classical analog; spin angular momentum is determined by the internal nuclear structure of the spin system. A classical limit is only approached in the case of orbital angular momentum and in the limit of large quantum numbers. Appendix 2 has some background introduction in quantum mechanics. This chapter presents the quantum mechanical description of the fundamental NMR concepts.

3.1 NUCLEAR MAGNETISM

A nuclear spin in quantum mechanical description is represented by a spin angular momentum operator **I**, which can be written in the usual Cartesian coordinate system as a dimensionless quantity

$$\boldsymbol{I} = I_x\boldsymbol{i} + I_y\boldsymbol{j} + I_z\boldsymbol{k}, \tag{3.1}$$

Essential Concepts in MRI: Physics, Instrumentation, Spectroscopy, and Imaging, First Edition. Yang Xia.
© 2022 John Wiley & Sons Ltd. Published 2022 by John Wiley & Sons Ltd.

where I_x, I_y, and I_z are the spin operators representing the x, y, z components of the spin operator $\boldsymbol{I}$.

The magnetic moment $\boldsymbol{\mu}$ is proportional to its spin angular momentum,

$$\boldsymbol{\mu} = \gamma\hbar\boldsymbol{I}, \tag{3.2}$$

where γ is a proportionality constant (called the gyromagnetic ratio), different for different nuclear species. This equation is identical to that in the classical description [Eq. (2.1)], except the spin $\boldsymbol{I}$ is now an operator.

A *single* nucleus in an external magnetic field ($\boldsymbol{B}_0 = B_0\boldsymbol{k}$) experiences the nuclear Zeeman interaction[1] with the field. The evolution of a spin system ψ is governed by the time-dependent Schrödinger equation,

$$i\hbar\frac{\partial}{\partial t}|\psi(t)> = \mathcal{H}|\psi(t)>, \tag{3.3}$$

where $\mathcal{H}$ is a Hamiltonian. (This equation plays a similar role as Eq. (2.3) in classical treatment of NMR in Chapter 2, where the Newton's second law was used.) If $\mathcal{H}$ is considered time-independent, the evolution of the spin system can be derived from the above equation as

$$|\psi(t)> = U(t)|\psi(t_0)>, \tag{3.4}$$

where $U(t)$ is the evolution operator. This equation effectively separates the time-independent part $|\psi(t_0)>$ from the time-dependent part $U(t)$.

Since the Hamiltonian operator for the case of $\boldsymbol{B}_0 = B_0\boldsymbol{k}$ is given by the Zeeman Hamiltonian, we can write down the operator $\mathcal{H}$ as

$$\mathcal{H} = -\boldsymbol{\mu} \cdot \boldsymbol{B}_0 = -\gamma\hbar\boldsymbol{I} \cdot \boldsymbol{B}_0 = -\gamma B_0\hbar I_z. \tag{3.5}$$

Note that only the I_z component is present in the last part of Eq. (3.5), which is due to the properties of the dot product (cf. Appendix A1.1) since $\boldsymbol{B}_0 = B_0\boldsymbol{k}$.

As in the classical description where $\boldsymbol{I}$ is the spin angular momentum (a vector) and the half-integer or integer values of $\boldsymbol{I}$ are called spin quantum number I, the spin operator I_z has m possible values (the eigenvalues), ranging from $-I$, $-I + 1$, ..., I, where m is the azimuthal quantum number.

Therefore, the evolution operator $U(t)$ can be written as

$$U(t) = \exp(-\frac{i}{\hbar}\mathcal{H}t) = \exp(-i\theta I_z) \tag{3.6}$$

where the second step considers the fact that $\omega_0 = \gamma B_0$ and $\theta = \omega_0 t$. $U(t)$ is hence just a rotation operator [recall that $\exp(i\theta) = \cos\theta + i\sin\theta$, also in Appendix A1.1], which corresponds to a rotation of the spin state $|\psi>$ about the z axis with an angular frequency ω_0, known as the Larmor precession frequency:

$$\omega_0 = \gamma B_0. \tag{3.7}$$

This equation is identical to the equation that we have derived in the classical description [Eq. (2.5)].

---

[1] The Zeeman effect is named after Dutch physicist Pieter Zeeman for his 1896 observation of splitting of the optical spectral lines in a magnetic field.

3.2   ENERGY DIFFERENCE

From the energy eigenvalue equation,

$$\mathcal{H}\,|\psi\rangle = E(m)|\psi\rangle,$$

(3.8)

one can obtain the energy eigenvalues of the Zeeman Hamiltonian $\mathcal{H}$, which are the energy levels (called the Zeeman levels):

$$E(m) = -m\hbar\gamma B_0.$$

(3.9)

Therefore, the energy difference ΔE between any two adjacent eigenstates of a spin system, known as the Zeeman splitting, is

$$|\Delta E| = \hbar\gamma B_0 = \hbar\omega_0.$$

(3.10)

As indicated in Eq. (3.9), a spin-1/2 system ($I = 1/2$) has only two eigenstates, corresponding to $m = +1/2$ (spin-up) and $m = -1/2$ (spin-down) states. Its two energy levels are therefore given by

$$E(+1/2) = -(1/2)\,\hbar\gamma B_0,$$

(3.11a)

$$E(-1/2) = +(1/2)\,\hbar\gamma B_0.$$

(3.11b)

Schematically, these two energy levels, which were shown once before briefly in Figure 2.2a, are now shown more completely in Figure 3.1. Note that for the spin-1/2 system, the "spin-up" and "spin-down" states are stationary states of the system and exist only in a time-independent magnetic field, that is, when $\mathcal{H}$ is time-independent. Note also that the schematic in Figure 3.1, where $E(-1/2)$ has a higher energy than $E(+1/2)$, is based on a positive γ.

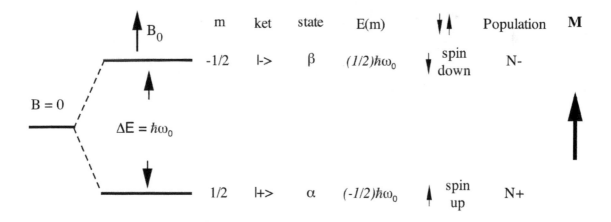

Figure 3.1   The quantities in a spin-1/2 system. The application of an external magnetic field B_0 introduces two energy levels for the spin-1/2 system. The population difference between the two states, determined by the Boltzmann distribution, results in a macroscopic magnetization, M, pointing along the same direction as the external magnetic field.

The terms "spin-up" and "spin-down" refer to the z-component $m\hbar$ of the angular momentum. The actual spin vector has a magnitude of $\hbar\sqrt{I(I+1)}$, which is greater than $m\hbar$. Hence, a spin vector I cannot lie graphically along any fixed axis in space. This is the reason that the precessional motion of a nucleus spin in a classical description is tilted at a fixed angle (Figure 2.3b). Since I_x, I_y, and I_z do not commute, we cannot specify or measure any two quantities simultaneously. Only the z-component I_z and the magnitude of I are known with certainty as $I_z = 1/2$ and $I^2 = 3/4$, which can be used to determine the fixed angle of the cone in Figure 2.3b and Figure 2.4b.

Instead of visualizing a vector $\boldsymbol{\mu}$ precessing on the surface of a cone, a spin vector in the stationary state can be thought of as uniformly smeared out over the surface of a cone, similar to the advanced concept in modern physics that visualizes an electron as a cloud around a nucleus instead of a point charge in an orbit around the nucleus. In addition, quantum mechanically, a nuclear spin in a stationary state does *not* precess, since the probability density and expectation values are independent of time. Since I is uniformly smeared out as described and cannot be specified to lie completely along any axis, we should only be concerned with the components of I, not I itself. Therefore, the spin-up state can be thought of as a spin vector lying along the z axis, parallel with the field direction.

3.3   MACROSCOPIC MAGNETIZATION

Any practical sample, no matter how small, contains an enormous number of nuclei. It is the macroscopic ensemble average of the observable quantities in which we are interested. In these ensembles, different nuclei may occupy different states $|\psi>$. We use the concept of sub-ensemble in which all nuclei are in identical states $|\psi(t)>$. The sum over all sub-ensembles, each with a classical probability p_ψ, gives the observable ensemble average in which we are interested. The averaged expectation value, by definition, is

$$\overline{<\psi|I_z|\psi>} = \sum_\psi p_\psi <\psi|I_z|\psi>, \tag{3.12}$$

where the bar refers to the statistical ensemble average, and the pair of arrow brackets, $<\ >$, represents the quantum mechanical expectation value.

Now consider spin-1/2 particles. In NMR, the dominant interaction of a spin with its environment is always via the Zeeman interaction [as in Eq. (3.5)]. This means that the natural eigenstates are those whose quantum numbers are eigenvalues of I_z, that is, $|1/2>$ and $|-1/2>$.

In general, we can express any state in this basis using the Pauli's spin matrices formalism (cf. Appendix A2.5), as

$$|\psi> = a_{1/2}|1/2> + a_{-1/2}|-1/2> = a_{1/2}\begin{bmatrix}1\\0\end{bmatrix} + a_{-1/2}\begin{bmatrix}0\\1\end{bmatrix}. \tag{3.13}$$

Since $I_z = \dfrac{1}{2}\begin{bmatrix}1 & 0\\0 & -1\end{bmatrix}$, the ensemble averaged expectation value of I_z is determined by the difference in populations between the upper and lower energy levels, according to the Boltzmann distribution. This distribution describes the polarization of the ensemble in thermal equilibrium. We can calculate the normalized population at thermal equilibrium as

$$\overline{|a_{\pm 1/2}|^2} = \frac{\exp\left(\pm\hbar\gamma B_0/2k_BT\right)}{\exp\left(\hbar\gamma B_0/2k_BT\right) + \exp\left(-\hbar\gamma B_0/2k_BT\right)}, \tag{3.14}$$

where the numerators are individual populations and the denominator is the total population.

Note that $k_B T$ is the Boltzmann energy and $\hbar\gamma B_0$ is the Zeeman energy difference. For example, for protons in a magnetic field of strength $B_0 = 7$ Tesla and at room temperature ($T = 300$ K), we have

$$k_B T = 1.38 \times 10^{-23} \times 300 = 4.41 \times 10^{-21} \, \text{J}$$

and $\quad \hbar\gamma B_0 = \dfrac{6.626 \times 10^{-34}}{2\pi} \times 2.675 \times 10^{8} \times 7 = 1.97 \times 10^{-25} \, \text{J}.$

Since $k_B T$ is over four orders of magnitude bigger than $\hbar\gamma B_0$, the ratio $\hbar\gamma B_0 / k_B T$ is tiny. This is a good news and bad news situation: the good news is that the exponentials in Eq. (3.14) can be simplified using the Taylor expansion since the high-order terms would be very small, while the bad news means that our signal will be very small, since the signal is proportional to the population difference.

Due to this tiny ratio between the Zeeman energy and the Boltzmann energy, we can apply the Taylor expansion (Appendix A1.1.4) to simplify the expression in Eq. (3.14) by keeping only the first two terms, as

$$\overline{|a_{\pm 1/2}|^2} = \frac{1}{2}\left(1 \pm \frac{\hbar\gamma B_0}{2k_B T}\right). \tag{3.15}$$

This equation at the room temperature and a 7 Tesla B_0 equals approximately $0.5 \pm 1.12 \times 10^{-5}$, which is almost a half and half situation between the two populations. This approximation is known as the "high-temperature approximation" in the NMR literature, except the "high temperature" in this estimation actually means the room temperature.

3.4 MEASUREMENT OF THE X COMPONENT OF ANGULAR MOMENTUM

In Appendix A2.6, we introduce the concept of density matrix operator ρ, which relates the expectation value of any operator $\boldsymbol{A}$ to the trace of the matrix product $\boldsymbol{A}\rho$, via Eq. (A2.33). Since $I_x = \dfrac{1}{2}\begin{bmatrix} 0 & 1 \\ 1 & 0 \end{bmatrix}$ in the formalism of Pauli's spin matrices (cf. Appendix A2.5), we have, from Eq. (3.12),

$$\begin{aligned}
\overline{<\psi|I_x|\psi>} &= \sum_\psi p_\psi <\psi|I_x|\psi> \\
&= \text{Tr}\left(I_x \rho\right) \\
&= \frac{1}{2}\left(\overline{a_{1/2}{}^* a_{-1/2}} + \overline{a_{1/2} a_{-1/2}{}^*}\right),
\end{aligned} \tag{3.16}$$

where Tr represents the trace of the matrix (the sum of the diagonal elements) – see Eqs. (A2.32) and (A2.33), and the two examples at the end of Appendix A2.6. The term in brackets describes the degree of "single quantum coherence" of the ensemble, while the average (the bars) reflects the phase coherence between the +1/2 and −1/2 states. At thermal equilibrium, both $\overline{a_{1/2}{}^* \, a_{-1/2}}$ and $\overline{a_{1/2} a_{-1/2}{}^*}$ are zero (no transverse component), while the term $(|a_{1/2}|^2 - |a_{-1/2}|^2)$ has a stable value (the longitudinal magnetization along the z axis).

3.5 MACROSCOPIC MAGNETIZATION FOR SPIN 1/2

In the current context, the observable quantity is just the (macroscopic) magnetization **M**, given by

$$\boldsymbol{M} = \overline{< N\gamma\hbar\boldsymbol{I} >} \tag{3.17a}$$

$$\text{or} \quad \boldsymbol{M} = N\gamma\hbar\left(\overline{< I_x >}\boldsymbol{i} + \overline{< I_y >}\boldsymbol{j} + \overline{< I_z >}\boldsymbol{k}\right), \tag{3.17b}$$

where N is the number of spins, and **i**, **j**, and **k** are the unit vectors in the Cartesian coordinates. Equation (3.17) is important because it may be shown that any state of the density matrix (defined in Appendix A2.6) for an ensemble of non-interacting spin-1/2 particles can be described using the macroscopic magnetization defined in this manner, thus permitting a classical description of simple spin systems.

In the absence of an external magnetic field, the ensemble average of the magnetization vector should be zero due to the random directions of the magnetic dipoles of the nuclei.

If a sample is immersed in an external field and in thermal equilibrium, the density operator associated with this magnetization vector is given by

$$\rho = \frac{\exp\left(-\mathcal{H}/k_\mathrm{B}T\right)}{\mathrm{Tr}\left[\exp\left(\mathcal{H}/k_\mathrm{B}T\right)\right]}. \tag{3.18}$$

The transverse component of **M** is zero due to the even distribution of the azimuthal phase angles of the precessing nuclei in the transverse plane. This corresponds to phase incoherence leading to the zero value of the off-diagonal elements of ρ,

$$\overline{a_{1/2}{}^* \, a_{-1/2}} = \overline{a_{1/2} \, a_{-1/2}{}^*} = 0. \tag{3.19}$$

The z component of the magnetization **M** arises from the difference in populations between the upper and lower energy states. At room temperature, the magnitude of this magnetization in the equilibrium state, M_0, can be derived as

$$M_0 = <M_z> = N\gamma\hbar <I_z> = N\gamma\hbar\,\mathrm{Tr}(\rho I_z) = \frac{N(\gamma\hbar)^2 B_0}{4k_\mathrm{B}T}. \tag{3.20}$$

For a spin-1/2 system at room temperature, the population difference between the spin-up ($m = +1/2$) state and the spin-down ($m = -1/2$) state can be calculated from the diagonal elements of ρ, as

$$\overline{|a_{1/2}|^2} - \overline{|a_{-1/2}|^2} = \frac{\hbar\gamma B_0}{2k_\mathrm{B}T}. \tag{3.21}$$

For protons at $B_0 = 1.4$ T (60 MHz), it is equal to about 5×10^{-6}, a small value resulting from the small value of $\gamma\hbar B_0$ (Zeeman splitting) compared to $k_\mathrm{B}T$ (Boltzmann energy). It is this small magnetization of nuclei at room temperature that limits NMR detection sensitivity and leads to resolution limitations in MRI experiments [7, 8].

3.6   RESONANT EXCITATION

When both B_0 and $B_1(t)$ are present and perpendicular to each other (B_1 in the transverse plane), we can write down the Hamiltonian in the laboratory frame as

$$\mathcal{H}_{\text{lab}} = -\hbar\gamma B_0 I_z - 2\hbar\gamma B_1 \cos(\omega t) I_x. \tag{3.22}$$

In the rotating frame, the Hamiltonian becomes

$$\mathcal{H}_{\text{rotating}} = -\hbar\gamma (B_0 - \omega/\gamma) I_z - \hbar\gamma B_1 I_x. \tag{3.23}$$

At $\omega = \omega_0$, we have

$$\mathcal{H}_{\text{rotating}} = -\hbar\gamma B_1 I_x. \tag{3.24}$$

Since $I_x = \frac{1}{2}(I_+ + I_-)$, where I_+ and I_- are the raising and lowering operators defined in Appendix A2.4, the time evolution of the spin system corresponds to an inter-conversion of each spin between |1/2> and |–1/2> at a rate of γB_1 (an oscillation).

3.7   MECHANISMS OF SPIN RELAXATION

Spin relaxation is truly fundamental and central to the theory of NMR and MRI; the influence of spin relaxation on both NMR and MRI measurements is wide, deep, and quite often subtle. It is therefore worth taking some time to learn to appreciate the subtleties of spin relaxation.

A nucleus in a liquid experiences a fluctuating field, due to the magnetic moments of nuclei in other molecules as they undergo thermal motions. (This experience of a fluctuating field is actually true for all environments, not just a nucleus in a liquid.) This fluctuating field may be resolved by Fourier analysis into a series of terms that are oscillating at different frequencies, which may be further divided into components parallel to B_0 and perpendicular to B_0. The component parallel to B_0 could influence the steadiness of the static field B_0, while the components perpendicular to the static field at the Larmor frequency could induce transitions between the levels in a similar way to B_1. These influences give rise to a non-adiabatic (or non-secular) contribution to relaxation of both the longitudinal and transverse components of M.

If the fluctuating field manages to alter the populations of the states, the populations would evolve immediately until they reach the values predicted by the Boltzmann equations for the temperature of the Brownian motion (lattice temperature). This process is described by T_1 and results in the relaxation of the longitudinal component of M. The contribution of the fluctuating field to T_2 can be seen from the following argument. According to Eq. (3.9)–Eq. (3.11), the Zeeman energy levels are known precisely (Figure 3.1), which implies the resonance frequency associated with the transition between two neighboring Zeeman levels should have a unique value, that is, a delta function at a singular ω_0 (Figure 3.2a). In reality, however, the resonant peak even in simple liquids is broadened due to the fluctuation of the Zeeman levels (Figure 3.2b), caused by the distributions of local interactions in their environment experienced by the nuclear spins. The line width of the resonant peak (excluding the effect of inhomogeneity in B_0), which is inversely proportional to T_2, is a measure of the uncertainty in the energies between two

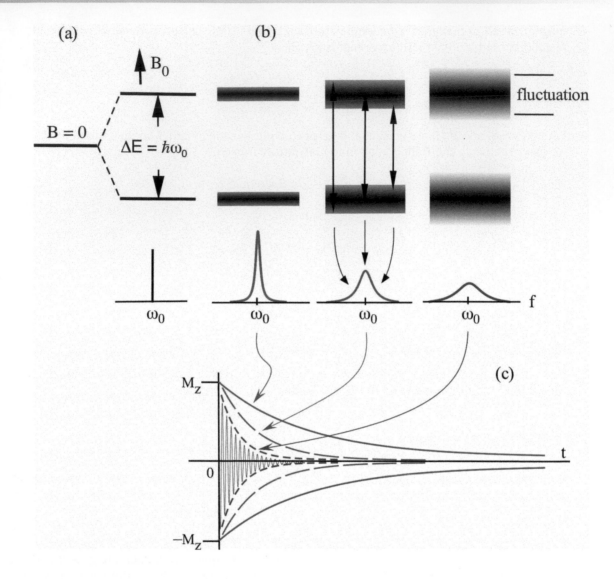

Figure 3.2 (a) A precise value of the Zeeman energy difference between the two states in a spin-½ system should imply a single value in the transition, hence a delta function in the frequency distribution. (b) In reality, a wider line shape such as a Lorentzian or Gaussian suggests an uncertainty in the difference between the energy levels. For simple liquids, the line shape is a Lorentzian in the frequency domain, which corresponds to the exponential decay in the time-domain FID, shown in (c). A fast decay of the FID (e.g., short blue dash) implies a short T_2 and a wide line shape; a slow decay (e.g., red solid line) implies a long T_2 and a narrow line shape. A precise value of the energy difference as in (a) would imply a sinusoidal oscillation without any decay in the time domain (as shown in Figure 2.15d).

neighboring Zeeman levels. This uncertainty can also be traced back to the fluctuating fields due to the magnetic moments of nuclei in other molecules as they undergo thermal motions.

The Bloch equation [Eq. (2.18)] contains a phenomenological term leading to exponential relaxation. This classical description is quite accurate for spins in rapidly tumbling molecules but breaks down when the motions of molecules become slow or complex, such as in the case of internal motion in macromolecules. In the following sections, we first explain the relaxation mechanisms in terms of quantum transitions between eigenstates of operators I_x, I_y, and I_z; then, we briefly describe the results of the random field model of relaxation.

3.7.1   Relaxation Mechanism in Terms of Quantum Transitions

For spin-1/2 particles, the relaxation mechanism can be understood with a set of equations and analysis in terms of quantum transition [9, 10]. In this approach, the spin populations (the occupancies of the eigenstates of I_z with eigenvalues $m = \pm 1/2$) are defined as

$$N_+ = N_0 < |a_+|^2 >$$

(3.25a)

$$\text{and} \quad N_- = N_0 < |a_-|^2 > .$$

(3.25b)

We also define the total population N_0 and the population difference n as

$$N_0 = N_+ + N_-,$$

(3.26a)

$$n = \Delta N = N_+ - N_-$$

(3.26b)

Hence, the macroscopic magnetization M is proportional to the population difference n. Using Eq. (3.21), the z component of the magnetization at time = 0 can be written as

$$M_z = \frac{1}{2}\gamma\hbar N_0\left(< |a_+|^2 > - < |a_-|^2 >\right) = \frac{1}{2}\gamma\hbar n.$$

(3.27)

Since the population is at equilibrium with the environment according to the Boltzmann distribution, the population ratio is

$$N_+ / N_- = \exp\left(\frac{\hbar\gamma B_0}{k_B T}\right).$$

(3.28)

To consider the dynamics of the two populations, we define w_{+-} as the probability of transition of a spin from |+> state to |–> state per spin per second, and w_{-+} as the probability of transition of a spin from |–> state to |+> state per spin per second. At equilibrium, we have

$$w_{+-}N_+ = w_{-+}N_-.$$

(3.29)

Combining Eq. (3.28) and Eq. (3.29), we have

$$\frac{w_{-+}}{w_{+-}} = \frac{N_+}{N_-} = \exp\left(\frac{\hbar\gamma B_0}{k_B T}\right).$$

(3.30)

With this equation, the changes of the spins with time can be defined as

$$\frac{dN_+}{dt} = N_- w_{-+} - N_+ w_{+-}$$

(3.31a)

$$\text{and} \quad \frac{dN_-}{dt} = N_+ w_{+-} - N_- w_{-+}.$$

(3.31b)

Each equation in Eq. (3.31) has two terms, the increment term (the first term) and the reduction term (the second term). Given the fact that $w_{+-} \approx w_{-+}$, we define

$$w_0 \approx w_{+-} \approx w_{-+} \qquad (3.32a)$$

and $\quad w_0 = \frac{1}{2}(w_{+-} + w_{-+}). \qquad (3.32b)$

Note that w_0 can be considered as the probability of induced transitions, while the term $\hbar\gamma B_0 / k_B T$ can be considered as the probability of spontaneous transitions; their differences were distinguished first by Albert Einstein in 1916 when he published a paper on different processes occurring in the formation of an atomic spectral line in optical studies. And hence we can show that

$$\frac{dn}{dt} = -2w_0 (n - n_\infty), \qquad (3.33)$$

where dn/dt goes to zero as n (a variable) goes to n_∞ (a constant given by the population difference in the presence of $\boldsymbol{B}_0$ but in the absence of another rf field). If we define

$$\frac{1}{T_1} = 2w_0, \qquad (3.34)$$

and multiplying Eq. (3.33) with $(1/2)\gamma\hbar$, we can derive

$$\frac{dM_z}{dt} = -\frac{M_z - M_0}{T_1}, \qquad (3.35)$$

which has been defined previously as Eq. (2.15a). Therefore, we can interpret T_1 in terms of quantum transitions between the states.

The process of transverse relaxation may also be viewed as the result of quantum transitions. By defining the probability of transition between the states of operators $I_{x'}$ and $I_{y'}$ as $w_{0x'}$ and w_{0y}, and recognizing at $w_{0x'} = w_{0y'} = w_{0\perp}$, we can derive

$$\frac{d|M_\perp|}{dt} = -2w_{0\perp} |M_\perp|, \qquad (3.36)$$

where $|M_\perp| = (M_{x'}^2 + M_{y'}^2)^{1/2}$. Hence,

$$\frac{1}{T_2} = 2w_{0\perp}. \qquad (3.37)$$

3.7.2   Relaxation Mechanisms in the Random Field Model

More complicated explanation of relaxation mechanisms employs time-dependent Hamiltonians of the nuclear interactions. While in higher spin systems quadrupole interactions are significant, the dominant interaction in the spin-1/2 nuclei arises from the dipolar Hamiltonian, where the random field model of relaxation contains a very useful parameter (τ_c) that relates to the motional properties of the molecules. A full account of this model is too lengthy for this introductory book; here we quote some arguments and conclusions of this so-called Bloembergen–Purcell–Pound theory (BPP theory) [11].

The random field model of relaxation is simplified by restricting consideration to a pair of like spins under the influence of the dipolar interaction. The arguments for the spin-lattice relaxation in this model consider the fact that each spin is not completely isolated from the rest of the molecular ensemble (the lattice). There is an interaction between the spin and the lattice, due to the molecular motions, that enables the exchange of the thermal energy. Each nucleus is influenced by a number of nearby nuclei. These neighboring nuclei, in the same and other molecules, are in thermal motion with respect to the observed nucleus. This collection of motions gives rise to a fluctuating magnetic field, which disturbs the applied external field B_0. Consequently, the magnetic moment will also experience the fluctuating local fields of its neighbors. Given the random nature for the motions of each molecule and of its neighbors, the motion would have a broad distribution of the frequencies. When the fluctuating local fields have components in the direction of B_1 and at the precessing frequency ω_0, these field components will induce transitions between energy levels, just like a purposefully applied B_1 field.

The model defines a spectral density function $J_q(\omega)$, which is the Fourier transform of the auto-correlation function of the spatial tensor component $q(t)$. $J_q(\omega)$ represents the relative intensity of the motional frequency ω. The auto-correlation function has some characteristic time τ_c where the function goes to zero when $t \gg \tau_c$. Hence, $J_q(\omega)$ has a characteristic frequency (τ_c^{-1}), with which $J_q(\omega)$ goes to zero when $\omega \gg \tau_c^{-1}$. The correlation time τ_c is the characteristic time of the signal decay, which can be defined as the *average* time between molecular collisions for translational motion. The value of τ_c depends upon many factors of the sample, such as molecular size, molecular symmetry, and solution viscosity. For random molecular tumbling, τ_c corresponds approximately to the average time for a molecule to rotate through one radian. A shorter/longer τ_c corresponds to samples with more/less mobile molecules.

Figure 3.3 shows schematically the variations of the spectral density function $J_q(\omega)$ in three different cases. Since the area under each curve represents the total energy available for the motion, the area under any curve stays constant among the curves. If τ_c is long (case a in Figure 3.3), as in solids or viscous liquids, the low motional frequencies have a higher chance of occurring, and consequently $J_q(\omega)$ has weaker components at ω_0. At the other extreme, if τ_c is short (case c in Figure 3.3), as in liquids of low viscosity, molecular motions are distributed over a wide frequency range, and all motional frequencies within that range have equally probable

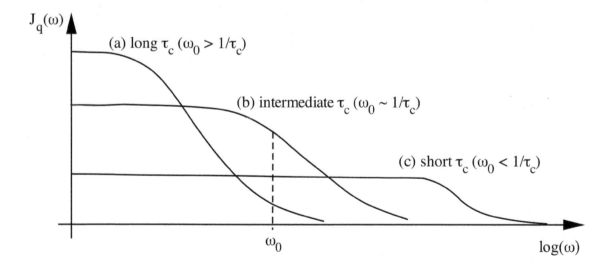

Figure 3.3 Schematic power spectra of the spectral density function $J_q(\omega)$, as the function of the frequency $\log(\omega)$. (Case a) Specimens having high viscosities or rigid lattice have long τ_c ($\omega_0 > 1/\tau_c$), which leads to weak $J_q(\omega)$ at ω_0. (Case b) Specimens having intermediate viscosities have shorter τ_c ($\omega_0 \sim 1/\tau_c$), which leads to a maximum $J_q(\omega)$ at ω_0, hence optimum relaxation. (Case c) Specimens having low viscosities have short τ_c ($\omega_0 < 1/\tau_c$), which leads to a broad spectrum of $J_q(\omega)$ with no component having a high value.

chance of occurring; however, no one component, in particular that at ω_0, can be very intense. An optimum efficiency for thermal relaxation of the spins can be expected when $J_q(\omega)$ reaches a maximum at ω_0, when $\omega\tau_c \sim 1$ (case b in Figure 3.3).

This model shows that the spin-lattice relaxation will depend on $J_q(\omega_0)$, which has a line shape of a Lorentzian. We can also show that, by similar arguments or using the density operator formalism, the spin-spin relaxation will depend on $J_q(0)$, because it involves no energy exchange. Figure 3.4 shows schematically the plots of relaxation times as the function of $\log(\tau_c)$. On the left of the condition when $\omega_0\tau_c = 1$, T_1 is approximately equal to T_2, a regime that is called the extreme narrowing regime. When the molecules become less mobile [towards the right along the $\log(\tau_c)$ axis], differences between T_1 and T_2 become significant, where T_1 becomes long (tens of minutes or much longer) while T_2 becomes short (tens of milliseconds or much shorter). This random field model is valid in the "weak collision" case where $\tau_c < T_2$; the theory will fail at the situation in relatively rigid materials. Some of the longest T_1 are commonly found in solids, on the order of tens of minutes; some of the shortest T_2 are on the order of 10 µs, which occurs in solids as well as well-organized connective tissues such as tendon and ligaments. Typical ranges of T_1 and T_2 for biological tissues are marked in Figure 3.4.

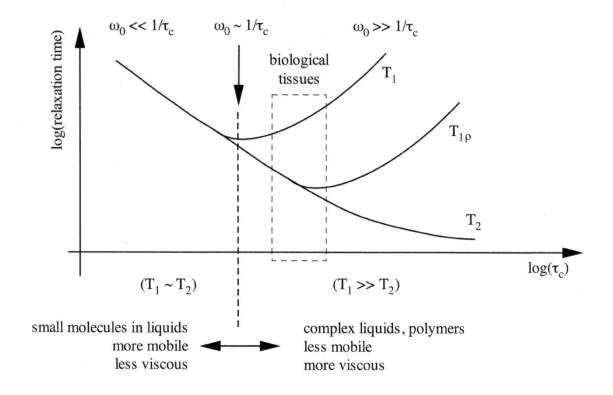

Figure 3.4 Schematic log/log trends of relaxation times on the correlation time τ_c, which also depend upon the resonance frequency and temperature (not shown). T_1 relaxation time has a minimum, occurring when $\omega \sim 1/\tau_c$. This minimum separates two regimes of relaxation. On the left of the minimum is the extreme narrowing regime ($\omega \ll 1/\tau_c$) where $T_1 \sim T_2$, where one can find the reduction of the homogeneous line width ($1/\pi T_2$); one commonly finds simple and less viscous liquids of small molecules in this regime. On the right of the minimum there is a regime ($\omega \gg 1/\tau_c$) where T_1 and T_2 diverge significantly; one can find in this regime viscous liquids and polymers of relatively large and complex molecules, as well as a variety of biological tissues.

References

1. Abragam A. *The Principles of Nuclear Magnetism*. Oxford: Clarendon; 1960.
2. Blum K. *Density Matrix Theory and Applications*. New York: Plenum; 1981.
3. Poole CPJ, Farach HA. *Theory of Magnetic Resonance*. 2nd ed. New York: John Wiley & Sons; 1987.
4. Ernst RR, Bodenhausen G, Wokaun A. *Principles of Nuclear Magnetic Resonance in One and Two Dimensions*. Oxford: Clarendon; 1987.
5. Callaghan PT. *Principles of Nuclear Magnetic Resonance Microscopy*. Oxford: Oxford University Press; 1991.
6. Slichter CP. *Principles of Magnetic Resonance*. 3rd ed. Berlin: Springer-Verlag; 1992.
7. Callaghan PT, Eccles CD. Sensitivity and Resolution in NMR Imaging. *J Magn Reson*. 1987;71:426–45.
8. Xia Y. Contrast in NMR Imaging and Microscopy. *Concepts in Magn Reson*. 1996;8(3):205–25.
9. Bovey FA. *Nuclear Magnetic Resonance Spectroscopy*. 2nd ed. San Diego, CA: Academic Press; 1988.
10. Hennel JW, Klinowski J. *Fundamentals of Nuclear Magnetic Resonance*. Essex: Longman Scientific & Technical; 1993.
11. Bloembergen N, Purcell EM, Pound RV. Relaxation Effects in Nuclear Magnetic Resonance Absorption. *Phys Rev*. 1948;73:679–712.

4

Nuclear Interactions

The nuclear interaction discussed in the previous chapters is called the Zeeman interaction, which is the interaction between the longitudinal magnetic field $\boldsymbol{B}_0$ and the nuclear ensemble (a large collection of nuclear spins). We showed that the difference between two neighboring energy levels is $\Delta E = \gamma \hbar B_0$. Even though the Zeeman energy is tiny compared to the Boltzmann thermal energy (Chapter 3.3), the Zeeman interaction is much stronger than other nuclear interactions that we will discuss in this chapter. A unique feature of NMR is that many weak nuclear interactions can be observed in the experiments despite the presence of a dominant Zeeman interaction. This feature is a consequence of the remarkable coherence time exhibited by spin ensembles, which is true especially in liquids.

In general, the spin Hamiltonian in NMR can include the following nuclear interactions,

$$\mathcal{H} = \mathcal{H}_Z + \mathcal{H}_D + \mathcal{H}_{CS} + \mathcal{H}_S + \mathcal{H}_Q + \text{etc.,} \tag{4.1}$$

where $\mathcal{H}_Z$ is the Hamiltonian for Zeeman interaction, which is the interaction between the longitudinal field $\boldsymbol{B}_0$ and the nuclei; $\mathcal{H}_D$ is the Hamiltonian for the dipolar interaction, which is the mutual interaction between the nuclear spins; $\mathcal{H}_{CS}$ is the Hamiltonian for the chemical shift interaction, arising from the electron orbital shielding that perturbs the Zeeman interaction; $\mathcal{H}_S$ is the Hamiltonian for the scalar interaction (spin-spin coupling or J coupling), which is the interaction mediated by the electronic framework of the molecule; and $\mathcal{H}_Q$ is the Hamiltonian for the quadrupole interaction when spin $I > 1/2$, which is the interaction between the nuclear quadrupole moment and the surrounding electric field gradient. The final term *etc.* contains Hamiltonians for other interactions that may be present in a spin system, for example, the effect of a radio-frequency field, $\mathcal{H}_{rf}$. All of these nuclear interactions can be treated properly using quantum mechanics [1, 2]. In this chapter, we describe some of these interactions with a minimal use of advanced theories.

4.1 DIPOLAR INTERACTION

Dipolar interaction is a *direct through-space* interaction between the nuclear spins. Since NMR-sensitive nuclei possess an angular momentum $\boldsymbol{I}$ and magnetic dipolar moment $\boldsymbol{\mu}$, a spinning nucleus generates a local magnetic field, behaving much like a tiny magnetic dipole, with a dipolar field that extends into the nearby space. One feature of the dipolar field is that at any field location, its magnitude and direction are different (Figure 4.1a). Any given nucleus in the nuclear ensemble will be affected by the dipolar fields of its local neighbors at the molecular

Essential Concepts in MRI: Physics, Instrumentation, Spectroscopy, and Imaging, First Edition. Yang Xia.
© 2022 John Wiley & Sons Ltd. Published 2022 by John Wiley & Sons Ltd.

level (Figure 4.1b). Using the knowledge of electromagnetism, the energy of interaction between two magnetic point dipoles μ_1 and μ_2 can be written as

$$E = \frac{\mu_0}{4\pi}\left[\frac{\mu_1 \cdot \mu_2}{r^3} - 3\frac{(\mu_1 \cdot r)(\mu_2 \cdot r)}{r^5}\right], \tag{4.2}$$

where μ_0 is the permeability of free space (which was until 2019 a physical constant, $4\pi \times 10^{-7}$ H/m, and is now based on some experimental measurement, $1.25663706212(19) \times 10^{-6}$ H/m); r is the distance that separates the two point-dipoles μ_1 and μ_2. By using $\mu = \gamma\hbar I$, the expression for the Hamiltonian can be derived from Eq. (4.2) as

$$\mathcal{H}_D = \frac{\mu_0}{4\pi}\gamma_1\gamma_2\hbar^2\left[\frac{I_1 \cdot I_2}{r^3} - 3\frac{(I_1 \cdot r)(I_2 \cdot r)}{r^5}\right], \tag{4.3}$$

where γ_1 and γ_2 are the gyromagnetic ratios for μ_1 and μ_2, respectively. In the NMR literature, it is common to express Eq. (4.3) more conveniently in spherical coordinates (Figure 4.1b) as

$$\mathcal{H}_D = \left(\frac{\mu_0}{4\pi}\frac{\gamma_1\gamma_2\hbar^2}{r^3}\right)\{A + B + C + D + E + F\}, \tag{4.4}$$

where

$$A = -I_{1z}\,I_{2z}\,(3\cos^2\theta - 1), \tag{4.5a}$$

$$B = \frac{1}{4}\left[I_{1+}I_{2-} + I_{1-}I_{2+}\right](3\cos^2\theta - 1), \tag{4.5b}$$

$$C = -\frac{3}{2}\left[I_{1z}I_{2+} + I_{1+}I_{2z}\right]\sin\theta\,\cos\theta\,\exp(-i\phi), \tag{4.5c}$$

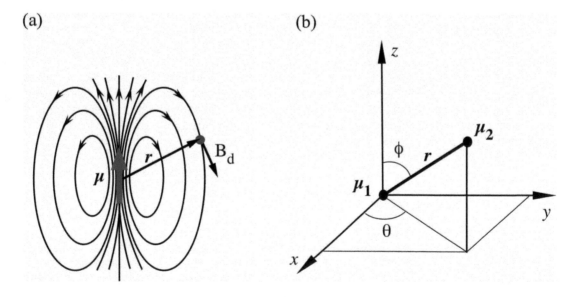

Figure 4.1   (a) A nuclear spin behaves like a tiny magnetic dipole, generating a 3D dipolar field around itself (the schematic is drawn in 2D, which is a slice in the 3D space). At any location nearby (at a distance r), the magnitude and direction of this dipolar field B_d is different. (b) Spherical coordinates relating two nuclear spins (μ_1, μ_2).

$$D = -\frac{3}{2}\left[I_{1z}I_{2-} + I_{1-}I_{2z}\right]\sin\theta\,\cos\theta\,\exp(i\phi),\tag{4.5d}$$

$$E = -\frac{3}{4}I_{1+}I_{2+}\,\sin^2\theta\,\exp(-2i\phi),\tag{4.5e}$$

$$F = -\frac{3}{4}I_{1-}I_{2-}\,\sin^2\theta\,\exp(2i\phi).\tag{4.5f}$$

[The subscripted + and − in Eq. (4.5a–f) indicate the raising and lowering operators; see Appendix A2.4.]

A few comments can be made to the above-mentioned equations:

1. Equations (4.4) and (4.5) contain a spin operator tensor of rank 2, where A and B are the rank 0 part of the tensor (i.e., scalars), C and D are the rank 1 part of the tensors (i.e., vectors, each with three components), and E and F are the rank 2 part of the tensors (i.e., 3×3 matrices, each with nine components). Each equation for Eq. (4.5a) to Eq. (4.5f) contains the spin operators I and the spherical harmonics.

2. The common header in Eq. (4.4), that is, $\left(\dfrac{\mu_0}{4\pi}\dfrac{\gamma_1\gamma_2\hbar^2}{r^3}\right)$, is referred to as the dipolar coupling constant R. For protons at $r = 1.5$ Å, the dipolar coupling constant R is around 100 kHz.

3. When the molecules tumble rapidly, which is the case for simple non-viscous liquids, the average over the space will cause all of the terms in Eq. (4.5a–f) to equal zero, that is, no dipolar interaction for any collection of isotropically tumbling molecules. (Pages 97–98 of

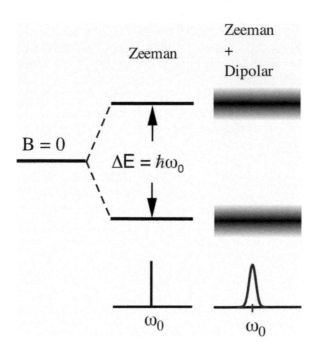

Figure 4.2   The application of the external field B_0 results in the formation of two Zeeman energy levels for a spin-1/2 nucleus. In the presence of the dipolar interaction, the uncertainties of the Zeeman energy levels lead to the broadening of the resonant line.

Harris [3] contain a proof of this statement.) This is a good news because much weaker interactions such as the scalar coupling can then be observed.

4. In contrast, for solid samples $\mathcal{H}_D$ does not average to zero and will play an important role in NMR. The dominant terms in the non-zero $\mathcal{H}_D$ are the rank 0 part of the spin operator tensor, which contains the geometric factor $(3\cos^2\theta - 1)$. This is an important factor in solid-state NMR and some biological tissues with organized macromolecules (e.g., collagen fibers in articular cartilage and tendon [4]), and we will discuss later how to utilize it.

5. When the dipolar interaction plays an important role in solid-state NMR, the line shape is approximately Gaussian, in contrast to the Lorentzian shape in liquids when the effect of T_2 relaxation is dominant.

6. In many liquids, there is still some residual effect of dipolar interaction on the relaxation (line width), which could be visualized conceptually as fluctuations of the Zeeman energy that lead to line broadening (Figure 4.2).

4.2   CHEMICAL SHIFT INTERACTION

If the frequency of an NMR transition for a specimen were entirely determined by the equation that we introduced in Chapter 2 ($\omega_0 = \gamma\boldsymbol{B}_0$), the NMR technique would have little application in chemistry. Fortunately, this is not true. The reason is that $\omega_0 = \gamma\boldsymbol{B}_0$ was determined for a *naked* nucleus; in atoms, the nucleus is surrounded by electrons that modify the resonance frequency of the nucleus. This modification can be determined by the chemical shift interaction [3, 5].

Depending upon the electronic structure of an atom or molecule, a material is called *paramagnetic* if there is an unpaired electron in the molecule, or *diamagnetic* if all electrons are paired. Paramagnetic materials are attracted by an external magnetic field, and their internally induced magnetic fields are parallel with the externally applied magnetic field, which adds to the external field. In contrast, diamagnetic materials are repelled by an external magnetic field, and their internally induced magnetic fields are antiparallel with the externally applied magnetic field, which opposes the external field. Since nearly all water-rich biological tissues are diamagnetic, *chemical shift is discussed in the rest of this book with the assumption of diamagnetic materials*. Diamagnetic materials have a negative bulk susceptibility χ and their chemical shifts vary with the number of electrons in the atoms (cf. Chapter 15.4 for additional discussion on magnetic susceptibility).

Since the internally induced magnetic field $\boldsymbol{B}_i$ of diamagnetic materials is antiparallel with the externally applied magnetic field $\boldsymbol{B}_0$, one can visualize the electronic orbital motion ω_i to be in the opposite sense to the proton precession described by ω_0. Since the circulation involves a motion of charge, there will be an associated induced magnetic moment $\boldsymbol{\mu}_i$, which points antiparallelly with the magnetic moment of the nuclear spin $\boldsymbol{\mu}$. Hence, instead of the straightforward equation ($\omega_0 = \gamma\boldsymbol{B}_0$), the local magnetic field $\boldsymbol{B}_{loc}$ becomes different from the applied magnetic field $\boldsymbol{B}_0$, as

$$\boldsymbol{B}_{loc} = \boldsymbol{B}_0(1 - \sigma) \qquad (4.6)$$

and

$$\omega_{loc} = \gamma\boldsymbol{B}_0(1 - \sigma), \qquad (4.7)$$

where σ is a dimensionless number called the shielding constant and often expressed in parts-per-million (1 ppm = 1×10^{-6}). Note that the shielding constant σ is a molecular parameter that does not depend on the magnetic field.

The chemical shift effect is characteristic of the local chemical environment of the molecule, which makes NMR a valuable tool in structural chemistry. This modification to the magnetic field $\boldsymbol{B}_0$ is equivalent to the reduction of the Zeeman energy for diamagnetic materials (Figure 4.3). *Please pay attention to the peculiarity of the axis labeling in NMR spectra.* By convention, a *reduced* Zeeman energy leads to a peak shift towards the *right* side of the spectrum plot, towards a *lower* frequency. Chapter 8 will have more discussion on the origin of this convention.

Since the disposition of electrons in a molecule is related to the orientation of chemical bonds and therefore is anisotropic, the shielding depends on the orientation of the molecule in the magnetic field. Consequently, the shielding constant should be in general expressed by a tensor. The treatment of the chemical shift tensor follows a similar approach as in the dipolar interaction in Section 4.1. For solids, solid-like specimens, and liquid crystals, the Hamiltonian for the chemical shift interaction can be written as

$$\mathcal{H}_{CS} = \hbar[-\sigma_i \omega_0 I_z - \frac{1}{2}(3\cos^2\beta - 1)(\sigma_{zz} - \sigma_i)\omega_0 I_z], \tag{4.8}$$

which has two terms, the first called the secular term ($\sigma_i \omega_0 I_z$) and the second the anisotropic term. In the equation, β is the polar angle between the polarizing field direction and the principal axis system of the chemical shift tensor, and σ_i is the isotropic component of the chemical shift.

For liquids, the rapid molecular tumbling causes the geometric factor ($3\cos^2\beta - 1$) to vanish, an effect known as "isotropic averaging" (pages 162–163 of Hennel and Klinowski contain a

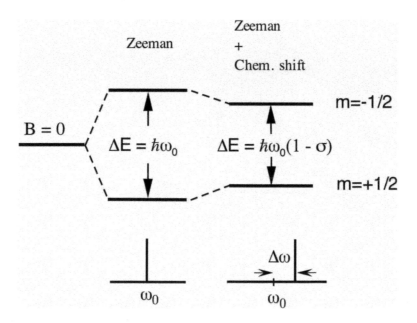

Figure 4.3 The introduction of the chemical shift interaction in diamagnetic materials leads to the reduction of the Zeeman energy, which causes a shift of the resonant peak to a lower frequency (i.e., towards the right side of the ω_0 axis).

proof of this statement [6]). Hence, the anisotropic term in Eq. (4.8) disappears in liquids, which simplifies the chemical shift Hamiltonian for liquids to

$$\mathcal{H}_{cs} = -\sigma_i \omega_0 \hbar I_z. \tag{4.9}$$

Comparing this expression with the equation for the Zeeman Hamiltonian Eq. (3.5), we note that Eq. (4.9) is simply $\sigma_i \mathcal{H}_Z$.

Since the difference between $\boldsymbol{B}_0$ and $\boldsymbol{B}_0(1 - \sigma)$ is small and the value of a magnetic field is difficult to measure accurately, the chemical shift is commonly measured in experiments by comparing the resonance frequency of an unknown specimen with that of a known reference. The usual reference for ^{1}H, ^{13}C, and ^{29}Si is tetramethylsilane (TMS), which has the molecular formula of $Si(CH_3)_4$ (Figure 4.4). Since the 12 hydrogen atoms in TMS are chemically equivalent and both Si and C are spin zero (assuming the carbon is ^{12}C and the silicon is ^{28}Si), TMS gives a single narrow line occurring at a lower frequency than that of most other proton resonances. Therefore, we commonly use the δ *scale* with the unit of parts-per-million (ppm) in proton NMR,

$$\delta = 10^6(f_{sample}/f_{TMS} - 1), \tag{4.10}$$

where f is the temporal frequency ($f = \omega/2\pi$, where ω is the angular frequency), and the 10^6 factor is used to convert the frequency ratio into ppm.

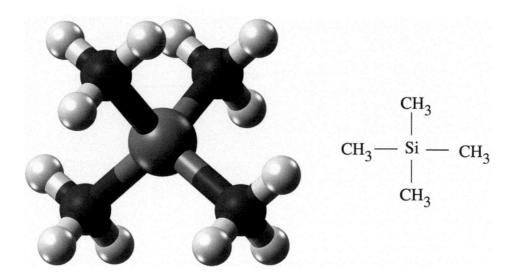

Figure 4.4 TMS is commonly used as the chemical shift reference in NMR spectroscopy. The main drawing is the 3D ball-and-stick model of TMS (the dark green object at the center is silicon-28; the black objects are carbon-12, and the silver objects are hydrogen-1); the small insert is the common 2D schematic of the same molecule. Source: The main drawing is open source - accessed 27 February 2021 from https://commons.wikimedia.org/wiki/File:Tetramethylsilane_molecule_ball_from_xtal.png.

The advantages of using the δ scale can be illustrated from the following example. If a ^{1}H spectrum measured on a 60-MHz spectrometer (in a magnet of 1.4 Tesla) has two peaks with a separation of 60 Hz, that separation is 1 ppm. Now compare this 60-MHz spectrum with a spectrum measured on a 300-MHz spectrometer (in a magnet of 7 Tesla). Without using the δ scale, we would have to know both the 60-MHz B_0 and 60-Hz separation. Using the δ scale, a peak separation of 300 Hz can be determined instantly at the new resonant frequency, since a 300-Hz separation equals to 1 ppm at 300-MHz B_0. Therefore δ is a molecular parameter that depends only on the sample conditions (solvent, concentration, temperature, etc.) and not on the spectrometer's resonant frequency.

In practical NMR spectroscopy, Eq. (4.10) is often written into a more convenient form, as the following. Given two resonant peaks, a unknown peak $\omega_i = \gamma B_0 (1 - \sigma_i)$ and a reference peak $\omega_{ref} = \gamma B_0 (1 - \sigma_{ref})$, we have $\omega_i - \omega_{ref} = \gamma B_0 (\sigma_{ref} - \sigma_i)$, which can be organized into

$$10^{-6}\delta = \frac{\omega_i - \omega_{ref}}{\omega_{ref}} = (\sigma_{ref} - \sigma_i), \tag{4.11}$$

which is valid because σ_i is much less than 1. Equations (4.10) and (4.11) are the fundamental equations in NMR spectroscopy, which will be discussed and used in detail in Chapters 8 and 10.

4.3   SCALAR INTERACTION

The scalar interaction (or the spin-spin interaction, or indirect interaction, or J coupling) is an indirect interaction between nuclei that arises via the mediation of electrons in the molecular orbital [1, 6]. The nuclear spin causes a slight electron polarization, which can be transmitted to neighboring nuclei (Figure 4.5). The transmission of this polarization will cause the resonant peak to split into multiple lines, due to its sense of its neighbors. Since this interaction requires a molecular orbit, it acts only through the medium of covalent bonds, that is, within the same molecule. Hence it is an intramolecular interaction, which is in contrast to the dipolar interaction that acts through space. The scalar interaction is also a much weaker interaction than the dipolar interaction.

Assuming I and S are the two interacting spins, the Hamiltonian of the scalar interaction can be written as

$$\mathcal{H}_S = 2\pi\hbar J I \cdot S \tag{4.12}$$

where J is the coupling constant, which for protons has the typical values from 0 to 20 Hz; and 2π is used to convert J to the angular frequency unit. Since J is a scalar, J coupling is rotationally invariant. Since Eq. (4.12) lacks the B_0 term, $\mathcal{H}_S$ is independent of the external magnetic field. Hence J coupling is always expressed in Hz, never in ppm.

The consequence of the scalar interaction is to split the resonant line from a single line to multiple lines. In general, the resonance of spin I is split into $(2S + 1)$ lines and that of spin S is split into $(2I + 1)$ lines. Considering the fact that $\mathcal{H}_{CS}$ is dependent on the external magnetic field [Eq. (4.9)] while $\mathcal{H}_S$ is independent of the external magnetic field [Eq. (4.12)], several general conclusions can be summed up from the equations in the two interactions, as shown in Figure 4.6:

(a) When the chemical shift difference δ is equal to zero, an NMR spectrum does not show the scalar coupling, because the two spins I and S are equivalent in their environment (e.g., two protons in a water molecule). Hence there is only a single line (Figure 4.6a).

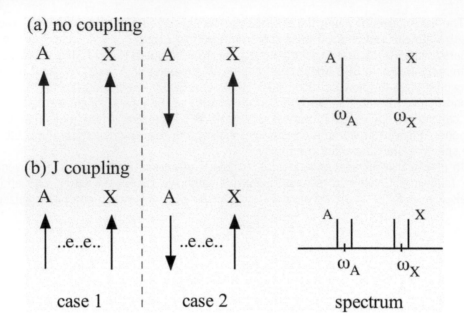

Figure 4.5 The scalar interaction arises via the mediation of electrons in a molecular orbit. (a) Without the mediation of the electrons, spins A and X do not sense the polarity of the other (i.e., one spin could not sense whether the other is spin up or spin down). Hence A and X each has one resonant line. (b) In the presence of electron mediation in a molecular orbital (indicated by ".e."), spin A or X can sense the polarity of the other due to a slight electron polarization, which causes the split of the resonance peaks. Case 1, where both A and X are in parallel with each other, has a higher energy, while case 2 has a lower energy because A and X are antiparallel with each other.

(b) When chemical shift difference δ between the two spins is not zero but the coupling constant J is equal to zero, there will be no scalar coupling. There are two lines separated by its chemical shift difference, where each line represents one nuclear spin (Figure 4.6b).

(c) When $|\omega_I - \omega_S| \gg 2\pi|J|$, the interaction is called weak scalar coupling (the two spins I and S are far away, but each can sense the spin state of the other). The split lines have *equal* intensity (Figure 4.6c).

(d) When $|\omega_I - \omega_S|$ is the same size or even smaller than $2\pi|J|$, the situation is called strong scalar coupling. The split lines have *unequal* intensities, where in the simplest cases the *inside* peaks of the line group have *higher* intensities than the outside peaks of the same line group (Figure 4.6d).

Examine closely the scalar coupling for a spin-1/2 system with two spins I and S. Given

$$I \cdot S = I_z S_z + I_x S_x + I_y S_y = I_z S_z + \frac{1}{2}(I_+ S_- + I_- S_+), \qquad (4.13)$$

where the subscripted $+$ and $-$ indicate the raising and lowering operators (Appendix A2.4). The spin Hamiltonian for this two-spin system can be written as

$$\mathcal{H}_s = \hbar\omega_I I_z + \hbar\omega_S S_z + \pi\hbar J(I_+ S_- + I_- S_+ + 2I_z S_z). \qquad (4.14)$$

When $|\alpha\alpha\rangle$, $|\alpha\beta\rangle$, $|\beta\alpha\rangle$, and $|\beta\beta\rangle$ (or $|++\rangle$, $|+-\rangle$, $|-+\rangle$, and $|--\rangle$) are used as the four basis kets for this two-spin system (in which the first and second letters in the ket represent I_z and S_z), the four eigenvalues for the associated eigenlevels are

(a)

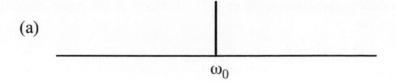

ω_0

(b)

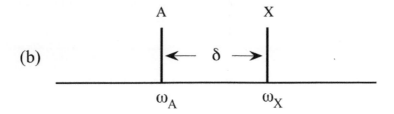

A   X

$\leftarrow \delta \rightarrow$

ω_A ω_X

(c)

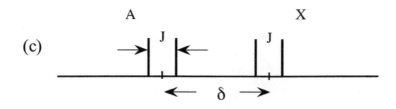

A   X

$\rightarrow$ J $\leftarrow$ J

$\leftarrow \delta \rightarrow$

(d)

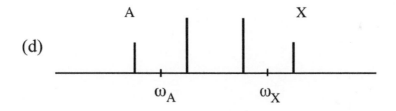

A   X

ω_A ω_X

Figure 4.6 The influences of the chemical shift and scalar interactions on the patterns of the resonance lines for a two-spin system (A, X). (a) The chemical shift difference δ is equal to zero. (b) The chemical shift difference δ is not zero but the coupling constant J is zero. (c) The weak scalar coupling case, when $|\omega_I - \omega_S| \gg 2\pi|J|$. (d) The strong scalar coupling case, when $|\omega_I - \omega_S| \sim 2\pi|J|$ or $|\omega_I - \omega_S| \ll 2\pi|J|$.

$$|\alpha\alpha> = (\hbar/2)(\omega_I + \omega_s + \pi J), \tag{4.15a}$$

$$|\alpha\beta> = (\hbar/2)(\omega_I - \omega_s - \pi J), \tag{4.15b}$$

$$|\beta\alpha> = (\hbar/2)(-\omega_I + \omega_s - \pi J), \tag{4.15c}$$

$$|\beta\beta> = (\hbar/2)(-\omega_I - \omega_s + \pi J), \tag{4.15d}$$

which are shown graphically in Figure 4.7. The matrix of the scalar coupling Hamiltonian can then be filled as

$$
\mathcal{H}_S = \frac{\hbar}{2} \begin{bmatrix} (\omega_I + \omega_S) + \pi J & 0 & 0 & 0 \\ 0 & (\omega_I - \omega_S) - \pi J & 2\pi J & 0 \\ 0 & 2\pi J & -(\omega_I - \omega_S) - \pi J & 0 \\ 0 & 0 & 0 & -(\omega_I + \omega_S) + \pi J \end{bmatrix}. \tag{4.16}
$$

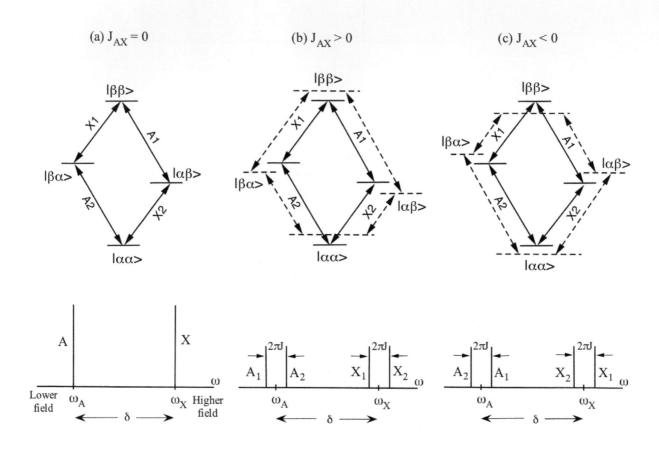

Figure 4.7 The effect of the *J*-coupling constant on the Zeeman energy levels for a two-spin system (labeled as A and X), when (a) *J* = 0 (two non-coupling spins), (b) *J* > 0 (when the antiparallel alignments (αβ and βα) reduce the energy levels), and (c) *J* < 0 (when the parallel alignments (ββ and αα) reduce the energy levels). The solid horizontal lines are the Zeeman levels without *J* coupling; the dashed horizontal lines are the modified Zeeman levels due to the *J* coupling. Since a reduced Zeeman energy difference leads to a peak shift to the right (to a lower frequency) while an increased Zeeman energy difference leads to a peak shift to the left (to a higher frequency), A₁ line in (b) shifts to the left of ω_A and A₂ line in (b) shifts to the right of ω_A. Note that for homonuclear systems (i.e., two spins are of the same species, such as proton and proton), the energies of the αβ and βα states are similar but very different from the other two states. The reductions and increments of the Zeeman energy levels under the influence of *J* in the schematics are grossly exaggerated for illustration (i.e., not drawn in proportion) – the Zeeman levels correspond to the frequency in hundreds of MHz while the *J* coupling changes the Zeeman levels by merely several Hz.

If the coupling is weak, i.e., $|\omega_I - \omega_S| \gg 2\pi|J|$, the two non-diagonal terms $2\pi J$ can be set to zero (the weak coupling limit). The four diagonal terms are then approximately equal to the eigenvalues. There are four allowed transitions at $\Delta m = \pm 1$, that is, there will be two doublets with a splitting $\Delta f = J$. Since the probabilities of the transitions are equal, the four spectral lines have equal intensity (Figure 4.6c). Note that discarding the non-diagonal terms is equivalent to ignoring the I_+S_- and I_-S_+ terms in Eq. (4.14), which leads to a truncated Hamiltonian.

Since the weak scalar coupling represents a major simplification in interpreting NMR spectra, it is possible to satisfy the condition of the weak coupling experimentally by increasing the external field (i.e., increase ω_0). That explains in part the race to higher B_0 fields in modern NMR spectroscopy. We will have more discussion of the peak splitting in Chapter 8.

If $|\omega_I - \omega_S|$ is of the same order of magnitude as $2\pi|J|$, we cannot make any approximation in $\mathcal{H}_S$, that is, the $2\pi|J|$ terms in Eq. (4.16) [or the I_+S_- and I_-S_+ terms in Eq. (4.14)] cannot be ignored, and the analysis is much more complicated. The spectrum still contains four lines but the intensities of the lines are unequal (Figure 4.6d) [3].

Note that the scalar coupling constant J can be either positive or negative, depending on the energy reduction of one spin from the parallelism of the coupling spin. J is negative if a parallel alignment of the two spins lowers the energy and positive for an antiparallel alignment between the two coupling spins. Recall that a reduced Zeeman energy leads to a peak shift to the right of the NMR spectrum (Figure 4.3), towards a lower frequency, while an increased Zeeman energy leads to a peak shift to the left, towards a higher frequency. Figure 4.7 shows the changes of the Zeeman energy levels when $J = 0$, $J > 0$, and $J < 0$.

4.4 QUADRUPOLE INTERACTION

When spin $I > 1/2$, a simple vector description of NMR is no longer possible and quantum mechanical analysis must be used [3, 7]. This is because the spin states at $I = 1$ or higher will no longer have two basic states (spin up and spin down). For spin $= 1$ that has three energy levels (Figure 2.2b), its Hamiltonian will contain electric quadrupole terms that are quadratic in the spin operations. [A dipole can be visualized as two point-charges (one positive and one negative) separated by a small distance along a line; a quadrupole can be visualized as four point-charges (two positive and two negative) located at the four corners of a square.] Quadrupole terms arise from an interaction between the nuclear quadrupole moment and the surrounding electric field gradient (dE/dr), due to their non-uniform charge distributions. The quadrupole Hamiltonian for a single spin I is generally written in the form of $I \cdot Q \cdot I$, where Q is the nuclear quadrupole moment tensor. We will not go into the details of the quadrupole interaction, except to say that it is useful for determining the electrostatic potential in solids and molecules, which can be used to study electron distributions, intermolecular and intramolecular binding, molecular motions, phase transitions, and other properties of solids and molecules [3].

4.5 SUMMARY OF NUCLEAR INTERACTIONS

Figure 4.8 summarizes graphically the consequences of nuclear interactions for NMR resonance frequencies. For molecules having large and complicated scalar interactions, the scalar line patterns can overlap. When this happens, one way to simplify the pattern recognition is to make the Zeeman difference larger, which implies the use of a higher magnetic field in NMR spectroscopy.

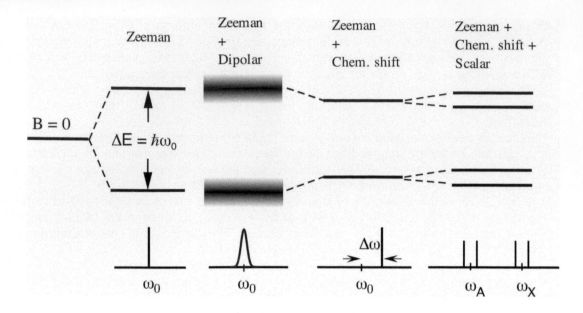

Figure 4.8 A schematic summary of the modification to the Zeeman energy levels under various nuclear interactions, and the spectra under each set of the interactions.

The following two tables (Tables 4.1 and 4.2) summarize the typical magnitudes (ranges) and orientational properties of nuclear interactions for three selected spin-1/2 nuclei at B_0 = 2 Tesla [3, 6]. The higher the atomic number of the nucleus, the wider the range of possible chemical shift interactions, because of the large number of electrons surrounding the nucleus. We note also that the larger γ (as in proton) is, the larger the dipolar interaction will be.

Table 4.1   Magnitudes of nuclear interactions.

Nucleus	γ (10^7 rad T^{-1} S^{-1})	Zeeman (kHz)	Dipolar (kHz)	Chemical shift (kHz)	Scalar (kHz)
^{1}H	26.7520	85,200	90	2	0.020
^{13}C	6.7283	21,400	20	10	0.100
^{31}P	10.841	34,500	35	57	0.150

Table 4.2   Orientational properties of nuclear interactions in solutions and solids.

Sample	Zeeman	Dipolar	Chemical shift	Scalar
Solutions		zero	Isotropic	Isotropic
Solids		$3\cos^2\theta - 1$	Anisotropic	Often too small to be seen

References

1. Callaghan PT. *Principles of Nuclear Magnetic Resonance Microscopy*. Oxford: Oxford University Press; 1991.
2. Slichter CP. *Principles of Magnetic Resonance*. 3rd ed. Berlin: Springer-Verlag; 1992.
3. Harris RK. *Nuclear Magnetic Resonance Spectroscopy - A Physicochemical View*. Essex: Longman Scientific & Technical; 1983.
4. Xia Y. Relaxation Anisotropy in Cartilage by NMR Mcroscopy (µMRI) at 14 µm Resolution. *Magn Reson Med.* 1998;39(6):941–9.
5. Callaghan PT. Pulsed Field Gradient Nuclear Magnetic Resonance as a Probe of Liquid State Molecular Organization. *Austr J Phys.* 1984;37:359–87.
6. Hennel JW, Klinowski J. *Fundamentals of Nuclear Magnetic Resonance*. Essex: Longman Scientific & Technical; 1993.
7. Poole CPJ, Farach HA. *Theory of Magnetic Resonance*. 2nd ed. New York: John Wiley & Sons; 1987.

Part II

Essential Concepts in NMR Instrumentation

5

Instrumentation

This chapter discusses the instrumentation that facilitates the experiments in (high-resolution) NMR spectroscopy. A complete NMR system contains four essential parts: a magnet that provides the B_0 field, a computer that provides the human machine interface, a spectrometer that includes all hardware and software, and a probe that hosts an rf coil and the specimen (it is possible to use the same coil to apply a B_1 field and to pick up the NMR signal). Figure 5.1 shows a conceptual diagram for such an NMR system. (Note that the essential components in this chapter are part of any NMR machine, including MRI. The chapter focuses on a spectroscopy machine; additional hardware for MRI will be discussed in Chapter 13.)

5.1 MAGNETS

The heart of any NMR spectrometer is its magnet. Since all NMR specimens need a very uniform magnetic field, the design and quality of the magnet is critically important and needs to be tailored to the particular NMR or MRI experiment. In general, there are three fundamental types of magnets, namely an electromagnet, a superconducting magnet, and a permanent magnet, which provide different orientations of the B_0 field (Figure 1.2).

5.1.1 Electromagnet

An electromagnet was used in the early days of NMR (both spectroscopy and imaging) but has become obsolete in modern NMR systems because of its poor stability and low field strength. The electromagnet gets its magnetic field by passing an electric current continuously through its coil, which is commonly wound around an iron core. The Biot–Savart law in electromagnetism can be used to determine the magnetic field generated by an electric current I in a loop or coil,

$$\boldsymbol{B(r)} = \frac{\mu_0}{4\pi} \int \frac{I d\boldsymbol{l} \times \hat{r}'}{|\,r'\,|^2},\tag{5.1}$$

where μ_0 is the permeability of free space, $d\boldsymbol{l}$ is a vector wire element along the path (the wire), and $\boldsymbol{r}'$ is the displacement vector with $\hat{r}'$ as its unit vector.

The highest stable magnetic field of an electromagnet is around 2 Tesla, beyond which the iron core materials become saturated and any further increase of electric current won't increase the field significantly. In addition, since the current runs through the resistive coil continuously, water cooling is essential to remove the heat generated by the coils.

Essential Concepts in MRI: Physics, Instrumentation, Spectroscopy, and Imaging, First Edition. Yang Xia.
© 2022 John Wiley & Sons Ltd. Published 2022 by John Wiley & Sons Ltd.

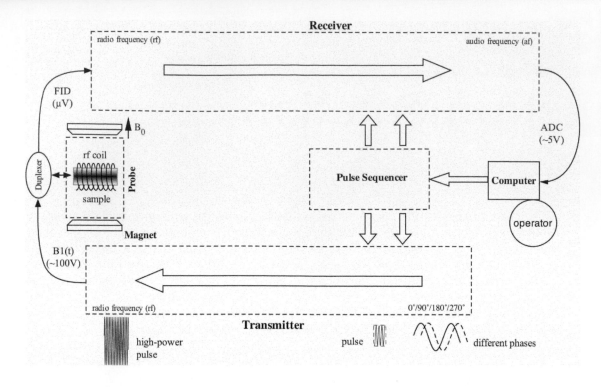

Figure 5.1 A block diagram for an NMR spectrometer, which has six major components: two amplifier channels, a transmitter channel to convert small but precise oscillations to high-powered rf pulses, and a receiver channel to amplify the small FID signal; two computers, one that interfaces with the operator and one that interfaces with the electronics to apply the pulse sequences and to control the data acquisition; the magnet; and the probe that hosts an rf coil inside.

In the conventional orientation of the electromagnet (Figure 1.2c), the B_0 direction in an electromagnet is horizontal, which could be advantageous in some imaging experiments (since one can use simple solenoid coils as the rf coil with a convenient access to the center of the magnetic field, cf. Section 5.2) [1, 2]. The main disadvantage of an electromagnet, apart from its limited field strength, is the instability of its field, which is subject to the fluctuations of the current supply as well as the temperature. The short-term (transient) stabilization of the B_0 field can be provided by a flux stabilizer, and the long-term field stabilization can be provided by a deuterium NMR lock system in the probe. The lock system is actually a small continuous-wave NMR spectrometer with a drift detector and feed back-loop.

5.1.2   Superconducting Magnet

The superconducting magnet is the magnet of choice in a modern NMR instrument. There are two different designs for a superconducting magnet: a vertical bore (Figure 1.2a) and a horizontal bore (Figure 1.2d). The vertical bore magnet has a small bore size and is commonly used for high-resolution NMR spectroscopy and NMR microscopy, while the horizontal bore magnet has a large bore size and is designed specifically for human imaging in the clinic because of the easy access to the center of the field. In addition to superconducting magnets for whole-body human MRI, there are smaller-sized horizontal superconducting magnets for animal research. Chapter 12.10 discusses the imaging resolution for each of these types of magnets.

A superconducting magnet consists of multiple concentric cryogenic dewars, with the innermost one having a superconducting coil sitting in a liquid helium bath (at a temperature of 4 K or −269 °C). There is a liquid nitrogen dewar (at a temperature of 77 K or −196 °C) surrounding the helium dewar. (A dewar is a double-walled flask of metal or silvered glass with a vacuum between the walls, which insulates the content of the inner container from the outside temperature. It can be used to store either hot or cold liquids.) There are also several radiation shields (e.g., aluminum layer) between the helium dewar and nitrogen dewar, between the helium dewar and the room temperature bore, and between the nitrogen dewar and the metal shell of the magnet, which lowers the liquid cryogen boil-off rate. Figure 5.2 shows a cut-open high-field superconducting magnet.

In the superconducting coil, a large electric current (hundreds of amperes) circulates continuously. Since the coil is kept in the superconducting condition, the coil has *zero* resistance and hence does not cause any loss to the electric current. Once the desired current is established in

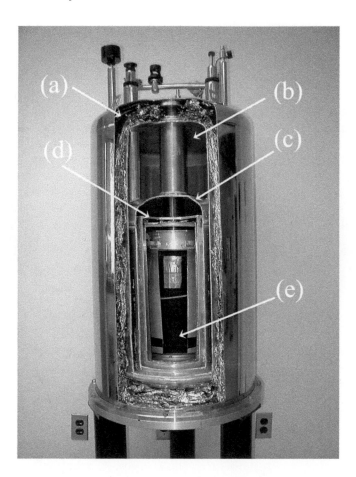

Figure 5.2   A cut-open vertical-bore superconducting magnet (original manufacturer: Oxford NMR Instruments UK; central field 6.34 Tesla, which would be 270 MHz for ^{1}H; room-temperature bore diameter 54 mm; electric current at field 34.735 A). (a) An outside vacuum chamber. (b) The liquid nitrogen vessel (at 77 K). (c) A 20 K radiation shield (to lower the liquid helium boil-off rate). (d) The liquid helium vessel (at 4.2 K). (e) The superconducting coil, which contains approximately 12 miles (19 km) of superconducting wire. (Source: Reproduced with permission from JEOL USA, Inc., https://www.jeolusa.com/RESOURCES/Analytical-Instruments/NMR-Magnet-Destruction; accessed on 12 December 2020).

the coil, this current should in principle circulate through the coil forever. Furthermore, because of the zero resistance of the coil, there will be no heat generated. Since there is no need to have a power supply connected to the superconducting magnet, its field can be very stable. The field strength of modern magnets can reach as high as 10–20 Tesla, which is critical for high-resolution NMR spectroscopy. The major disadvantages of the superconducting system include the high capital cost and high running cost (liquid helium and liquid nitrogen are needed to top the cryogen dewars periodically).

5.1.3 Permanent Magnet

A permanent magnet can be made by magnetizing certain materials that are ferromagnetic. A permanent magnet (such as a horseshoe magnet) was never seriously considered for an NMR spectrometer, not even in the early days of NMR. When compared to an electromagnet, a permanent magnet is low in magnetic field strength and unstable because it is *very* sensitive to the environmental temperature.

In the recent years, the use of a special type of permanent magnet, the Halbach arrangement (Figure 5.3), has become a hot topic in NMR and MRI. The Halbach arrangement capitalizes on the special magnetic properties of the oriented rare-earth material, which can be precisely machined and specifically treated so that its magnetization is very strong and permanent along a preferred axis [3]. One can design various arrangements using small pieces of permanent magnets (each piece has a fixed magnetization that generates a dipolar field, as in Figure 4.1a, and is oriented at a particular angle in the arrangement), so that the magnetic field of the arrangement is augmented on one side or inside of the magnets [4]. Examples include 1-dimensional (1D) Halbach linear arrays (which have a stronger magnetic field on one side of the bar than the other side), 2D Halbach cylinders (which confine a magnetic field within the cylinder), and 3D Halbach spheres (which confine a magnetic field in the center of

(a) Halbach bar

(b) Halbach cylinder

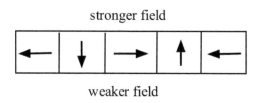

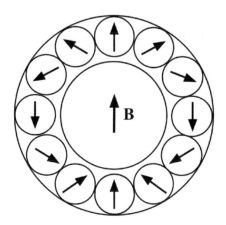

Figure 5.3   Halbach magnet configurations. (a) A 1D Halbach bar has many individual magnet blocks – each block has the magnetic field in the direction of its arrow. The magnetic field of the bar is stronger on one side. (b) A 2D Halbach cylinder where the field is limited to the inside of the cylinder. The circular shapes are individual magnet rods – each rod has the magnetic field in the direction of its arrow. The central space of the cylinder is empty where the arrow represents the direction of the magnetic field.

the arrangement). The feature of these magnets having a strong field on one side or inside of the magnet itself is a major advantage for any practical NMR and MRI instrument. The motivation of these Halbach magnets in NMR and MRI is to reduce the major capital cost of the NMR and MRI hardware (its magnet cost), to reduce the major running cost of the NMR and MRI system (its cryogen cost), and to reduce the weight of the instrument (so it becomes potentially mobile). Although the field strength of a Halbach magnet is still weak compared to a high-field superconducting magnet, there is an increasing number of applications of Halbach magnets in NMR and MRI. For example, several commercial companies currently are able to offer desktop cryogen-free NMR instruments with a ^{1}H resonance frequency up to 80 MHz [5].

5.1.4 Ways to Make the Magnetic Field Uniform

Despite advanced designs and optimized manufacturing, none of the magnets described above can provide a magnetic field that is sufficiently homogeneous for high-resolution NMR spectroscopy. In addition, the sample itself can distort the magnetic field. Two additional techniques are used in NMR experiments to improve the field homogeneity after each new sample is loaded into the magnet.

The first technique is to use a set of correction coils to generate non-uniform magnetic fields that can compensate the non-uniformity of the original field. This set of correction coils is called the shim coils [6]. Each of the shim coils produces independently a well-controlled field profile (Figure 5.4), such as the zero-order deviation of the main field (to increase or decrease the main field slightly), the first-order deviation of the main field (the field gradients in different directions), and higher-order deviations of the main field (different curvatures), as shown in Table 5.1.

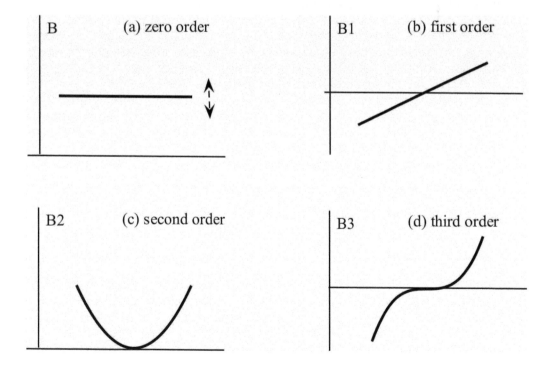

Figure 5.4 Shimming patterns that make the magnetic field uniform. (a) The zero-order adjustment of the main field, (b) the first-order adjustment of the main field (i.e., the field gradients in different directions), (c) and (d) the second- and third-order adjustments of the main field (different curvatures).

Table 5.1 Shimming of magnetic field.

Order	Shim name	Function	Comment
0	Z_0	ΔB_0	main field
1	Z_1	$\partial B_z/\partial z$	z gradient
1	X	$\partial B_z/\partial x$	x gradient
1	Y	$\partial B_z/\partial y$	y gradient
2	Z_2	$2\partial^2 B_z/\partial z^2 - (\partial^2 B_z/\partial x^2 + \partial^2 B_z/\partial x^2$	Curvature
2	ZX	$\partial^2 B_z/\partial z\partial x$	Curvature
2	ZY	$\partial^2 B_z/\partial z\partial y$	Curvature
2	XY	$\partial^2 B_z/\partial x\partial y$	Curvature
2	$X_2 - Y_2$	$\partial^2 B_z/\partial x^2 - \partial^2 B_z/\partial y^2$	Curvature
3	...		

The process of achieving an optimal field homogeneity is called shimming, which is done by adjusting the currents to different shim coils individually in several iterations to achieve a maximal signal with a minimal peak width. It used to be a time-consuming task when done manually. Modern spectrometers all have automatic shimming programs.

The second technique to improve the field homogeneity in NMR spectroscopy is to rotate the sample tube along its axis so that the field inhomogeneities (especially the inhomogeneities for the x and y components) can be averaged out, a process called sample spinning. A higher spinning rate results in better averaging. However, there are two consequences of sample spinning. One is the formation of a vortex on the top surface of a liquid sample that is being rotated, and the other is the formation of spinning sidebands in the NMR spectrum. Spinning sidebands are some small peaks that appear at multiples of the spinning frequency on both sides of the main spectral peak (cf. Chapter 8), which complicates the interpretation of the spectrum when J coupling is present. The higher the sample spinning speed and the bigger the sample tube, the easier the formation of a vortex and the worse effect on the spectrum.

5.2   RADIO-FREQUENCY COIL, ITS RESONANT CIRCUITRY, AND THE PROBE

At least one coil operating at a radio frequency (rf) is needed to provide two distinct functions: to deliver the time-dependent transverse field $\boldsymbol{B}_1(t)$ to the sample via the transmitter channel of the spectrometer, and to detect the induced FID signal via the receiver channel of the spectrometer. In NMR spectroscopy and small-size imaging systems, in order to minimize the relaxation loss during the time of pulsing, the duration of the rf pulses is often reduced to a minimum, down to the order of several microseconds. (This need is greater for multi-pulse sequences or for semi-solid samples.) Hence the peak voltage of the rf pulse needs to be very high, up to hundreds of volts. At the same time, the receiving mode of the rf coil needs to detect the induced FID signal, whose magnitude is often on the order of several microvolts or less. Although these two functions are very different in nature, one can use a single rf coil for both receiving and transmitting, via some clever switching of the resonant circuit. *(To appreciate the function of the circuit, please review essential concepts in rf electronics in Appendix A3.2.)*

5.2.1   Rf Coil Configurations

There are many configurations for an rf coil. For volume coils (i.e., the sample is inserted into the middle of the coil), two traditional and easy-to-make configurations are the solenoid coil (Figure 5.5a) and the saddle coil (Figure 5.5b). Both the saddle coil in a cylindrical-bore superconducting magnet (Figure 1.2a, d) and the solenoid coil in an electromagnet (Figure 1.2b, c)

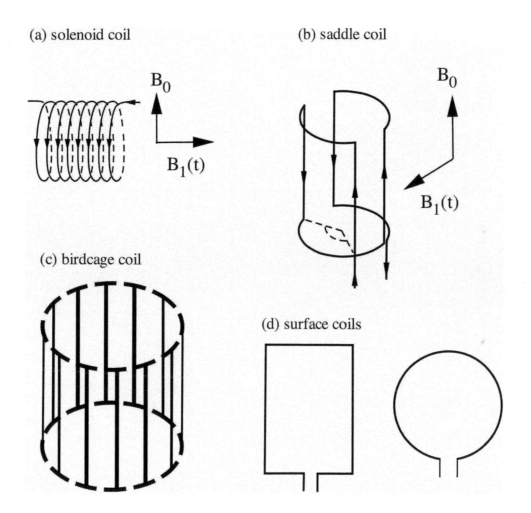

(a) solenoid coil

(b) saddle coil

(c) birdcage coil

(d) surface coils

Figure 5.5   Basic configurations of rf coils, where the field direction can be determined with the aid of the right-hand rule in electromagnetism. (a) A solenoid coil, which can provide a uniform magnetic field inside the central portion of the coil. This configuration is convenient for the coil used in an electromagnet. (b) A saddle coil and (c) a birdcage coil (where the gaps between the end segments are connected with capacitors), both of which can generate a directional B_1 field in the transverse plane of the coils. These configurations are useful in a superconducting magnet. (d) Surface coils, which are semi-flexible and used to detect the signal from the part of the specimen close to the coil.

can provide easy access to the center of the coil [2], while the employment of a solenoid coil in cylindrical-bore superconducting magnets requires a side entry for the sample. In terms of sensitivity, however, a solenoid coil is about three times better than a saddle coil [7]. For good performance at high magnetic field, other coil configurations can be used, such as different types of rf resonators (Figure 5.5c) [8].

Another type of rf coils is the surface coil (Figure 5.5d), which is used to detect the signal from a small region that is close to the surface on a large object (e.g., a human body). Surface coils (often semi-flexible to bend) improve SNR in two ways, by getting the tissue of interest closer to the coil and by receiving less sample-generated noise. An optimal SNR can be achieved when the radius of the coil equals the depth/height of the sample volume interested.

5.2.2   Rf Resonant Circuit and the Switching Between Transmitter and Receiver

The simplest version of an rf resonant circuit (also called an LC resonant circuit, a tank circuit) consists of two variable capacitors and an rf coil (Figure 5.6). One of the capacitors is the tuning capacitor, which is to tune the resonant circuit to resonate at ω_0; the other capacitor is the

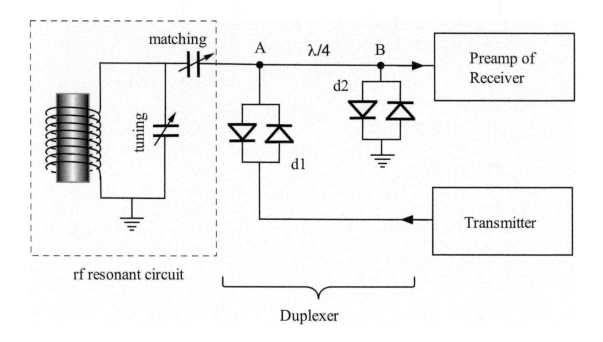

Figure 5.6   The rf coil and two capacitors form the basic rf resonant circuitry. The coil is switched between the transmitter mode and receiver mode by the duplexer, which has a quarter-wave cable, located between points A and B, and two sets of crossed diodes. In the transmitting mode, both crossed diodes become conducting at the high voltage. The conducting of the d2 diodes effectively grounds point B. Because of this short termination, Z_{in} of the quarter-wave cable goes to infinity (at point A), which blocks the transmitting energy from the receiver. Hence all energy is delivered to the coil. In the receiving mode, the FID voltage is too small to make the crossed diodes d1 conducting, effectively cutting off the connection to the transmitter. At the same time, the crossed diodes d2 are also not conducting, giving a clear path for the FID signal to be delivered to the preamplifier of the receiver.

matching capacitor, which transforms the impedance of the resonant circuit to 50 ohms. At a pure resistive 50 ohms impedance, all of the rf power (the B_1 field) can be delivered to the coil without any loss by a coaxial cable that typically has a characteristic impedance of 50 ohms.

The rf resonant circuitry is capable of transmitting high-power rf pulses to the rf coil through a cable while detecting the delicate FID signal between pulses using the same cable. In order to prevent any damage to the sensitive amplifiers in the receiver by the leakage of the high-power pulses, it is necessary to turn off the high-power route completely while the signal is being detected. This type of fast switching between the transmitter and receiver can be accomplished by a device called the duplexer. Modern spectrometers use digital electronics to switch. One of the original switching operations was accomplished by utilizing the characteristic features of the quarter-wave cable and the cross diodes, as shown in Figure 5.6.

During the transmission phase, both sets of the cross diodes conduct since the voltage of the rf pulse is much higher than the conducting threshold of the cross diodes (~0.7 V). The conduction of the cross diodes at the receiver side shorts the far end of the quarter-wave cable (point B), thus producing a perfect reflection of the incoming high voltage (cf. Appendix A3.2.3), so it effectively removes the receiver from the circuit. The rf resonant circuit, which is tuned to 50 ohms at ω_0, on the other hand, presents a perfect match to the impedance of the quarter-wave cable, so all of the power will be delivered to the rf coil. In the reception mode, both sets of diodes are not conductive due to the small voltage of the FID signal. Therefore, there is a clear path of the signal to the receiver.

5.2.3 Probe

An important hardware piece of any NMR spectrometer is called the probe. It is effectively a box that is situated between the magnet pole faces of a permanent or electromagnet, or a tube that fits inside the bore of a superconducting magnet. The purpose of the probe is to host the sample, the rf coil, and its tuning circuits, the necessary hardware to adjust the temperature of the sample, and to spin the sample. In a high-resolution probe, a separate rf coil and its tuning circuit could be included in the probe, for decoupling and other purposes. In imaging probes, a set of gradient coils is fitted either into the probe or outside the rf part of the probe. If the probe is used in electromagnets, it also hosts the external lock system used to provide long-term stability for the main field (the external lock is in contrast with the internal lock, where the lock material is added into the sample tube).

Probe design is an art, in which one tries to fit everything – often more than the space permits – into the box or tube and to optimize the performance of the rf and gradient system for a particular type of experiment. Some examples can be found in published papers [2].

5.3   FREQUENCY MANAGEMENT

Figure 5.7 fills out the block diagram in Figure 5.1 with the essential components. Among the numerous hardware and software components that are integrated into a working spectrometer, the frequency management is critically important, due to the need to precisely identify the frequency and phase in all oscillations in NMR experiments. In modern NMR spectrometers all frequencies are generated by one common oscillator, the so-called master clock. This highly precise and stable oscillation source (i.e., a crystal oscillator) is then used by the frequency synthesizer to produce oscillations at different frequencies and with different phase delays. For example, the dashed oscillation in Figure 5.7 (under the "variable phase" in the transmitter channel) is 90° behind the solid oscillation. To change the frequency of the oscillation, one uses the mixer, or phase sensitive detector, which is a circuit that accepts two

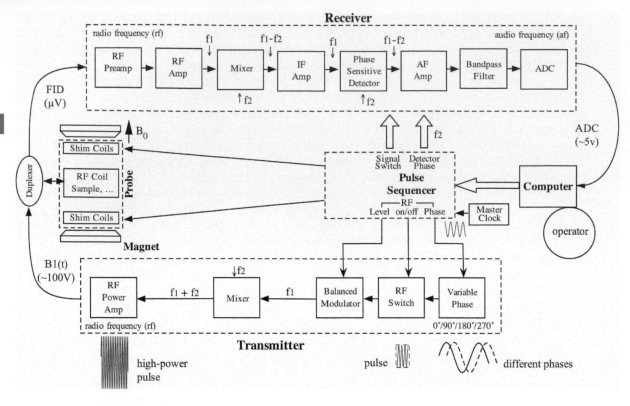

Figure 5.7 An NMR spectrometer, where all six major components are filled with some essential parts. The frequency of a signal can be changed by the use of a mixer (phase sensitive detector), to be the sum or difference of the two input frequencies. In the transmitter, the frequency of B_1 is converted to the radio frequency. In the receiver, the FID at radio frequency is converted to an intermediate frequency (if). After further amplification, this if signal is then fed into another mixer, producing an audio frequency signal. One can do the analog-to-digital conversion either at the end or in the middle of the receiver amplification.

signals (f_1, f_2) and forms an output signal that has the sum or difference of the two frequencies $(f_1 + f_2, f_1 - f_2)$. Various designs are available in modern electronic circuitry.

5.4 TRANSMITTER

The transmitter in an NMR spectrometer consists of a series of amplifiers. Briefly, the low-voltage continuous oscillation from the master clock is amplified in magnitude and chopped into rf pulses (i.e., with precise durations) by an rf switch. The rf pulses can be envelope modulated by a modulator to the desired pulse form (such as a sinc pulse used for the slice selection in MRI; see Chapter 13), frequency adjusted by a number of mixers to produce the rf signal at the desired frequency, and finally amplified by a broad-band rf power amplifier. (The broad-band amplifier is required because the signal it processed is of the form of short pulses.) The output pulses of the transmitter are sent to the rf coil that surrounds the sample via the rf resonant circuit. The final stage of the transmitter can have a power of several hundreds or even thousands of watts.

One fundamental parameter in the electronic amplifier is its rise time, which measures the response of the amplifier to fast input signals. In electronic circuitry, the rise time is the time that the circuit takes to rise from 10% to 90% of the output step function. Figure 5.8 shows the same square pulse of 5 μs duration being visualized on an (analog) oscilloscope at different time resolutions. As the time resolution (the horizontal axis) improves, the imperfection of the amplifier can be easily visualized and measured. In this case, the rise time was about 7 ns (Figure 5.8c). One can also see the leading edge of the ideal pulse over-shot the desirable level, which took about 30 ns to settle down (Figure 5.8c). In addition to the imperfection of the leading edge, the falling edge of the same pulse also had imperfections, which did not drop to zero quickly and cleanly (Figure 5.8d). For a resistive circuit, the values of the rise time are mostly limited by the bandwidth of the electronic circuitry, which is further distorted by the stray and residual capacitance and inductance in the circuitry.

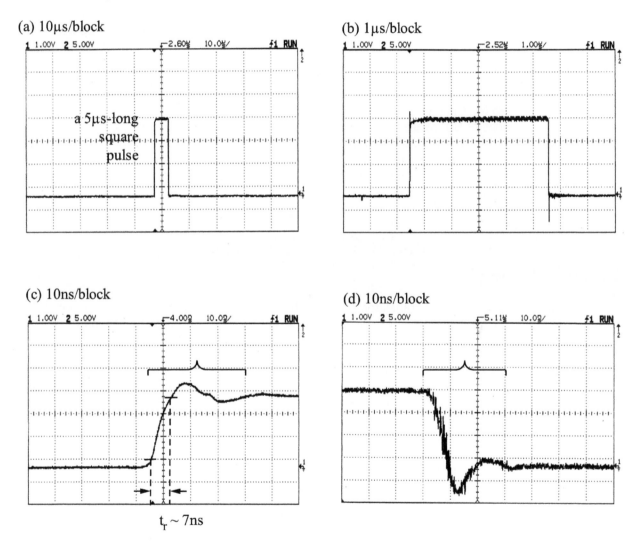

(a) 10μs/block

(b) 1μs/block

(c) 10ns/block

(d) 10ns/block

$t_r \sim 7\text{ns}$

Figure 5.8 The imperfections in the leading and falling edges of an *ideal* square pulse (a), visualized at different time resolutions on a *real* analog oscilloscope. The rise time characterizes how long it takes for the circuit to rise from 10% to 90% of the output step function. Additional imperfections can also be seen in the square pulses, in both the rising edge (c) and falling edge (d).

5.5   RECEIVER

The other series of amplifiers in an NMR spectrometer is in the receiver channel, which has two functions. First, a number of amplifiers are needed to amplify the tiny FID signal (on the order of several microvolts) to a final amplitude of several volts (which is needed to drive the analog-to-digital converter, ADC). The first of these amplifiers is called the preamplifier, which sits as close to the probe as possible to minimize the signal loss and noise inclusion, since any noise picked up by the preamplifier will be further amplified by the subsequent amplifiers. Second, just like in our daily life when we enjoy the audio components of an FM radio, the receiver needs to remove the base/carrier frequency (e.g., 300 MHz), leaving only the offset information of the signal (e.g., one peak is 20 Hz away from the reference). Hence, a number of mixers and phase sensitive detectors are used in the receiver channel to reduce and remove the carrier frequency.

There are at least three requirements for the receiver. First, it needs to have a good linearity in its gain, that is, the same amplification for strong and weak signals, as well as for high- and low-frequency signals. The second requirement is an equal delay time for high- and low-frequency signals. If the signals travel through the receiver at different speeds, a variable phase shift that is introduced will be very difficult to correct (cf. Chapter 2.13). The third requirement is for the receiver to sample the signal with a minimum delay; otherwise a first-order phase distortion will be introduced (cf. Chapter 6.10). Also, modern spectrometers have two phase-sensitive detectors with references of 90° phase shift to generate the in-phase and quadrature signals (either from the same coil or two coils), which are respectively proportional to M_x and M_y and represent the real and imaginary parts of the complex time-domain NMR data. When the receiver has more than one channel, the performance characteristics (e.g., rise time) and consistency in the electronics become critically important.

5.6   PULSE PROGRAMMER AND COMPUTER

A pulse programmer is a translator, responsible for the generation of a list of commands (cf. Chapter 6.3), based on the text or graphical instruction of the operator, to control various parts of the NMR instrumentation. Once the operator finishes the experimental planning and hits the run button, the pulse programmer takes over the tasks and runs the experiments. The pulse programmer has its own memory to store the commands for various operations, such as the real-time controlling of pulse timing and phase, the collection and averaging of data, initial analysis of data including apodization filtering, integration, and fast Fourier transformation. The signal from the master clock is used to control the timing of all pulses and actions.

A modern NMR spectrometer typically uses a commercial personal computer as the front interface between the human and machine. On the computer, one can write programs, design and compile pulse sequences, and select experimental parameters. The run button sends the complied commands and parameters to the pulse programmer for execution. During the signal collection, the FID data are sent to the computer in real time. The computer then carries out all post-acquisition operations and data archiving.

5.7   OTHER COMPONENTS

There are other components in a modern NMR spectrometer. For example, a decoupler is required to provide a decoupling signal to the rf coil (cf. Chapter 9.1), which often requires a separate channel for its own operation. One may need more than one decoupler in modern NMR experiments. The decoupler can be homonuclear (working at the same nucleus) or heteronuclear (decouple one nucleus and detecting another nucleus). A good temperature control unit is another requirement for some NMR studies.

References

1. Xia Y, Callaghan PT. Study of Shear Thinning in High Polymer Solution Using Dynamic NMR Microscopy. *Macro Mol*. 1991;24(17):4777–86.
2. Xia Y, Jeffrey KR, Callaghan PT. Purpose-designed Probe for Dynamic NMR Microscopy in an Electromagnet. *Magn Reson Imaging*. 1992;10:411–26.
3. Halbach K. Design of Permanent Multipole Magnets with Oriented Rare Earth Cobalt Materials. *Nuclear Instruments and Methods*. 1980;169(1):1–10.
4. Raich H, Blümler P. Design Construction of a Dipolar Halbach Array with a Homogeneous Field from Identical Bar Magnets: NMR Mandhalas. *Concepts Magn Reson Part B*. 2004;23B:16–25.
5. Blümich B, Singh K. Desktop NMR and Its Applications from Materials Science to Organic Chemistry. *Angew Chem*. 2018;57(24):6996–7010.
6. Anderson WA. Electrical Current Shims for Correcting Magnetic Fields. *Rev Sci Instrum*. 1961;32(3):241–50.
7. Hoult DI, Richards RE. Critical Factors in the Design of Sensitive High Resolution Nuclear Magnetic Resonance Spectrometers. *Proc R Soc Lond A*. 1975;344:311–40.
8. Webb AG, editor. *Magnetic Resonance Technology*. Cambridge: Royal Society of Chemistry; 2016.

6

NMR Experimental

This chapter discusses some of the experimental issues in NMR spectroscopy. Although they are covered in terms of NMR terminology, many discussions are equally applicable to other branches of modern imaging technology that involve signal acquisition and digital processing. We will look at them one by one, roughly in the order that one does an experiment.

6.1 SHIMMING

Shimming refers to a process in which one optimizes the homogeneity of the magnetic field (cf. Chapter 5.1). The name "shimming" came from the early days of NMR when a permanent magnet or electromagnet was used. By mechanical alignment of the magnet pole surfaces (cf. Figure 1.2c) to be as parallel as possible, the homogeneity of the magnetic field could be optimized. This mechanical alignment is done by placing thin wedges of brass (called shim stock) between the magnet and pole pieces, a tedious process that gained the name "shimming." In modern NMR with a superconducting magnet, the process of optimizing the field homogeneity is still called "shimming," but it is done electronically by adjusting the currents through various "shim" coils that generate small local magnetic fields with profiles that approximate the spherical harmonic functions (Figure 5.4). The lowest order of correction is Z_0, which changes the strength of the main magnetic field $\boldsymbol{B}_0$. The three first-order shims are the first-order deviations of the main field; they are the linear gradient of the main field. The second-order shims are called the curvature, a name that comes from the field profiles of their spherical harmonics [1]. Since many of these shim functions interact with each other, multiple iterations through the different shims are necessary to optimize the field homogeneity.

A good practice in solution NMR spectroscopy is, regardless of what chemical you are going to measure, to first shim on a tube of water, approximately the same volume as or larger than the volume of the unknown liquid specimen. The goal is to make the magnetic field within the sample volume as uniform as possible. Since the two hydrogen atoms in a water molecule have the same chemical shift, a single exponential decay is expected in the FID signal from water, which is easy to judge visually. If one shims on an unknown sample with multiple resonant peaks, the FID would not be a single exponential, so it is harder to judge visually the quality of the shimming. In shimming, it is also useful to set the FID off-resonance (Figure 6.1), which gives the FID a nice envelope in the exponential decay, which is much easier to optimize than an on-resonant decay.

Essential Concepts in MRI: Physics, Instrumentation, Spectroscopy, and Imaging, First Edition. Yang Xia.
© 2022 John Wiley & Sons Ltd. Published 2022 by John Wiley & Sons Ltd.

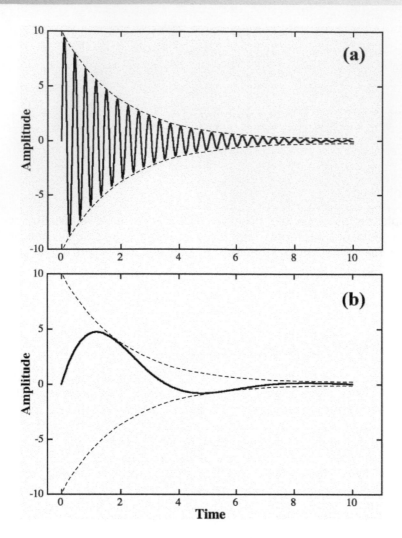

Figure 6.1   (a) A slightly off-resonance FID signal has a recognizable envelope of exponential decay (the dashed curve), which is useful in the monitoring of the shimming process. (b) An on-resonance FID signal does not have an obvious envelope of exponential decay. The dashed envelope in (b) is the same as in (a). If this dashed envelope in (b) is absent, it is harder to judge that the FID in (b) has the same shimming as in (a). Note that NMR signal is complex, with both real and imaginary components (shown previously in Figures 2.14a and 2.14b). In most figures of FID, it is common to show just one component of the complex signal.

6.2   PREPARING SAMPLES

The samples for NMR spectroscopy can range from soluble materials dissolved in an appropriate solvent (e.g., water) to biological tissue to solids. Consequently, there are a multitude of sample tubes/cells and sample preparation methods, each optimized for the particular material to be studied. Most high-resolution NMR work is done with samples in the liquid phase. It is therefore desirable to consider suitable solvents, which can be water or organic solvents. An ideal solvent should not contain the isotope whose resonance is being studied or have a solvent peak at the same chemical shift of the solute. However, it should be noted that the use of a solvent can change the chemical environment of the chemicals to be measured. This is because the chemical shift of a given molecule or chemical depends upon its molecular environment, which includes the particular solvent in which the chemical is dissolved, the concentration of the chemical solution, and other experimental parameters such as the measurement temperature and so forth. Many books on NMR spectroscopy contain extensive lists of common solvents and discussion on what constitutes a *good* solution [2, 3].

6.3   PULSE SEQUENCES AND FID

A pulse sequence in a modern spectrometer is usually a text file containing what the operator intends to do in an experiment, often with a graphical representation (Figure 6.2). When you read a graphical pulse sequence, no matter how complicated it is (how many pulses, how many channels, and how many loops), you read a pulse sequence from left to right, following the arrow of time (the horizontal axis). Figure 6.2 is a simplest possible pulse sequence, which has a single rf pulse for the excitation followed by a signal acquisition. Different manufacturers always have different codes and formats, even for the same operations. This sequence is translated in the pulse programmer into a sequential list of machine-level commands, in order of time. The execution of a pulse sequence manipulates the spin system, and its response to the pulse sequence provides an opportunity to observe specific features of the spin Hamiltonian. There are numerous sequences for spin manipulation. The next chapter, "*Spin Manipulations* by Pulse Sequences," will discuss several individual pulse sequences, each with a specific aim. For the moment, we know that after a sample is loaded into the magnet and an NMR experiment is carried out under the direction of a particular pulse sequence, an FID similar to Figure 6.1 is picked up by the receiver coil. Note as we have discussed earlier in Chapter 2.11 that the NMR signal is complex and has the real and imaginary components, which are detected in two quadrature channels and shown in Figure 2.14a and b. The complex nature of the NMR signal means that we can measure not only the amplitude of the signal but also its phase (i.e., the orientation of the magnetization in the transverse plane, cf. Chapter 2.13).

A common experimental parameter in NMR is the repetition time (TR). Since all pulse sequences in NMR are repeatable (to average several FIDs in order to achieve a better signal-to-noise ratio, see Section 6.11), TR is the amount of time between two successive executions of the same pulse sequence. Many considerations can go into the setting of TR in a practical experiment. If the desire is to have the full magnetization for each signal acquisition, for example, TR should be approximately $5 \times T_1$. More commonly, TR is much shorter than $5 \times T_1$.

We introduced a descriptive term for the rf pulses in Chapter 2.10, which labeled an rf pulse as *hard* or *soft*. This term refers to the excitation range of an rf pulse in frequency. A soft rf pulse has a long duration in time, which results in a narrow range of excitation frequencies. In contrast, a hard rf pulse has a short duration in time, which results in a wide range of excitation frequencies. Since the amount of spin rotation, ϕ, is given by the time integral of the amplitude of the rf pulse [the area under a pulse, as in Eqs. (2.26) and (2.27)], a hard pulse requires a high-voltage rf amplifier.

A useful comparison can be made between the $\boldsymbol{B}_0$ field strength and $\boldsymbol{B}_1$ field strength. In modern high-field spectrometers, a typical 90° rf pulse is on the order of 10 µs in duration. This suggests that it would take 4×10 µs = 40 µs for the $\boldsymbol{B}_1$ field to rotate the magnetization by 360°. The rate of the rotation by the $\boldsymbol{B}_1$ field is therefore 1/40 µs = 25 kHz. This rotation rate can be considered as the $\boldsymbol{B}_1$ field strength, which is about 1/10,000 to 1/20,000 of a $\boldsymbol{B}_0$ field (250 MHz to 500 MHz).

(a) A text file for a 1D sequence

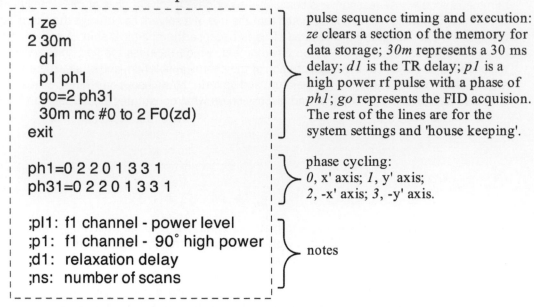

1 ze
2 30m
　d1
　p1 ph1
　go=2 ph31
　30m mc #0 to 2 F0(zd)
exit

ph1=0 2 2 0 1 3 3 1
ph31=0 2 2 0 1 3 3 1

;pl1: f1 channel - power level
;p1: f1 channel - 90° high power
;d1: relaxation delay
;ns: number of scans

pulse sequence timing and execution:
ze clears a section of the memory for
data storage; *30m* represents a 30 ms
delay; *d1* is the TR delay; *p1* is a
high power rf pulse with a phase of
ph1; *go* represents the FID acquision.
The rest of the lines are for the
system settings and 'house keeping'.

phase cycling:
0, x' axis; *1*, y' axis;
2, -x' axis; *3*, -y' axis.

notes

(b) A graphical representation for the actions of the sequence

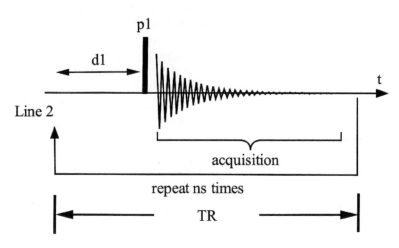

Figure 6.2 (a) A text file for a simple pulse sequence, which has one hard rf pulse, followed by the signal acquisition. (b) The graphical representation for the actions of the pulse sequence. TR stands for the repetition time.

6.4 DIGITIZATION RATE AND DIGITAL RESOLUTION

The FID signal from the rf coil is an analog signal, which is commonly plotted by using the time on the horizontal axis and the amplitude on the vertical axis (Figure 6.1). This analog signal is digitized (i.e., converted to the digital format) in the receiver channel by electronic circuitry called an analog-to-digital convertor (ADC). The digitization rate describes how fast the analog FID signal is sampled along the time axis. According to the sampling theorem, a sine wave must be sampled at least twice during a single cycle in order to be properly represented, which is known as the Nyquist rate (Figure 6.3). Since an NMR signal is not a simple sine wave but a mixture of waves at various frequencies, sufficient sampling is an important parameter in NMR signal acquisition. An under-sampled signal could miss the fine features of the signal and appear to have a lower frequency (aliasing).[1]

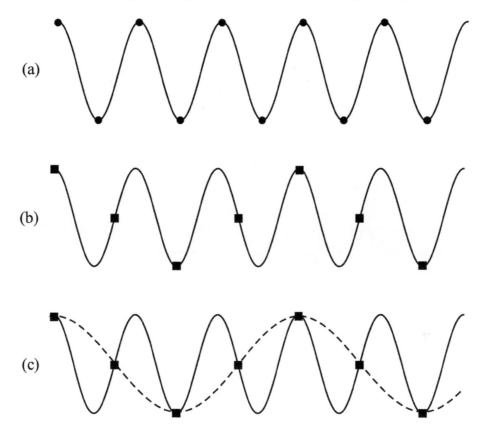

Figure 6.3 Consequences of under-sampling. (a) Sampled at the Nyquist rate, twice per cycle. (b) Under-sampled (4/3 times per cycle). (c) Incorrect determination of the frequency due to the under-sampling (dashed waveform), known as aliasing.

---

[1] When the wheels of a wagon start to speed up in a movie, the wheels would appear to accelerate normally in low speeds, then become stationary at a certain moment, then appear to move backwards. This odd appearance is caused by the fixed shutter speed of the camera (a fixed sampling rate). When the time required to move one spoke on the wheel from one position to the next position is more than the shutter speed (slow speed wheels), the wheel appears normal; when the shutter is at the precise speed to sample one spoke moving from one position to the next, the wheel seems stationary; when the spoke moves more than one position during the shutter time, the viewer will lose the tracking of position and follow the reversal motion, which is an alias due to insufficient sampling.

The time interval between any two sampling points is called the dwell time (DW) in NMR, which is given by the following equation:

$$DW = \frac{1}{2f_N},$$ (6.1)

where f_N is the Nyquist frequency. If f_N is 2 kHz, then DW will be 250 µs (1/4000).

The digital resolution (DR) in NMR is then determined by the spectral width divided by the total data points that are used to acquire the signal:

$$DR = \frac{2f_N}{SI},$$ (6.2)

where SI is the total number of data points. If SI = 8192 points and f_N = 2 kHz, DR will be 0.49 Hz/point.

Figure 6.4 shows the deterioration in the line shape of a triplet, which is a part of the high-resolution spectrum of ethanol (in Chapter 8), due to insufficient digital resolution. Note that the minimum digital resolution is determined by the features of the spectrum, typically from a few hertz to tens of kilohertz. In other words, we are only interested in the differences among the peaks

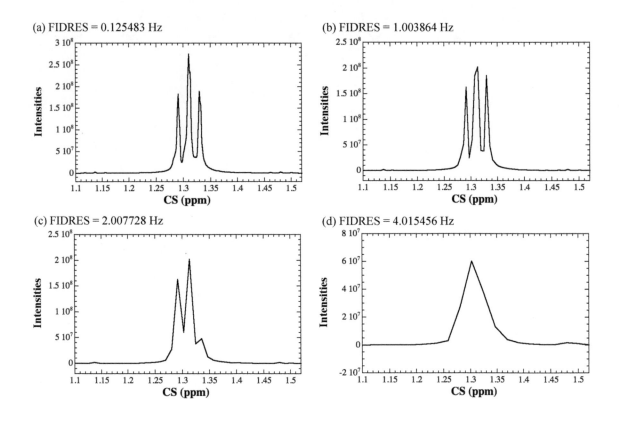

Figure 6.4   A spectrum of the ethanol's triplet. As the FID resolution is reduced from 0.125483 Hz/pixel (a) to 1.003864 Hz/pixel (b), 2.007728 Hz/pixel (c), and 4.105456 Hz/pixel (d), the appearance of the NMR spectra deteriorates significantly. (Due to the use of 32-bit or 64-bit data in signal acquisition, NMR data can have many significant figures.) The full spectrum of the ethanol is explained in Chapter 8.5. (CS, chemical shift).

but not their absolute frequencies. If we were to detect the absolute frequencies of the resonant peaks in NMR, we would have to sample the signal at the rf frequencies (hundreds of megahertz). Finally, the acquisition time (AQ), the time that is needed for the signal detection, will be

$$AQ = \frac{1}{DR} = \frac{SI}{2f_N} = SI \times DW. \tag{6.3}$$

When DW = 250 μs, the AQ will be 2.05 seconds (8192 points x 250 μs). The duration of AQ contributes significantly to the minimum echo time, which could become critically important in some spectroscopic and imaging experiments.

6.5   DYNAMIC RANGE

For the same NMR signal, the dynamic range describes how well the amplitude of the FID is being recorded, that is, the digital resolution in the amplitude (vertical) axis, as shown in Figure 6.5. The basic consideration in the dynamic range is that the strongest part of the signal must be represented without overflowing the computer digital size/depth (i.e., smaller than the largest number that can be represented), and the weakest part of the signal must be represented by the smallest number in the dynamic range. In digital electronics, the data are stored in the form of binary numbers. So the dynamic range of any receiver/digitizer will depend upon the number of digital possibilities when converting the analog signal to the digital measurement,[2] which can be seen from the following list:

# of Bits (n)	Possibilities (2^n)
1 bit	2 (= 2^1)
8 bits (1 byte)	256 (= 2^8)
16 bits (2 bytes)	65,536 (= 2^{16})
32 bits (4 bytes)	4,294,967,296 (= 2^{32})

For example, an 8-bit dynamic range can accommodate 256 different possibilities (either using an unsigned representation from 0 to 255 or a signed representation from −128 to +127). That is, the amplitude in Figure 6.5 is divided into 256 possible levels. Hence, the resolution of the amplitude is 1/256 = 0.39%. If the maximum signal is 10 volts, then any signal less than 39 mV (= 1/256) will be lost, since it would be considered as 0. To detect smaller components, one can increase the dynamic range of the ADC to 16 bits (which has 65,536 possible levels) or 32 bits (which has 4.29×10^9 possible levels) per data point. Modern spectrometers typically have at least 32 bits in the dynamic range of their detectors.

The concept of the dynamic range described here is identical to the digital depth in consumer electronics. Most consumer digital cameras have 8-bit pixel depth, which allows for up to 256 possible values for the variation of a pixel's brightness. These images are saved as JPG format,

---

[2] An interesting question to think about is, "What are the maximum number of different values that one's 10 fingers can represent?" The common answer is 10, one for each finger. However, if we consider 10 fingers as 10 sequential digits, where, for example, sticking out counts as 1 and bending down counts as 0, then 1024 different numbers can be represented by your 10 fingers.

88

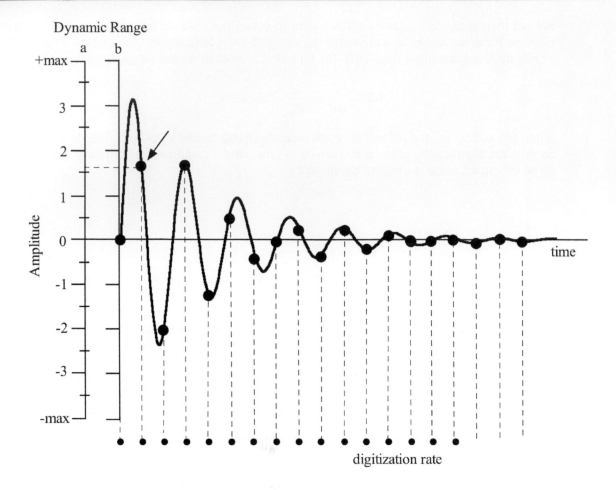

Figure 6.5 The issue of dynamic range in signal acquisition, where the digitization rate (the row of small dots) marks the sampling points along the axis of time. During the signal acquisition, the dynamic range determines how fine the amplitude of the signal can be represented/converted, with 0 (or below a certain value) as OFF and 1 (above a certain value) as ON. Two different dynamic ranges are given schematically. The range *a* can be determined to a half-integer level, while the range *b* can only be determined to an integer level. So the second point (arrow), for example, can get a more realistic value of 1.5 on the range *a* but gets rounded up to 2.0 on the range *b*.

which only allows for 8-bit data. High-end digital cameras can record the images in raw format with the bit depth from 10 bits (up to 1024 levels) to 14 bits (up to 16,384 levels). If you save the images in the JPG format, the computer chip in the camera processes the raw data and condenses it to 8-bit before saving the image.

The digital size/depth is determined by the quality of the ADC in the spectrometer hardware. There are several standard algorithms to carry out analog-to-digital conversion, such as ramp, successive approximation, and parallel encoder [4]. In general, higher bit depth leads to slower conversion, which also takes longer to save into the memory. An alternative approach in meeting the demand for dynamic range occurs when the strongest signal in an FID is from the solvent (e.g., water). In these situations, the dynamic range problem can be solved by selectively suppressing the non-essential large water or solvent peak during the signal acquisition so that the small peaks will not be lost [5].

6.6 PHASE CYCLING

As we discussed earlier in Chapter 2.13, the phase of the NMR signal is critically important. Since practical systems are never perfect, it is important to distinguish the NMR signal from any non-NMR interference such as noise. This can be done by using a practical procedure called phase cycling. This procedure relies on the fact that both the magnitude and orientation (i.e., phase) of the transverse magnetization ($M_\perp$) can be determined by the NMR acquisition method (i.e., the pulse sequence). During signal detection, the phase of the transverse magnetization is a relative property of the signal, relative to the phase of the reference signal of the NMR spectrometer (Figure 2.15). In a modern instrument, the phase of the reference signal can be adjusted in steps to any angle difference by a digital process. In consequence, the transmitter's pulse $B_1(t)$ field in the rotating frame can be set to any orientation (cf. Chapter 2.12).

When we adjust the phase of the excitation pulse in the transmitter and the detection phase in the receiver, the components of the transverse magnetization will be affected accordingly (Figure 6.6). First (Figure 6.6a), shifting the excitation phase by 180° will result in the inversion of signal while any non-NMR interference signal will be unaffected (i.e., inverting by 180° changes a +ve signal to a −ve signal). Second (Figure 6.6b), inverting the detection phase synchronously with the excitation phase leads to NMR signal addition and interference cancellation (i.e., there is no real change in this situation since we also change the reference synchronously, which is illustrated in Figure 6.6b with the exchange of the +ve and −ve axes). Third (Figure 6.6c), changing the excitation phase by 90° will interchange the real and imaginary signals in the receiver channels (i.e., swapping $M_{x'}$ and $M_{y'}$).

Based on these relationships, a very useful rf phase cycling sequence is called the cyclically ordered phase sequence (CYCLOPS), which is shown in Figure 6.7. It is a four-phase cycle that steps the four quadrants in 90° steps and is commonly used in modern dual-channel spectrometers. It can be used to solve a few practical issues in electronics, such as minor differences in the amplification (illustrated in the figure as the length of the arrows), phase (the orientation of the arrow in the transverse plane), and offset in the amplifiers between the two channels. Even just implementing the first two-phase cycle would improve the quality of the signal. In NMR

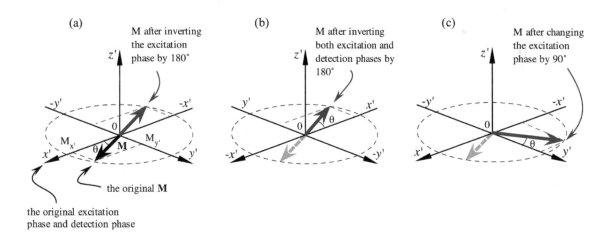

Figure 6.6 Basic relationships in phase cycling, where the original magnetization [the black arrow in (a) and the light gray arrows in (b) and (c)] is similar to the black arrow in Figure 2.15c, with an additional phase θ. (a) Inverting the excitation phase by 180°, (b) inverting the receiver phase synchronously with the excitation phase, and (c) changing the excitation phase by 90°.

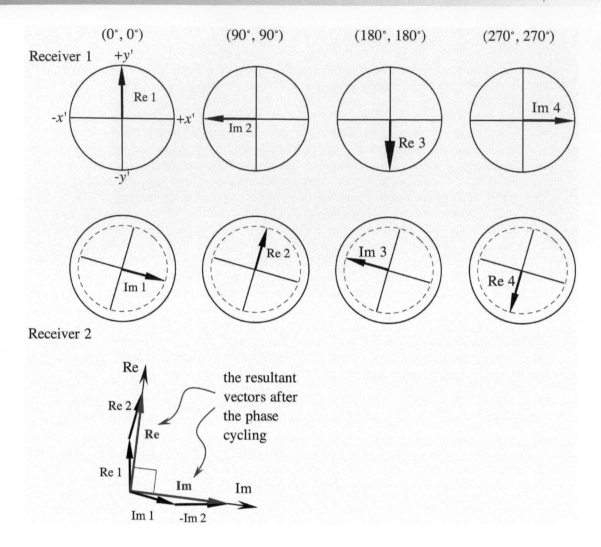

Figure 6.7 CYCLOPS phase cycling. The size of the circle and the length of the arrow represent the amplification, so that Receiver 2 has smaller amplification (the dashed circle) when compared with Receiver 1 (the solid circle). The orientation of the *xy* axis represents the phase of the receiver, so that Receiver 2 has a slightly different phase than Receiver 1. Even with a two-step phase cycling, one can eliminate the differences in the amplification and phase between the two receivers – by summing up the four imperfect individual vectors, one can obtain two resultant vectors (in red in the lower drawing) that are 90° with each other and have the same length (i.e., amplification).

spectroscopy and imaging sequences, there are numerous schemes for phase cycling, all with the goal to add up the desired signal and cancel out the undesired signal.

6.7   DATA ACCUMULATION

Data accumulation refers to the co-addition of the FID signal. Because the NMR signal is small and noise is everywhere, it is a common practice in NMR to acquire and co-add several or many identical spectra to improve the signal-to-noise ratio (SNR). This practice is based on the fact that the signal voltage adds linearly with the number of accumulations, whereas the noise power increases at a slower rate (due to the random nature of the noise). The SNR after n accumulations therefore increases as the square root of n

$$\text{SNR}_n = (n)^{1/2}\,\text{SNR}_1, \tag{6.4}$$

where SNR_1 refers to the SNR of a single scan. To double the SNR, one needs to repeat the signal acquisition four times. Figure 6.8 shows an example of SNR improvement due to the data accumulation.

According to the solution to the Bloch equation when a 90° pulse is used [Eq. (2.16a)], 99% of the magnetization will return to thermal equilibrium after waiting $5 \times T_1$. However, waiting $5 \times T_1$ to obtain the max signal for any single acquisition means one can repeat less frequently during *a given amount of total experimental time*, which is often the limiting factor in real life. By some trivial analysis, one can find out that the maximum SNR for *a given amount of time* can be obtained by setting the repetition time to ~ $1.26 \times T_1$.

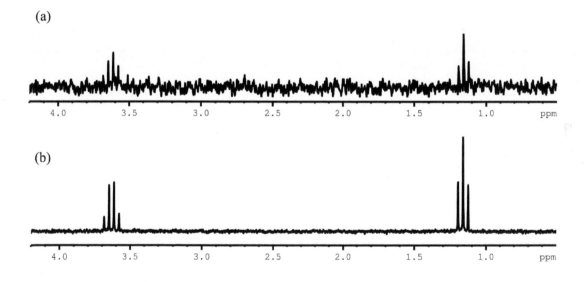

(a)

(b)

Figure 6.8   Improvement of SNR by data accumulation. (a) Signal acquisition one time, (b) signal acquisition 100 times. (The sample was a trace amount of pure ethanol in D_2O; the signals were acquired on a 200-MHz NMR spectrometer. Refer to Figure 8.6 for the full spectrum and peak assignments.)

6.8   PRE-FFT PROCESSING TECHNIQUES

Now we have the FIDs in the computer memory ready to be processed. Before the common oper-ation to fast Fourier transform (FFT) of the FID into a spectrum, there are several pre-FFT operations that may be helpful to prepare the raw data. Note that any operation before the Fourier transformation could have an impact on the spectrum, since according to the Fourier theorem, the FT of a product of two functions is the convolution of the two transformed functions in the conjugate space (cf. Appendix A1.2). Note also that these pre-FFT operations *cannot* increase the actual information in the data beyond what is already there; they can only enhance some of the information at the expense of other information or additional artifact.

Baseline correction: This is an operation that removes any residual DC offset in the two quad-rature channels (Figure 6.9). The Fourier transform of a DC offset is a spike (a delta function) in the frequency domain (cf. Chapter 2.8.1). Since the spike likely has a high amplitude, it com-presses the dynamic range of the useful signal, which is likely to be small in amplitude. Baseline correction can therefore improve the quality of the spectrum, especially the minor peaks in the spectrum.

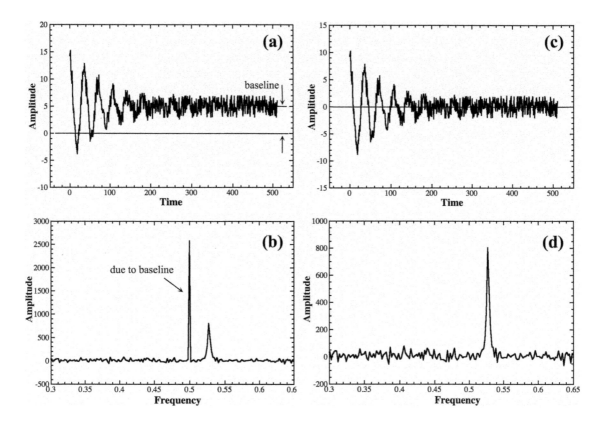

Figure 6.9   (a) The FID has a finite and constant baseline, indicated by the two arrows. (b) Since the FT of this baseline, which is a DC offset, would be a delta function, the FT of the FID with the baseline would have a spike at the center of the frequency axis. (c) It is simple enough to do a baseline correction to the FID data by subtracting a constant value, which returns a clean spectrum (d) with its peak better solved.

FID filters: Several methods fall into the general approach called FID filtering. The most commonly used is exponential filtering, which multiplies the FID with an exponential, exp(-t/k), where k is an adjustable parameter (Figure 6.10). This filter gradually reduces the amplitude of the final part of the FID or even replaces it with the baseline numbers (or even lots of zeros). The resultant spectrum can have an increased SNR since the contribution of the noise to the spectrum has been reduced. Since the FT of an exponential is a Lorentzian in the frequency domain, the application of this exponential filter to the FID will increase the line width of the spectrum, by about $1/(\pi k)$. If the parameters t and k are chosen so that the exponential reaches zero after any "visible" signal reaches zero, the line-broadening effect will be small. This process is sometimes called *sensitivity enhancement*. In contrast, if a function is chosen so that it stays flat for the initial part of the FID signal but gradually increases the amplitude of the middle to final parts of the FID when the signal is approaching the noise level, that is, to increase the influence of the high-frequency information that is buried inside the noises, the resultant spectrum would be noisier but may better detect the high-frequency components. This process is called *resolution enhancement*.

93

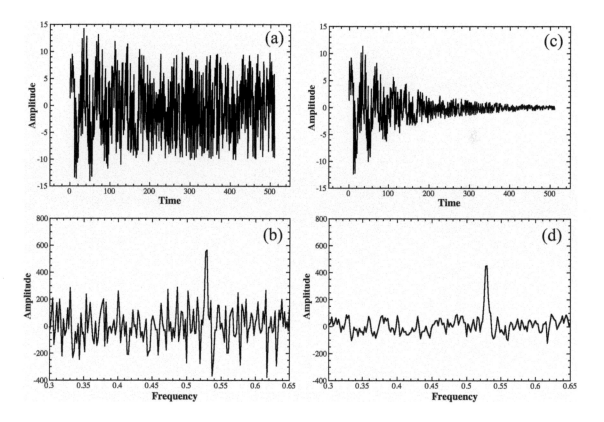

Figure 6.10 Sensitivity enhancement by filtering the FID before the Fourier transform. (a) A very noisy FID. (b) The FT of the noisy FID shown in (a). (c) Application of an exponential filter to the noisy FID in (a), which keeps the initial signal but substantially suppresses the latter part of the signal. (d) The FT of the exponentially decayed FID in (c), which has a much-reduced noise and an enhanced signal peak (as well as a much better SNR). A careful comparison can show that the FWHM of the (d) peak is 25% wider than that of the (b) peak, which comes from the convolution of the exponential filter upon the Fourier transform (FT of an exponential is a Lorentzian).

Zero-filling: The main purpose of the zero-filling (i.e., adding 0 beyond the end of the acquired data points) is to increase the number of the data points to the next 2^N or a large 2^N for Fourier transform, which can improve the digital accuracy of the results (the use of zero-filling in the time domain is equivalent to data interpolation in the frequency domain) [6]. Figure 6.11 shows the effect of zero-filling. Since zero-filling corresponds to the multiplication of the time-domain signal by a step function, a sinc convolution is introduced in the frequency space, which results in extra broadening of the peaks. But this broadening may be insignificant if the peak is sufficiently broader than the sinc function, a condition equivalent to requiring that the signal amplitude has significantly decayed before the onset of zero-filling. In addition, the SNR ratio of the spectrum can also be improved by the zero-filling since the additional zeros carry no noise.

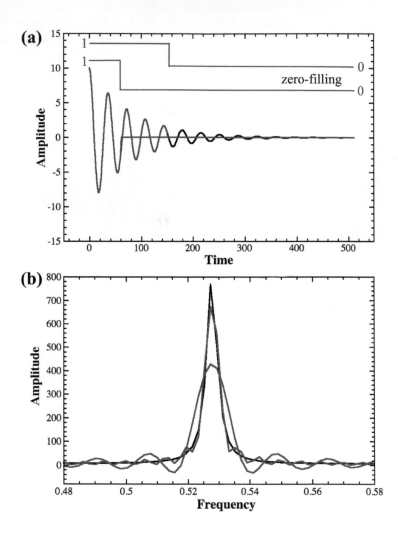

Figure 6.11   Zero-filling the time-domain data (a), which corresponds to the multiplication of the time-domain signal by a step function. The black peak in (b) is the FT of the un-truncated FID in (a). A sinc convolution will broaden the peak in the frequency domain (b). The blue step function does not cause a significant peak broadening since it is applied when the signal amplitude is sufficiently decayed. The red step function causes a significant peak broadening since it is applied when the signal amplitude is about ½ of the maximum. The red peak also shows clearly the oscillations due to the convolution of the sinc function.

6.9 FAST FOURIER TRANSFORM

The Fourier transform (FT), invented by Jean-Baptiste Joseph Fourier in 1822, decomposes a function of time into its constituent frequencies (Appendix A1.2). This is a very powerful process but could be time-consuming. The fast FT algorithm, developed in the mid-1960s by James Cooley and John Tukey, can perform the Fourier transform much faster. A price to pay to be processed by the FFT is that the number of the discrete points N in the function must be a power of two (i.e., 2^N = 16, 32, 64, 128, 256, 512, 1024, 2048, ...). When the original data are not equal to 2^N, one can always add a number of zeros to the end of the real data to lengthen it to 2^N (i.e., zero-filling – cf. Section 6.8) or interpolate the original data points into a new set of data points that meet the 2^N requirement (i.e., inserting new points based on the trend of the neighboring acquired data points – cf. Chapter 12.5.3). The time saving in FFT over FT is significant, since only $Nlog_2N$ operations are needed in an FFT, instead of N^2 operations in an FT. When N = 256, an FT would need 65,536 operations, while an FFT would need only 2048, a mere 3% of the operations.

6.10 POST-FFT PROCESSING

After the FFT, the NMR spectrum is ready to be further processed and displayed. The most common post-FFT processing is the phasing of the spectrum. Experimentally, most spectrometers use two phase parameters to describe the phase correction process. One phase parameter is called the zeroth-order phase shift ($P0$), which is the same for all peaks across the spectrum. The other phase parameter is called the first-order phase shift ($P1$), which varies linearly with frequency.

The zeroth-order phase correction: The zeroth-order phase shift (Chapter 2.12) means that the real and imaginary signals are not parallel with the x' and y' axes, which could be caused by a number of factors, including instrument imperfection where the signal gains an additional phase shift during the acquisition. Since it is the same shift for all peaks in the spectrum, it is easy to remove. When u_i and v_i represent the dispersion and absorption signals, respectively, and Re_i and Im_i represent the real and imaginary spectra, respectively, a rotation of the NMR signal can be accomplished by the application of a 2×2 matrix, as

$$Re_i = v_i \cos\theta_i + u_i \sin\theta_i \qquad (6.5a)$$

and

$$Im_i = -v_i \sin\theta_i + u_i \cos\theta_i, \qquad (6.5b)$$

where θ_i is the extra phase shift introduced by the spectrometer. Figure 6.12 shows the effect of this rotation matrix on the appearance of the spectrum. When $\theta_i = 0$ (no extra phase shift), we have

$$Re_i = v_i \qquad (6.6a)$$

and

$$Im_i = u_i, \qquad (6.6b)$$

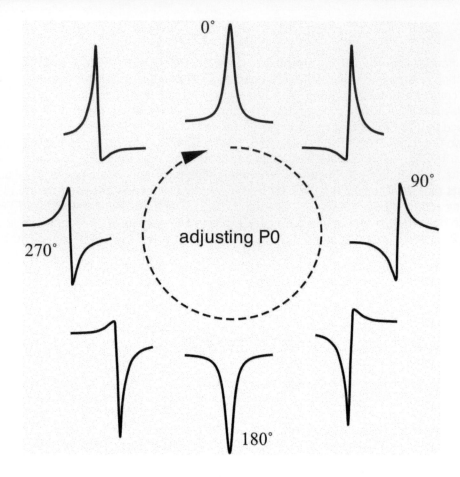

Figure 6.12 The effect of the zeroth-order phase shift ($P0$) on an NMR spectrum, which is equivalent to a rotation of the NMR spectrum by the application of a 2×2 rotation matrix [Eqs. (6.5a and 6.5b)] over a 360° angular space.

where v_i has been shown before in Eq. (2.24) as a Lorentzian line centered at ω_0 with a line width at half height of $1/(\pi T_2)$. A pure absorption spectrum can be obtained by taking combinations of the real and imaginary values as

$$v_i = Im_i \sin\theta_i + Re_i \cos\theta_i. \tag{6.7}$$

The correct value of θ_i can be chosen by either inspecting the spectrum from the computer display or by automation that maximizes the area under the absorption spectrum.

The first-order phase correction: The first-order phase shift ($P1$) is caused by the sampling not starting at the time origin. In an NMR experiment, the time origin ($t = 0$) is right after the application of the pulses. Since the electronic devices at this time are still recovering from the highly intense transmitter pulses (cf. Figure 5.8d), it is customary to delay the data acquisition for a short time, say 10 to 100 µs. During this time period, the individual magnetization components will precess away from each other in the rotating frame by an amount that varies with their individual frequencies. This is assuming that the spectrum has two or more major peaks, each having a different resonant frequency. Hence the peaks in the spectrum will bear a first-order

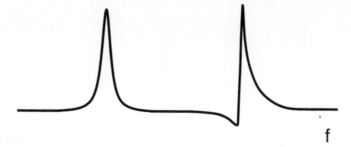

f

Figure 6.13   The first-order phase shift (*P*1) on an NMR spectrum, which has different phase offsets for different peaks hence is difficult to correct for all peaks.

phase shift, as shown in Figure 6.13. This first-order phase shift is difficult to correct perfectly for all peaks, since each peak needs a different amount of phasing. If the delay is not long or the frequency differences are not large, the *P*1 phase shift can be treated as a linear function. Sometimes one can circumvent the problem by displaying either the magnitude spectrum [i.e., $(u^2 + v^2)^{1/2}$] or the power spectrum [i.e., $(u^2 + v^2)$], neither of which has a phase dependence; one also gains SNR in both display modes. The price to pay is the loss of the phase information in the spectra.

6.11   SIGNAL-TO-NOISE RATIO

The population difference between the up and down spin states at thermal equilibrium is tiny (Chapter 3.3). For protons at 20 °C and $B_0 = 1$ Tesla ($f_0 = 42.86$ MHz) and 7 Tesla ($f_0 = 300$ MHz), for examples, the population differences equal 3.5×10^{-6} and 24×10^{-6}, respectively. These tiny population differences are due to the small value of $\hbar\gamma B_0$ (Zeeman energy) compared to $k_B T$ (Boltzmann energy). This rather weak signal is superimposed on the noise arising from the thermal motion of electrons in the receiver coil (the Johnson noise), together with other sources such as environmental noise. It is the available SNR that is important in any NMR experiment.

Commonly and practically, the SNR of a spectrum or an image profile can be estimated as

$$\text{SNR} = 2.5 \, \frac{\text{Signal}_p}{\text{Noise}_{pp}}, \tag{6.8}$$

where Signal_p is the peak signal and Noise_{pp} is the peak-to-peak noise (Figure 6.14). This equation is more commonly used in practice than the equation using the root-mean-square (rms) noise level in the spectrum. In noise theory, the rms noise has significance but the peak-to-peak noise does not. In practice, the peak values in the signal and noise are straightforward to estimate and measure from the spectrum on a computer monitor, but the former is not. The conversion factor, 2.5, comes from the statistical theory showing that the probability of the peak-to-peak noise equaling or exceeding 2.5 times the rms noise is only 1%.

To see ways to improve the SNR, however, one has to consult the theoretical equation. Briefly, the signal is proportional to the volume of the sample. Sources of noise may be numerous. In a well-designed instrument where the external noise from the environment is not of concern, the

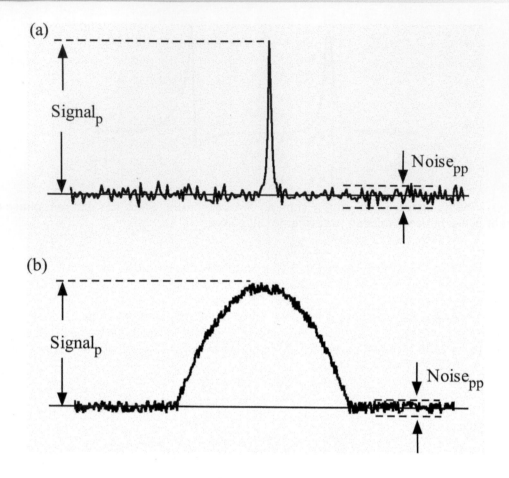

Figure 6.14   Estimation of the SNR from a spectrum (a) or a 1D image profile (b).

overall experimental noise in NMR is determined by the dielectric and inductive losses in a biological sample, the fundamental thermal noise in the rf receiver coil, and the noise in the preamplifier (the first amplifier in the receiver channel). The starting point for SNR analysis considers the fundamental signal due to the magnetization and the thermal noise in the instrument [7], which gives the ^{1}H NMR SNR expressed in terms of the peak signal over the rms noise as

$$\text{SNR} \propto \frac{B_{1xy} V_s N_s \gamma \omega_0^{7/4}}{T_s} \left(\frac{p}{T_c L F \Delta f} \right)^{1/2} \left(\frac{1}{\mu \rho_c} \right)^{1/4}, \tag{6.9}$$

where B_{1xy} accounts for the effective field over the sample volume produced by unit current flowing in the receiver coil, N_s is the number of spins per unit volume, V_s is the sample volume, T_s is the sample temperature, T_c is the coil temperature, p is the perimeter of the coil, L is the length of the coil conductor, F is the noise figure of the preamplifier, Δf is the bandwidth of the receiver's band-pass filter, μ is the permeability of the material, and ρ_c is the resistivity of the coil conductor (which is of course a function of temperature). The proportional symbol in Eq. (6.9) comes from the omission of several constants in the expression, including k_B, $\hbar$, a factor taking account of the reduction in skin depth of the coil (Appendix A3.2), a numerical factor based on the coil's geometry, and so on.

Equation (6.9) provides the upper limit to the SNR. A major simplification in the derivation of Eq. (6.9) is the absence of the dielectric and magnetic losses in biological samples that contain conductive tissues [8, 9]. For a given sample in a well-designed system, the use of a scaling law [10] for SNR in NMR gives the following equation that provides some intuitive insights:

$$\text{SNR} = \frac{\text{Signal}}{\text{Noise}_{\text{total}}} \propto \frac{\omega_0^2 r^2}{(a\omega_0^2 r^3 + b\omega_0^{1/2} + cF)^{1/2}}, \tag{6.10}$$

where r is the linear scale dimension of the sample and receiver coil; and a, b, and c are empirical constants depending upon practical situations. The three terms in the total noise come from the sample, the receiver coil, and the preamplifier, where F is the noise factor of the amplifier. Now let us see how to maximize SNR.

Nucleus: The most sensitive nucleus is the proton in hydrogen atom, because it has the largest gyromagnetic ratio γ (= 2.6752×10^8 rad s^{-1} T^{-1}) and almost 100% isotopic abundance (Table 2.1). Water molecules also have the highest concentration of any nucleus, at least in tissue. The relative sensitivities for other nuclei can be found in most NMR references [2, 11, 12].

Resonant frequency or field strength: Because ω_0 is proportional to the strength of the magnet, a higher field will give a better sensitivity. It is therefore advantageous to use a high field magnet in the experiments. This statement is generally true for solution NMR. For imaging involving biological samples (as opposed to chemicals dissolved uniformly in solutions), however, experimental artifacts and distortions due to magnetic susceptibility inhomogeneities (see also Chapter 15.4.1) also increase with the field strength [13, 14]. For solid-state NMR, a low field magnet may also be advantageous when we consider the spinning sideband problem and the transmitter power requirement. Note that SNR is proportional to $\omega_0^{7/4}$ in Eq. (6.9), while to ω_0^2 in Eq. (6.10). The discrepancy arises because Eq. (6.9) allows for the reduction in skin depth as ω_0 increases (which attenuates the quality factor of the rf coil).

Thermal noise: All noise terms are temperature-dependent. The lower the temperature, the less the noise. In practice, it is often impractical to change the sample temperature due to the biology or chemistry of the system under investigation. Using the scaling laws [10] in Eq. (6.10), it can be shown that the noise voltage from the sample dominates the total experimental noise in clinical situations where the samples are large and the magnetic fields are low (e.g., the clinical MRI situation). For sample sizes in the range of 1 cm or less in a field of 7 T, the noise voltage from the receiver coil becomes dominant. In NMR microscopy, therefore, cooling the rf coil and probe circuit can significantly improve the SNR [15], because the NMR sensitivity is inversely proportional to the square root of the receiver coil temperature, a term reflecting the Johnson noise contributions. Because the noise figure F in the preamplifier dominates the instrumental noise in the receiver channel, any investment in a "quieter" preamplifier such as cooling or better electronics may offer additional SNR improvement in NMR spectroscopy and NMR microscopy.

Geometry of the rf coil: Due to the B_{1xy} term in Eq. (6.9), it is clear that the choice of the receiver coil's geometry is important in the optimization of the SNR. Two classical coil configurations are the solenoid coil and the saddle coil, given in Figure 5.5. Both the saddle coil in a cylindrical-bore superconducting magnet and the solenoid coil in an electromagnet provide an easy access to the center of the probe (i.e., the center of the magnetic field), while the employment of a solenoid coil in a cylindrical-bore magnet requires a side entry for the sample. In terms of sensitivity, however, a solenoid coil is about three times better than a saddle coil [7, 16].

References

1. Bovey FA. *Nuclear Magnetic Resonance Spectroscopy*. 2nd ed. San Diego, CA: Academic Press; 1988.
2. Harris RK. *Nuclear Magnetic Resonance Spectroscopy – A Physicochemical View*. Essex: Longman Scientific & Technical; 1983.
3. Canet D. *Nuclear Magnetic Resonance – Concepts and Methods*. Chichester: John Wiley & Sons; 1996.
4. Horowitz P, Hill W. *The Art of Electronics*. 2nd ed. New York: Cambridge University Press; 1989.
5. Xia Y, Jelinski LW. Imaging Low-concentration Metabolites in the Presence of a Large Background Signal. *J Magn Reson Ser B*. 1995;107:1–9.
6. Callaghan PT, Xia Y. Velocity and Diffusion Imaging in Dynamic NMR Microscopy. *J Magn Reson*. 1991;91:326–52.
7. Hoult DI, Richards RE. The Signal-to-noise Ratio of the Nuclear Magnetic Resonance Experiment. *J Magn Reson*. 1976;24:71–85.
8. Hoult DI, Lauterbur PC. The Sensitivity of the Zeugmatographic Experiment Involving Human Samples. *J Magn Reson*. 1979;34:425–33.
9. Chen C-N, Hoult D. *Biomedical Magnetic Resonance Technology*. Bristol: Adam Hilger; 1989. 2436–8.
10. Black RD, Early TA, Roemer PB, Mueller OM, Mogro-Campero A, Turner LG, et al. A High-temperature Superconducting Receiver for Nuclear Magnetic Resonance Microscopy. *Science*. 1993;259(5096):793–5.
11. Hennel JW, Klinowski J. *Fundamentals of Nuclear Magnetic Resonance*. Essex: Longman Scientific & Technical; 1993.
12. Harris RK, Becker ED, Cabral De Menezes SM, Goodfellow R, Granger P. NMR Nomenclature. Nuclear Spin Properties and Conventions for Chemical Shifts (IUPAC Recommendations 2001). *Pure Appl Chem*. 2001;73(11):1795–818.
13. Callaghan PT. Susceptibility-Limited Resolution in Nuclear Magnetic Resonance Microscopy. *J Magn Reson*. 1990;87:304–18.
14. Xia Y. Contrast in NMR Imaging and Microscopy. *Concepts in Magn Reson*. 1996;8(3):205–25.
15. McFarland EW, Mortara A. Three-dimensional NMR Microscopy: Improving SNR with Temperature and Microcoils. *Magn Reson Imaging*. 1992;10:279–88.
16. Hoult DI. The NMR Receiver: A Description and Analysis of Design. *Progress in NMR Spectroscopy*. 1978;12:41–77.

7

Spin Manipulations by Pulse Sequences

Spin manipulation by the means of a pulse sequence prepares the nuclear spins for evolution and detection. This chapter discusses several sequences that are *fundamental* in NMR spectroscopy (and later in imaging).

7.1   SINGLE PULSE: $90°I_x$, $90°I_y$, $90°I_{-x}$, $90°I_{-y}$

The simplest pulse sequence in NMR consists of a 90° rf pulse followed by a signal acquisition (Figure 7.1a). The function of a 90° pulse is to tip the magnetization along the z axis entirely into the transverse plane for signal detection. When a 90° pulse is applied along the x' axis, ***M*** will be tipped to the y' axis if the notation specified in Figure 1.3 is followed. A set of quadrature signals will be acquired (Figure 7.1b), where one is perfectly in phase (the real part of the signal) and one is perfectly out of phase (the imaginary part of the signal). If the 90° pulse is not on any axis in the rotating frame or if an extra phase is picked up during the experiment, the real and imaginary signals are not perfectly in phase and out of phase (Figure 7.1c), which can be corrected in the post-acquisition data analysis (Chapter 6.10). To simplify the specification, one commonly writes 90° or $90°I_x$ instead of $90°I_{x'}$, unless the phase of the B_1 pulse is consequential to the experiment.

By adjusting the phase of the 90° pulse, one can tip the magnetization to any direction in the transverse plane, as shown in Figure 7.2 (as well as Figure 2.15c and 2.15e). As we have noted before (Chapter 2.13), for a pulse sequence that has a single pulse it does not matter which of the four phases is used, since a zeroth-order phase correction can always phase back the absorption profile (cf. Figure 6.12). The phase of an rf pulse does matter if it is part of a sequence of pulses. *It is the relationship among the pulses that matters.*

Essential Concepts in MRI: Physics, Instrumentation, Spectroscopy, and Imaging, First Edition. Yang Xia.
© 2022 John Wiley & Sons Ltd. Published 2022 by John Wiley & Sons Ltd.

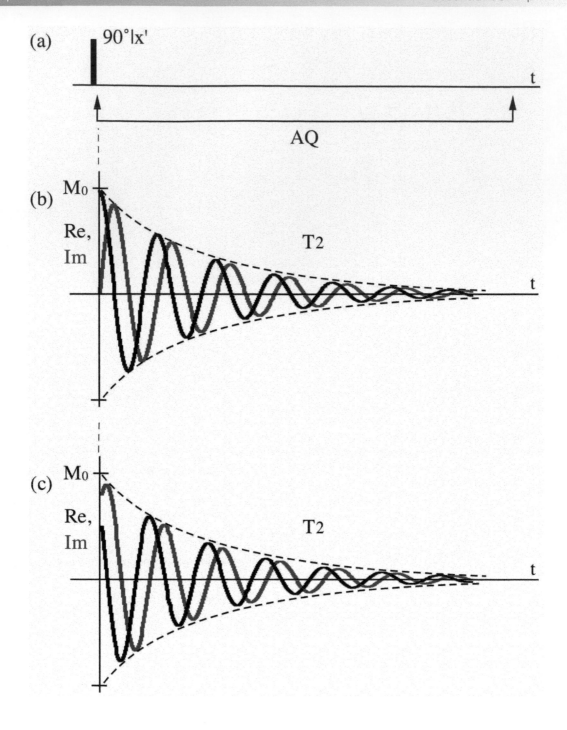

Figure 7.1 (a) A pulse sequence with a single 90° pulse. (b) The real (black) and imaginary (red) components of the signal by a 90° pulse, where the magnetization is on the axis of the rotating frame (so that the real component starts at the maximum and the imaginary component starts from zero). (c) The real (black) and imaginary (red) components of the signal by a 90° pulse, where the magnetization is not on the axis of the rotating frame (so that the real component does not start at the maximum and the imaginary component does not start from zero). Both signals in (b) and (c) have the same T_2 decay. AQ, signal acquisition.

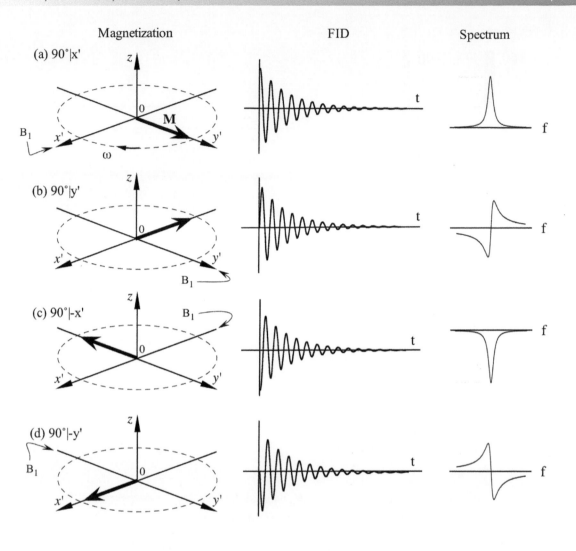

Figure 7.2 The magnetization vectors and the FID signals from a single 90° pulse, which can have four different on-axis phases. The spectra from the FID signals are also shown. Refer to Figure 2.15 and the related discussion for the phases of the magnetization.

7.2 INVERSION RECOVERY SEQUENCE, SATURATION RECOVERY SEQUENCE, AND T_1 RELAXATION

An inversion recovery (IR) sequence has two rf pulses, a leading 180° pulse and a subsequent 90° pulse after a delay τ (Figure 7.3). The 180° pulse inverts the magnetization M, making it align along the $-z$ axis (Figure 7.3b i). As soon as the 180° pulse ends, the inverted magnetization will relax back to the $+z$ axis, which is governed by the spin-lattice relaxation time, T_1. If τ is longer than $5 \times T_1$, the magnetization will return to its thermal equilibrium position (Figure 7.3b v). Of course, there will be a moment, at $0.693 \times T_1 (\ln 2 = 0.693)$, when the magnetization makes the zero-crossing (Figure 7.3b iii) from the negative M_z to the positive M_z.

(a) Pulse sequence

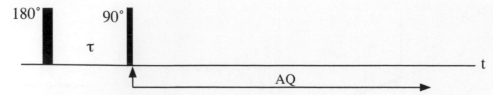

(b) Motion of M, after the 180° pulse

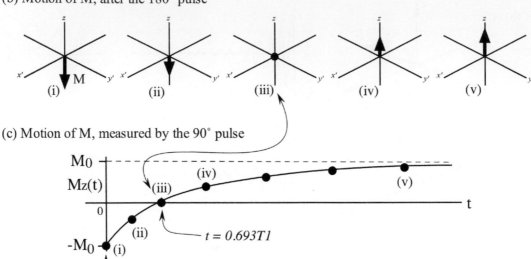

(c) Motion of M, measured by the 90° pulse

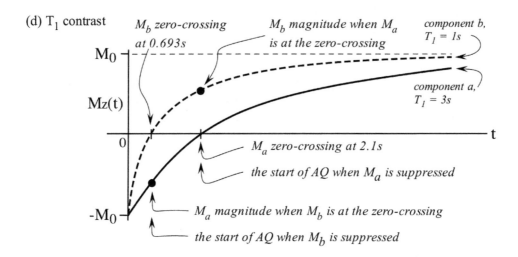

(d) T_1 contrast

Figure 7.3 (a) An inversion recovery pulse sequence. (b) The motion of the magnetization after the 180° pulse, which is along the *z* axis. (c) Changes in the longitudinal magnetization along the *z* axis, which can be measured by the use of the 90° pulse in the sequence. (d) The use of IR sequence for selective suppression, where the zero-crossing occurs at $0.693 T_1$.

If the 180° pulse is accurate, there will be no transverse component after the inversion. Since the motion of the magnetization is strictly along the z axis, no signal will be detected during τ – regardless which direction the magnetization is pointing, since the rf coils in NMR and MRI only detect the transversion magnetization. In order to measure the status of the spin-lattice relaxation at any particular moment, a second rf pulse is used in the sequence, the 90° pulse. The purpose of this 90° pulse is to tip the magnetization into the transverse plane for detection. If τ is as short as possible (i.e., the 90° pulse is applied directly after the 180° pulse, as in Figure 7.3b i), a full magnetization can be measured in the transverse plane (with careful analysis, this magnetization can be determined to be negative). If τ is $0.693 \times T_1$ (Figure 7.3b iii), no signal will be detected since the magnetization is at its zero-crossing. If τ is longer than $5 \times T_1$ (Figure 7.3b v), a full positive magnetization can again be measured in the transverse plane, which should result in a peak having the same magnitude but the opposite sign as in Figure 7.3b i.

There are two useful applications of the IR sequence. The first is to measure the bulk T_1 relaxation time of the specimen (Figure 7.3c). This can be done by repeating the IR sequence multiple times together with a series of different delay times τ. At each particular τ, one measures the peak signal (the solid dots in Figure 7.3c). By fitting the data points to

$$M(\tau) = M_0(1 - 2\exp(-\tau/T_1)), \tag{7.1}$$

a T_1 value of the specimen can be determined accurately. [Eq. (7.1) is the same as Eq. (2.16b), with the replacement of t with τ.] If the specimen only has one T_1, there is also a quicker way to estimate T_1. You can simply set the IR sequence on the real-time repeating mode and vary τ. When a particular τ results in a null signal, the T_1 of the specimen can be estimated using the equation $T_1 = \tau / 0.693$.

The second usage of the IR pulse sequence is to utilize the zero-crossing of the magnetization to selectively suppress an unwanted signal in the specimen, the so-called T_1-contrast approach (Figure 7.3d). Assuming that a sample contains two components with two different values of T_1, say $T_{1a} = 3$ s and $T_{1b} = 1$ s, by using $\tau = 2.1$s (which is 0.693×3 s) in the IR sequence, the FID will contain the contribution from the component b only because at the time of 90° pulse that tips the magnetization to the transverse plane, the component a is at its zero-crossing. A choice of $\tau = 0.69$ s will suppress the component b in a similar manner.

This approach works the best when the two T_1s are very different. The bigger the difference, the better the effect. The price to pay for using IR for selective suppression is the loss of signal for the wanted component by as much as

$$\frac{M(t)}{M_0} = \frac{M_0 - M_0(1 - 2\exp(-t/T_{1a}))}{M_0} = 2\exp(-t/T_{1a}) \tag{7.2}$$

because the signal acquisition starts at a later time τ, when the remaining magnetization is no longer at its maximum but a reduced value, indicated by the solid dots in Figure 7.3d. This approach can be implemented easily in spectroscopy and imaging to separate different tissues in the specimen, such as between cartilage and muscle [1].

One can use another pulse sequence that is even simpler than the inversion recovery sequence to measure T_1 relaxation, which is called the saturation recovery (SR) sequence (Figure 7.4). This sequence only has one 90° rf pulse and relies on a variable TR (the repetition time) to carry out the measurement. If TR is $5 \times T_1$ or longer, the acquired magnetization would be at the maximum value M_0. If TR is shorter than $5 \times T_1$, the acquired magnetization would be smaller than M_0. The maximum signal magnitude follows the curve in Figure 7.4b, and the equation,

$$M(TR) = M_0(1 - \exp(-TR/T_1)) \tag{7.3}$$

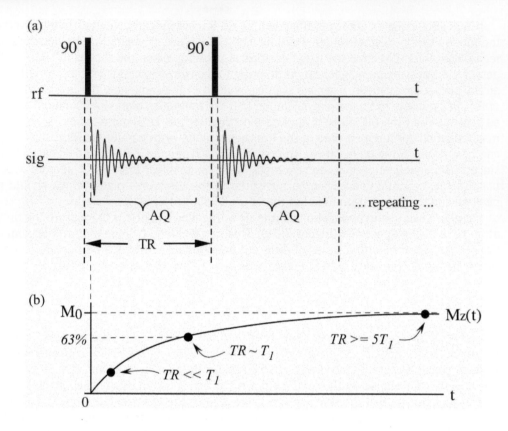

Figure 7.4 (a) A saturation recovery pulse sequence. (b) The motion of the magnetization after the 90° pulse, as the function of the repetition time TR.

[Eq. (7.3) is the same as Eq. (2.16a), with the replacement of t with TR.] To determine T_1 quantitatively, a series of SR experiments with a decreasing TR can be used. This SR sequence can also be used to impose a T_1 contrast to a specimen that has several different T_1 relaxation times. It should be noted that before each subsequent measurement, one needs to make sure there is no residual transverse magnetization. That means TR must be much longer than T_2; if it is not, one needs to apply some spoiler technique to destroy the residual transverse magnetization before any subsequent acquisition. A measurement of T_1 by the saturation recovery sequence can be made faster than by the inversion recovery sequence.

7.3 SPIN-ECHO SEQUENCE (HAHN ECHO) AND T_2 RELAXATION

The spin-echo (SE) sequence, invented by Erwin L. Hahn in 1950 [2], is the most influential sequence from the early days of NMR. It is widely used and incorporated in numerous sequences in both NMR spectroscopy and imaging. The spin-echo sequence has two hard rf pulses, first a 90° pulse and then a 180° pulse, with a delay before the 180° pulse and an equal delay after the 180° pulse until the echo appears.

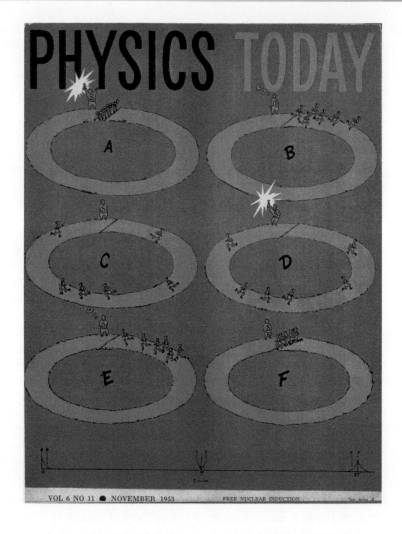

Figure 7.5 The cover of the November 1953 issue of *Physics Today,* which illustrated the principle of the spin-echo sequence. Source: Reproduced with the permission of the American Institute of Physics.

The principle of a spin-echo sequence was illustrated artfully on the cover of the November 1953 issue of *Physics Today* (Figure 7.5). Six runners (say a, b, c, d, e, f) are engaged in a race. After the signal to start (which corresponds to the 90° pulse), all runners race to the finish line based on their best abilities. Assume that at a certain time during the race *runner a* leads the pack while *runner f* lags behind. A normal race would let all runners reach the finish line and award the first prize to *runner a*. Now suppose we change the rules of the game and require the runners to turn around and run back towards the starting point *as soon as* they hear a second signal during their race (which corresponds to the 180° pulse). Since *runner a* is the fastest, *a* is also the farthest away from the starting point; in contrast, *runner f* is the slowest, but *f* is also the closest to the starting point. Assuming their running speeds stay constant, all runners should return to the starting point at the same time. Everyone would become the winner.

A similar situation happens on a microscopic scale to the motion of the magnetization in NMR. As we mentioned in Chapter 2.6, the NMR signal in the transversion plane decays with time, due to both homogeneous broadening and inhomogeneous broadening. The homogeneous broadening is due to spin relaxation, and its effect on the decay of the signal is inherently irreversible. The inhomogeneous broadening can come from a number of sources (e.g., the B_0 inhomogeneity, chemical shift distribution, or spin-spin interactions), which leads to a *faster* dephasing of the magnetization in the transverse plane, hence a *faster* decay of the signal. This additional signal decay due to the inhomogeneous broadening is potentially reversible, by the use of a spin-echo sequence, which has two basic forms, namely $90°|_{x'}$-τ-$180°|_{y'}$ and $90°|_{x'}$-τ-$180°|_x$. Both sequences are shown in Figure 7.6.

As shown in the vector drawings of Figure 7.6a, the initial magnetization is tipped by the 90° pulse to the transverse plane, along the y' axis, which precesses at a frequency ω in a rotating frame (which also has a frequency ω). Assume the bulk magnetization **M** in the specimen is the sum of five individual components, where the formation of these components is due to hetero-geneous differences in the molecular structures or nuclear interactions in the specimen. Each component is called *an isochromatic group* (i.e., packet), where the frequency spread of the nuclear spins within the group is negligible. These five isochromatic groups have five distinctly different precessional frequencies, represented by the five individual vectors in Figure 7.6a ii. The frequency differences among the isochromatic groups will dephase the magnetization vec-tors in time. Assuming the middle vector is in resonance in the rotating frame at the frequency ω, some isochromatic groups will precess faster than ω while others will be slower than ω. Consequently, the magnitude of the bulk magnetization (the sum of the five vectors) will decrease (become dephased) at a rate faster than that due to the transverse relaxation alone.

The application of the 180° at the time of τ in the spin-echo sequence inverts the magnetiza-tion vectors (Figure 7.6a iii), putting the faster-dephasing isochromatic groups behind ω and the slower-dephasing isochromatic groups ahead of ω. Since all isochromatic groups keep their own precessional frequencies (i.e., keep their own directions of frequency spreads), these individual magnetization vectors will be refocused at the time 2τ, which leads to the formation of a spin echo. Since the 180° inversion pulse can only refocus the effects due to the inhomogeneous broadening, the echo still bears the reduction of the magnetization due to the spin-spin relaxation, as

$$M_y(2\tau) = M_0 \exp(-2\tau/T_2). \tag{7.4}$$

Equation (7.4), which is essentially Eq. (2.17), actually points out a minor inaccuracy in the artful illustration of *Physics Today*. Similar to the signal loss due to T_2 relaxation in NMR, not all runners in the race should be able to return to the starting point at exactly the same time. If, for example, one of the runners gave up the race before returning to the starting point, then the *Physics Today* illustration would be both artistically beautiful and scientifically accurate.

Since the inhomogeneous broadening leads to a *faster* dephasing of the magnetization and a *faster* decay of the signal, the combined effect by both homogeneous broadening and inhomo-geneous broadening on the signal decay can be characterized by a T_2-like quantity, called T_2^* relaxation (Figure 7.6a). Typically, T_2^* relaxation time is much shorter than T_2 relaxation time. One can relate T_2 and T_2^* by the rate of the transverse relaxation,

$$\frac{1}{T_2^*} = \frac{1}{T_2} + \frac{1}{T_{\text{extra}}}, \tag{7.5}$$

where T_{extra} represents the extra transverse relaxation due to, for example, the inhomogeneity of the magnetic field. There are numerous factors that can lead to the field inhomogeneity, from inferior magnet or poor shimming to the material properties of the specimen and specimen holder and so on.

Since Eq. (7.4) states that the reduction in the magnetization amplitude is only due to T_2, the spin-echo sequence can be used to measure T_2 relaxation time experimentally. An example will be given at the end of this chapter, in Section 7.7. It is worth noting that a critical requirement for the spin-echo sequence to remain valid is that the spins have not moved from one site to another dur-ing the time 2τ. In other words, there is no translational motion of the molecules in the specimen.

(a) 90°|x' - τ - 180°|y' sequence

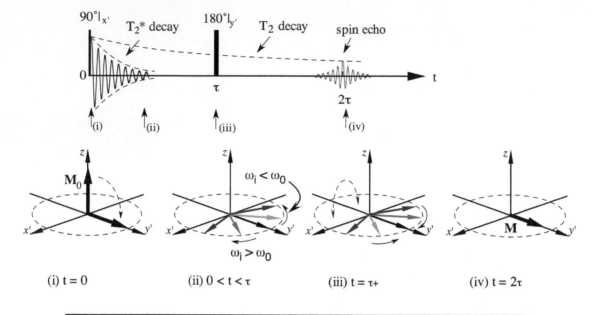

(i) t = 0　　　　(ii) 0 < t < τ　　　　(iii) t = τ+　　　　(iv) t = 2τ

(b) 90°|x' - τ - 180°|x' sequence

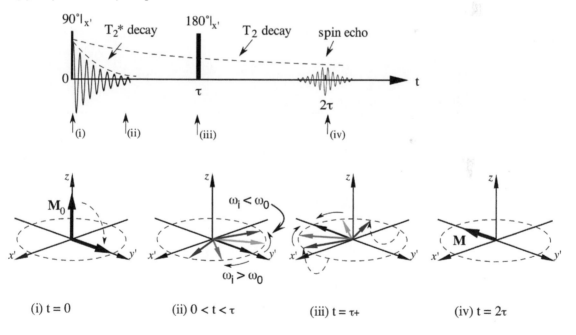

(i) t = 0　　　　(ii) 0 < t < τ　　　　(iii) t = τ+　　　　(iv) t = 2τ

Figure 7.6 Two versions of the spin-echo sequence, which can remove dephasing due to inhomogeneous broadening, chemical shift, and so forth. Each group of nuclear spins is represented by a small colored vector in events *ii* and *iii*, where the addition of all five small vectors form the bulk magnetization (black vector in events *i* and *iv*). The 180° pulse changes the positions of the individual spin groups but not the precessional directions of the individual spin groups. The signal loss due to T_2^* can be recovered ($T_2^* < T_2$).

7.4   CPMG ECHO TRAIN

In the spin-echo sequence, the phase coherence recovered in the spin echo is subsequently lost for $t \gg 2\tau$. Successive recoveries are possible if a train of additional 180° rf pulses is used, as suggested by Carr and Purcell [3] and modified by Meiboom and Gill [4]. This is because after the first 180° pulse, the second half of the first spin echo is fully equivalent to the FID, only attenuated by T_2 relaxation (Figure 7.7b). The spin-echo experiment can therefore continue with the addition of another 180° pulse, and another 180° pulse, ..., which forms a train of spin echoes. The envelope of the echoes in Carr–Purcell–Meiboom–Gill (CPMG) sequence is determined by T_2 decay alone, and it is therefore possible to determine the T_2 value in a single experiment. Either the $90°|_{x}$-τ-$180°|_{x}$ or the $90°|_{x}$-τ-$180°|_{y}$ pulse sequence could be extended to form a train of echoes, but the CPMG sequence uses $90°|_{x}$-τ-$180°|_{y}$ because it is less sensitive to small errors in the 180° pulses.

(a) CPMG sequence

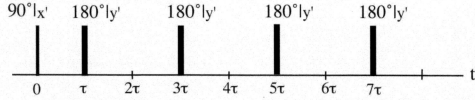

(b) Signals in CPMG

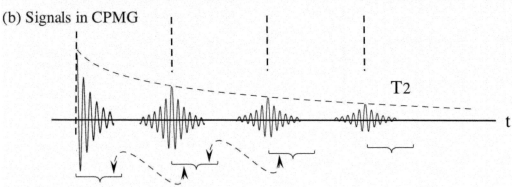

(c) A short form of CPMG sequence

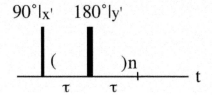

Figure 7.7   (a) CPMG sequence, which can have a number of $180°|_{y'}$ pulses; each can refocus inhomogeneous broadening of the nuclear spins and form a spin echo. (b) The peaks of all echoes decay as T_2 process while the individual FID decays as T_2^*. (c) A short form to represent CPMG sequence graphically in a pulse sequence.

7.5   STIMULATED ECHO SEQUENCE

For many specimens, T_1 relaxation time (the time for the spins returning to thermal equilibrium) is much longer than T_2 relaxation time (the time for the spin to dephase from coherence to disorder). For example, many biological samples can have T_1 on the order of 2–3 seconds and T_2 on the order of 50–100 ms. This relaxation characteristic of the specimen ($T_1 \gg T_2$) limits the window of time during which one can manipulate the magnetization; for example, applying a set of field gradients to measure the self-diffusion (cf. Chapter 15.1).

To maximize the time that one can use for spin manipulation, the concept of the stimulated echo [2] can be used, as shown in Figure 7.8. This pulse sequence looks like a spin-echo

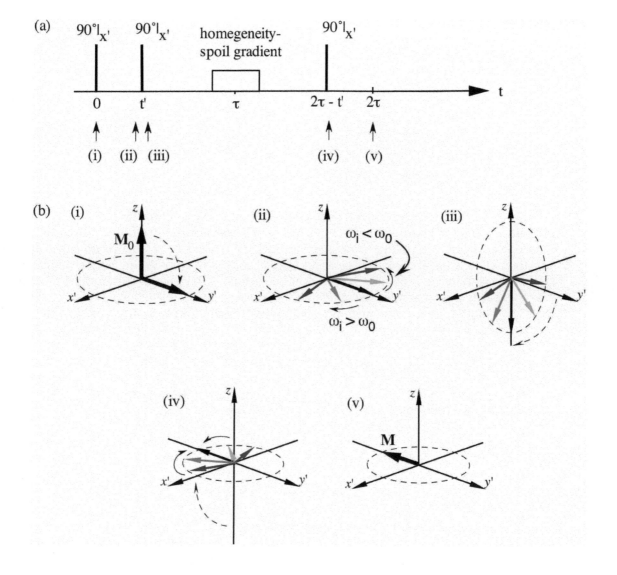

Figure 7.8   (a) Stimulated echo sequence. (b) The motion of the magnetization. Each group of nuclear spins is represented by a small colored vector in events *ii*, *iii*, and *iv*, where the addition of all five small vectors forms the bulk magnetization (black vector in events *i* and *v*). Note that only ½ of the magnetization can be recovered, which is reflected schematically as a smaller circle in *iv* and *v*.

sequence, except its 180° pulse is split into two 90° pulses. The initial 90° pulse establishes the transverse magnetization. The second 90° pulse tips the y' component of the transverse magnetization to the longitudinal axis so that it only undergoes T_1 relaxation. The x' component of the transverse magnetization is unaffected by this 90° pulse and so remains in the transverse plane to undergo T_2 decay. After a delay, the last 90° pulse brings back the longitudinal spin vector to the transverse plane for detection. Therefore, the maximum delay time of the transverse magnetization can be stretched to be on the order of T_1. This sequence has enabled the measurement of the vascular flow as low as 45 μm/s in living plants [5].

Two points are worthy of mention. First, only half the signal could possibly be recovered because only the y' component is restored to the longitudinal axis, which is a price paid for the long echo time. The maximum signal in a stimulated echo experiment can still be determined by Eq. (7.4), except the magnitude is reduced by 1/2. Second, more than one final signal could arise from the transverse magnetization established by the initial 90° pulse because each following 90° pulse could also produce a spin echo. Therefore, some sort of "cleaning" step is commonly used to destroy these unwanted transverse magnetizations (e.g., a homogeneity-spoil gradient pulse, as shown in Figure 7.8, which forces the transverse spins to precess at much faster rates, hence dephase very quickly).

7.6   SPIN-LOCKING AND $T_{1\rho}$ RELAXATION

In addition to T_1 and T_2 relaxation times, a different kind of spin relaxation can happen in NMR and MRI if the tipped magnetization is locked onto the transverse plane by a transverse $\boldsymbol{B}_1$ field, such as in Figure 7.9. In the experiment, the first 90° pulse has the usual job to tip the longitudinal magnetization M_0 to the transverse plane, in this case to the y' axis. Immediately after, another $\boldsymbol{B}_1$ pulse is applied, with a direction set along the y' axis. Without this $B_{1y'}$ pulse, the magnetization will immediately undergo the transverse spin relaxation as usual. When this $B_{1y'}$ field is applied and maintained, it becomes the only effective field in the rotating frame (cf. Chapter 2.5), which acts to keep (i.e., lock) the magnetization in the rotating frame, along the y' axis. Since the direction of the magnetization is in parallel with the only field in the rotation frame, there won't be any transverse spin relaxation to the magnetization. However, as discussed in Chapter 6.3, $|B_1|$ can be about 1/10,000 to 1/20,000 of a $\boldsymbol{B}_0$ field in modern instruments. Since the initial magnitude of the magnetization M_0 is proportional to the magnitude of $\boldsymbol{B}_0 (n \propto B_0)$, the current magnitude of the magnetization is far too large to be maintained by the B_1. Hence the magnitude of the magnetization will decay in time to the value of $(B_1/B_0) M_0$.

This decay is a new type of spin relaxation, termed as the spin-lattice relaxation in the rotating frame, or simply $T_{1\rho}$ relaxation. This relaxation can be determined by

$$M(\tau) = M_0 \exp(-\tau/T_{1\rho}). \tag{7.6}$$

Similar to the calculation of T_1 and T_2, repeating the $T_{1\rho}$ sequence at different τ can lead to the calculation of $T_{1\rho}$ quantitatively. $T_{1\rho}$ is a useful parameter in spin dynamics. Since the sensitivity of $T_{1\rho}$ relaxation to molecular motion is in a frequency range between T_1 and T_2 (cf. Figure 3.4), the value of $T_{1\rho}$ has been found to reflect a certain structural degradation in biological tissues (Chapter 15.2) [6].

(a)

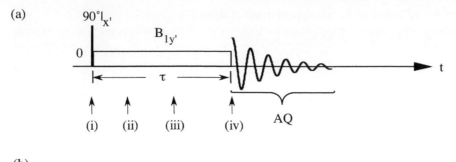

(b)

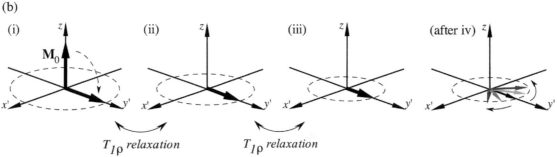

Figure 7.9   (a) The $T_{1\rho}$ sequence. (b) The motion of the magnetization. After being tipped to the y' axis, the magnetization is locked up immediately by the $B_{1y'}$ pulse. Because the magnitude of the $\boldsymbol{B_1}$ field is much smaller than B_0, $\boldsymbol{M}$ will relax along the y' axis and reduce its magnitude (i.e., graphically the size of the circle is getting smaller between events *i* and *ii* and between events *ii* and *iii*). The dephase of the magnetization starts after the end of the $B_{1y'}$ pulse, which starts the signal acquisition.

7.7   HOW TO SELECT THE DELAYS IN RELAXATION MEASUREMENT

All pulse sequences discussed in this chapter can be used to measure the relaxation characteristics of a specimen. Commonly, the same sequence is repeated multiple times, each time having a different delay time (either τ or 2τ in the sequence). Then the maximum signals from all measurements are fitted with the relaxation equation, which yields the relaxation time measurement. An immediate consideration in the experimental planning would be the selection of these delay times. Here, we use a T_2 measurement by the spin-echo sequence (Figure 7.6) as an example.

Let's suppose we plan to repeat the spin-echo sequence 10 times to measure the T_2 of a specimen, which we estimate to be 60 ms. Since T_2 decay is described by Eq. (7.4), an intuitive planning would be to step up the delay time (2τ) from 0 with a fixed step size (e.g., 15 ms). A series of delay times and their associated signal amplitudes are shown in Table 7.1a and plotted in Figure 7.10a. Since this is simulated data without noise, an exponential fit to Figure 7.10a yields precisely a T_2 of 60.0 ms with an R value (a measure of the goodness of fit) of 1.00 (a perfect fit).

Table 7.1   Two series of delay times in T_2 relaxation measurement.

(a) Equal time delays		(b) Equal amplitude decays	
2τ (ms)	Amplitude	2τ (ms)	Amplitude
0.0000	1.0000	0.0000	1.0000
15.000	0.77880	6.3216	0.90000
30.000	0.60653	13.389	0.80000
45.000	0.47237	21.400	0.70000
60.000	0.36788	30.650	0.60000
75.000	0.28650	41.589	0.50000
90.000	0.22313	54.977	0.40000
105.00	0.17377	72.238	0.30000
120.00	0.13534	96.566	0.20000
135.00	0.10540	138.16	0.10000

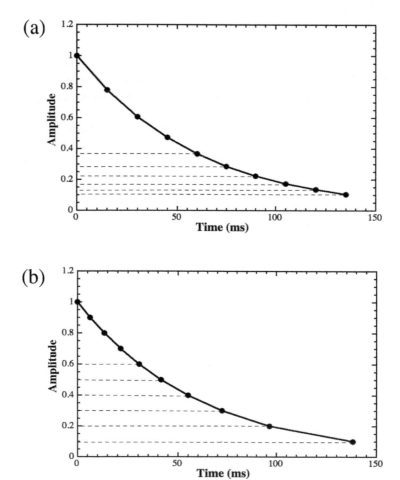

Figure 7.10   Two ways to sample an exponential decay: (a) equal time delay, (b) equal amplitude decay.

A close look at the plot in Figure 7.10a reveals that the second half of the data points in this equal-delay approach essentially catch the long tail of the exponential, which changes little. Since each individual experiment takes essentially the same amount of time, one does not really need to sample so densely the tail portion of the exponential. In practical situations, a certain amount of experimental noise is inevitable, which influences the low-amplitude data points (i.e., the tail of the exponential decay) more than the high-amplitude data points (the initial decays). It would therefore be beneficial to sample the amplitude of the exponential decay equally, which can be determined by inverting Eq. (7.4):

$$\ln(M_y/M_0) = -2\tau/T_2. \tag{7.7}$$

The determination of the delay times can be obtained by utilizing Eq. (7.7) with equal increments of M_y/M_0, which spreads the delay times logarithmically (in equal increments if plotted on semi-log graph paper), as shown in Table 7.1b and plotted in Figure 7.10b.

In any practical experiment, therefore, one should mentally select the sampling points logarithmically, placing them more densely for the early decays and gradually more sparsely for the later decays. This approach would reduce the influence of noise to the data fitting and also catch both short and long relaxation components if you suspect the specimen has multi-component relaxation (cf. Chapter 16.2). The influence of experimental noise on signal fitting is discussed further in Chapter 15.5.

References

1. Lambert RK, Pack RJ, Xia Y, Eccles CD, Callaghan PT. In Vitro Tracheal Mechanics by Nuclear Magnetic Resonance Imaging. *J Appl Physiol.* 1988;65(4):1872–9.
2. Hahn EL. Spin Echoes. *Phys Rev.* 1950;80:580–94.
3. Carr HY, Purcell EM. Effects of Diffusion on Free Precession in Nuclear Magnetic Resonance Experiments. *Phys Rev.* 1954;94(3):630–8.
4. Meiboom S, Gill D. Modified Spin-echo Method for Measuring Nuclear Relaxation Times. *Rev Sci Instrum.* 1958;29:688–91.
5. Xia Y, Sarafis V, Campbell EO, Callaghan PT. Non Invasive Imaging of Water Flow in Plants by NMR Microscopy. *Protoplasma.* 1993;173:170–6.
6. Xia Y. MRI of Articular Cartilage at Microscopic Resolution. *Bone and Joint Res.* 2013;2(1):9–17.

Part III

Essential Concepts in NMR Spectroscopy

8

First-order 1D Spectroscopy

This chapter deals with first-order spectra (i.e., the weak coupling cases in Chapter 4.3), where all differences in chemical shift $|\omega_j - \omega_k|$ are much greater than the corresponding coupling constant $|J_{jk}|$, that is, $|\omega_j - \omega_k| >> 2\pi |J_{jk}|$. In this "simplified" situation, spectral lines are split into multiple lines with equal intensities (i.e., only Figure 4.6c, not Figure 4.6d). For a given spectral line that has been split, the summation of its resultant multiple lines' intensities equals the original spectral line's intensity. It also includes the weak coupling of one spin to more than one equivalent spin (i.e., AX_N systems). Note that the term *line* is used here to represent a spectral peak in a schematic, where the line height provides a relative proton count.

This chapter first describes the nomenclature of the spin system in NMR spectroscopy, then three characteristic parameters that specify the environment of the specific nucleus in the sample (the peak shift, the peak area/intensity, and the peak splitting). The last section discusses spectra of selected small organic molecules.

8.1 NOMENCLATURE OF THE SPIN SYSTEM

In NMR spectroscopy, each nucleus in a system of neighboring nuclei is given a letter, for example, A, B, C, and so forth. If several nuclei have the same chemical shift, that is, they are magnetically equivalent nuclei, numerical subscripts are used, such as A_2, B_3, and so on. For example, the two hydrogen atoms in a water molecule (H_2O) are an A_2 system. Magnetic equivalence [1] could be due to structural symmetry, rapid rotation, or simply by coincidence.

If another group of nuclei has very different chemical shifts, the letters at the end of the alphabet are commonly used to name them, such as X, Y. (It may also be used as an approximation for nuclei of the same species where the coupling is much less than the chemical shift differences.) When there are more than two well-separated groups of nuclei, a middle letter of the alphabet can be used, for example, AMX implies the existence of three different nuclei with three different magnetic environments. Other nomenclatures of the spin system in NMR spectroscopy also exist [1–3].

Essential Concepts in MRI: Physics, Instrumentation, Spectroscopy, and Imaging, First Edition. Yang Xia.
© 2022 John Wiley & Sons Ltd. Published 2022 by John Wiley & Sons Ltd.

8.2 PEAK SHIFT – THE EFFECT OF CHEMICAL SHIFT

In Chapter 4.2, the equation for chemical shift was given as $f = \gamma B_0 (1 - \sigma) / 2\pi$, where σ is the shielding constant and f is the temporal/linear frequency in hertz; and chemical shift is discussed in this book with the assumption of *diamagnetic* materials. Practically, chemical shift of a resonance is compared to that of a reference, commonly tetramethylsilane (TMS) in ^{1}H and ^{13}C spectroscopy. In a typical plot of an NMR spectrum, the horizontal axis is the chemical shift in ppm and the vertical axis is the signal amplitude. The chemical shift of the nucleus is commonly quoted in the δ scale, given by

$$\delta = 10^6 \frac{f_{\text{sample}} - f_{\text{TMS}}}{f_{\text{TMS}}} = (\sigma_{\text{TMS}} - \sigma_{\text{sample}})10^6, \tag{8.1}$$

where the 10^6 factor converts the chemical shift from the frequency ratio to ppm [cf. Eq. (4.10)].

In the early days of NMR experiments, the electric current was slowly increased in the coil of an electromagnet to increase the external magnetic field B_0 [cf. Eq. (5.1)]. If a resonance occurred in the sample, an absorption peak would show up on the cathode-ray screen of an analog oscilloscope, which was recorded on paper by a chart recorder. If there was no resonance after a reasonable amount of time, one increased the electric current to produce a higher field. The left-hand side of the recording paper therefore represented the effect of the lower magnetic field and was called "downfield." The right-hand side of the recording paper contained the result of the higher magnetic field and was called "upfield." A larger shielding constant, σ, requires a higher field to reach the resonance; so a peak with a large σ would appear on the right side of the recording paper. Since $\omega = \gamma B_0 (1 - \sigma)$, the peak that has a larger σ would have a lower resonant frequency and consequently a smaller chemical shift.

The terms "upfield" and "downfield" used by the NMR pioneers remain in modern NMR spectroscopy, which can be confusing to beginners. Figure 8.1 shows schematically a two-peak system. Compared to the resonant peak B, peak A has a smaller shielding constant σ, a higher

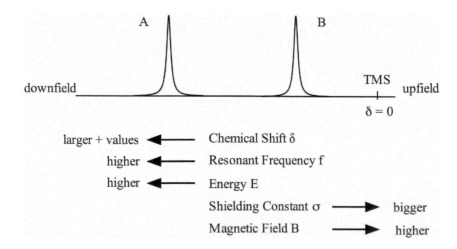

Figure 8.1 Fundamental relationships among the related parameters in chemical shift in ^{1}H NMR spectroscopy. The two peaks represent the two individual resonances, which are referenced to the 0 ppm chemical shift of TMS on the right-hand (upfield) side of the chemical shift axis. The horizontal axis can also be plotted as the resonance frequency f.

resonant frequency f, a resonance downfield (i.e., it requires a relatively lower magnetic field), larger positive values in chemical shift δ, and a higher energy. Note that according to Figure 4.3, a peak shift to the left, towards a higher frequency, implies an increased Zeeman energy.

Figure 8.2 shows the NMR spectrum of ethanol (CH_3CH_2OH), without the use of any solvent (an experimental condition sometimes called *neat*). This spectrum was deliberately shimmed improperly on a 400-MHz NMR spectrometer to illustrate poor resolution. It is comparable with the very first NMR spectrum of ethanol published in 1951 (Figure 1.4). Three resonant peaks in the main spectrum represent the protons in OH, CH_2, CH_3 of the ethanol molecule (ca. 5.48 ppm, 3.73 ppm, and 1.30 ppm, respectively). (Note that the spectrometer frequency was set on TMS = 0 ppm by a different sample earlier.) The little blip around 4.7 ppm was due to the protons in residual water contamination, including moisture in the air dissolved into the ethanol. Since no additional additive was used in the experiment, water resonated at its nominal frequency. If a deuterated solvent had been used, as is done commonly in modern NMR spectroscopic experiments, the chemical environment of the protons in water would have changed, and this water peak may appear at different frequencies in the measurement. In general, the chemical shifts of the peaks in a sample are not only known to be solvent-dependent, but also concentration-dependent and temperature-dependent [4].

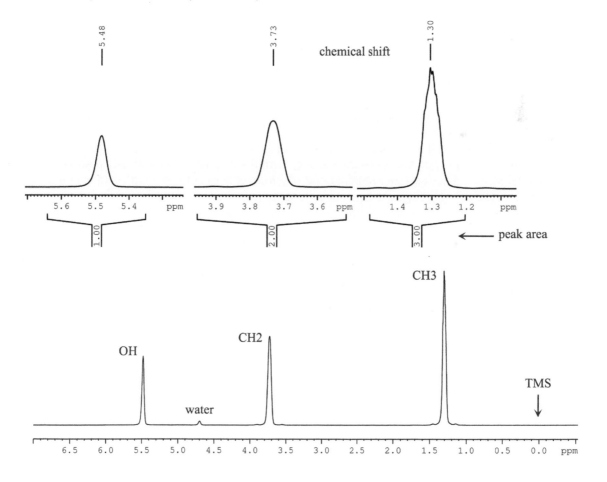

Figure 8.2   A low-resolution spectrum of ethanol (CH_3CH_2OH), without the use of any solvent. Due to the intentionally poor shim (hence poor field homogeneity), the peak splitting is not resolved. The chemical shifts and the ratios of the peak areas can be accurately determined from the spectrum. The small blip at 4.7 ppm is due to residual water contamination.

121

8.3 PEAK AREA – REFLECTING THE NUMBER OF PROTONS

NMR spectroscopy is a unique tool in quantitative analytical chemistry and industry. Peak area in NMR spectroscopy can be used to provide a relative proton count. (In simple schematics where a line is used to represent a resonant peak, the line height should be drawn to scale to provide a relative proton count.) In addition, the proportionality of peak area to concentration in NMR is valid for comparisons between different molecules, thus providing a rapid non-destructive method of obtaining concentration. It is straightforward to obtain absolute concentrations of a chemical in solution if another chemical, called a standard, is added in a known concentration. In the spectra shown in Figure 1.4 and Figure 8.2, the area ratios of the three peaks are 1:2:3, matching the numbers of protons in each of the three unique groups of protons (OH, CH_2, CH_3) in the ethanol molecule. This proportionality among the spectral peaks is of great importance for structure determinations.

8.4 PEAK SPLITTING – THE CONSEQUENCE OF *J* COUPLING

When there are only two nuclei with non-zero spin in the molecule under consideration (*I* and *S*), the Hamiltonian of the scalar interaction is given by Eq. (4.12). The consequence of the scalar interaction is to split the resonance peak of spin *I* into $(2S + 1)$ lines of equal intensity and that of spin *S* into $(2I + 1)$ lines also of equal intensity, with the line separations all being equal. For a pair of nuclei with $I = 1/2$ (e.g., two protons), the spectrum consists of two doublets (as in Figure 4.6c), where the doublet splittings are equal to *J* in hertz.

In general, for a system consisting of *n* spin-1/2 nuclei (none of which are equivalent, and all of which are interacting with their neighbors), each nucleus gives rise in principle to 2^{n-1} lines for the neighbors. The total number of lines in the spectrum is $n2^{n-1}$. In practice, the total number of lines can be less than $n2^{n-1}$, due to (a) some split lines may overlap and cause an increased intensity at the frequency location; (b) sometimes coupling constants are too small to be observed; and (c) some groups can be insulated by some part of the molecule; for example, the methyl group (CH_3) in an ethyl acetate molecule is insulated by the carbonyl and ether oxygen, hence only has a single peak (cf. Section 8.5.3).

8.4.1 Characteristics of *J* Coupling Constant and the Coupled Spectrum

1. *J* depends on the chemical environment, not on the operating frequency of the spectrometer [i.e., since there is no B_0 term in Eq. (4.12)]. Hence *J* is always quoted in hertz, not in ppm.
2. Individual absorption lines are not assigned δ values in ppm, since their positions are dependent of the spectrometer's operating frequency. Individual absorption lines are commonly given their resonance frequencies, and from these the *J* coupling constants can be determined. The δ scale in ppm should be reserved for true chemical shift positions.

3. J coupling is only effective between nuclei in the same molecule. When there are more than two nuclei in the molecule, coupling still can occur between pairs of nuclei, and the coupling constants normally differ in their magnitude. In some cases, nuclei with similar chemical environments will have equivalent coupling constants. The pattern of lines for a given nucleus may be explained by the method of successive splitting (see the examples later in this section, e.g., Section 8.4.4).

4. The magnitude of coupling constant tends to decrease as the number of chemical bonds separating the coupled nuclei increases. This is because the most common mechanism for coupling involves the transmission of the coupling effect through the bonding electrons. The magnitude of this effect is attenuated by each successive bond.

5. For saturated hydrocarbons (i.e., any organic compound consisting only of hydrogen and carbon), the coupling constant between protons separated by more than three bonds is often negligibly small. Hence, in most cases, spectral splitting can be attributed to groups of protons on adjacent carbons.

8.4.2 Basis Product Functions

As we see in Figure 3.1, the two energy states for spin-1/2 nuclei are labeled as α (spin-up, ↑) for $m_I = +1/2$ and β (spin-down, ↓) for $m_I = -1/2$, where α and β are the eigenkets of the spin system and the values of the azimuthal quantum number m_I are the eigenvalues of the spin operator I_z. In this notation, any overall spin state may be designated by the product of the wave functions of an individual nucleus. For example, one state in a three-spin AMX system can be labeled as $\alpha_A \beta_M \beta_X$, or simply $\alpha\beta\beta$ (i.e., ↑↓↓), where the order of the three letters or arrows matters. We call functions such as $\alpha\beta\beta$ "basis product functions," where the word "basis" means that α or β is a basis ket as defined in quantum mechanics (cf. Appendix 2). Since there are two possibilities for the wave function of each spin-1/2 nucleus, there are 2^n basis product functions for a spin system of n spin-1/2 nuclei. A two-spin system would have $2^2 = 4$ spin states (Table 8.1), while a three-spin system (none equivalent) would have $2^3 = 8$ spin states (Table 8.2). m_T in the two tables represents the total quantum number m_I for the system.

Table 8.1 Basis product functions for a two-spin AX system.

Spin states for AX	Sum of eigenvalues m_T	Product of $m_A m_X$
$\beta\beta$	−1	+1/4
$\alpha\beta, \beta\alpha$	0	−1/4
$\alpha\alpha$	+1	+1/4

Table 8.2 Basis product functions for a three-spin AMX system.

Spin states for AMX	Sum of the eigenvalues m_T	Product of $m_A m_M m_X$
$\beta\beta\beta$	−3/2	−1/8
$\beta\beta\alpha, \beta\alpha\beta, \alpha\beta\beta$	−1/2	+1/8
$\beta\alpha\alpha, \alpha\beta\alpha, \alpha\alpha\beta$	+1/2	−1/8
$\alpha\alpha\alpha$	+3/2	+1/8

123

124

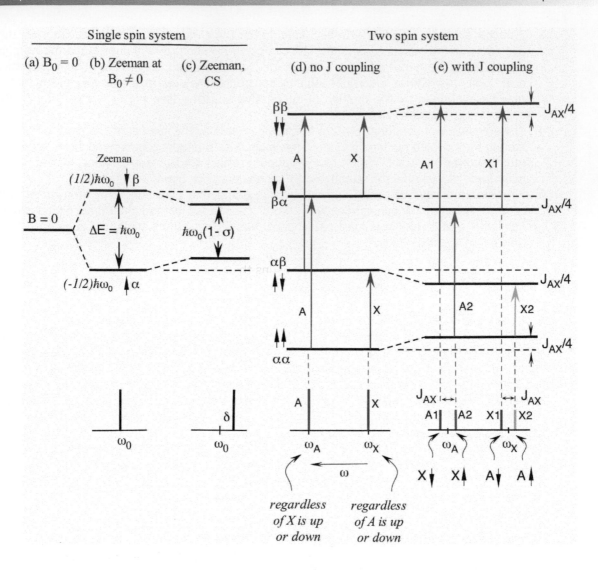

Figure 8.3 Schematic line patterns of a two spin-1/2 AX system when $J_{AX} > 0$, with the energy levels as in Figure 4.7b. Note again that for homonuclear systems, the energies of the $\alpha\beta$ and $\beta\alpha$ states are similar but very different from the other two states ($\alpha\alpha$ and $\beta\beta$). In the schematic drawings, the reductions and increments of the Zeeman energies under the influence of J coupling are grossly exaggerated (i.e., not drawn in proportion), for the purpose of visualization. The Zeeman levels correspond to frequencies of hundreds of megahertz, while the modifications to the Zeeman levels by J coupling are only several hertz.

8.4.3 Two-spin AX Systems

Figure 8.3 shows the schematics for a two-spin AX system (where both A and X are spin-1/2), which has four basis product functions (Table 8.1). There are four eigenvalues associated with the four individual states of the two spins, $m_A = \pm 1/2$, and $m_X = \pm 1/2$.

When the nuclei in this two-spin system do not couple (Figure 8.3d), any association of an energy level with neighboring levels of different energy does not affect the resulting spectrum.

For example, the α level of the A nucleus could be associated with either the α or β level of the X nucleus, hence having the spin states of $\alpha\alpha$ or $\alpha\beta$. However, since there is no coupling between A and X, the α level of the A nucleus remains the same, regardless of whether it is associated with the α or β level of the X nucleus. As a consequence, the spectrum contains two resonant lines (peaks) that are independent of each other at their own resonant frequencies ω_A and ω_X.

When two spins in the AX system do couple via J coupling (Figure 8.3e), whether the α level of the A nucleus is associated with the α or β level of the X nucleus does affect the spectrum. The energy levels of $\alpha\alpha$ and $\beta\beta$ both increase by $J_{AX}/4$ since $m_A m_X$ is positive in both cases. The energy levels of $\alpha\beta$ and $\beta\alpha$ both decrease by $J_{AX}/4$ since $m_A m_X$ is negative in both cases. This results in four unequal energy gaps, which leads to four resonant lines of equal-intensity. These four lines are collectively referred to as two doublets, where the two A lines form one doublet and two X lines form another doublet.

According to quantum mechanics, any direct observation of a spectral transition has to meet the selection rule

$$\Delta m_T = m_{initial} - m_{final} = \pm 1. \tag{8.2}$$

These allowed transitions are also called single-quantum transitions, which can be observed directly in NMR. Accordingly, the transitions of two A spins in a two-spin AX system have $\Delta m_A = \pm 1$ and $\Delta m_X = 0$; the transitions of two X spins have $\Delta m_A = 0$ and $\Delta m_X = \pm 1$. Any transition that does not meet the selection rule, for example, between $\alpha\alpha$ and $\beta\beta$ (which is called double quantum transition since $\Delta m_T = 2$) and between $\alpha\beta$ and $\beta\alpha$ (which is called zero quantum transition since $\Delta m_T = 0$), can only be observed indirectly in multidimensional NMR.

For A1 and X1 lines in the AX system in Figure 8.3e, their energy gaps increase, causing an increase in their Zeeman energies. Consequently, the resonant peaks, A1 and X1, are shifted by $J_{AX}/2$ to the left of ω_A and ω_X, respectively. In the same way, the A2 and X2 lines are shifted by $J_{AX}/2$ to the right of the ω_A and ω_X, respectively.

The magnitude of a line's frequency allows one to draw certain conclusions. The A1 line has the highest resonant frequency (i.e., the biggest value in chemical shift). So one can conclude that the A1 line also has the smallest shielding constant, the lowest magnetic field, and the highest energy. The gap between the two A lines as well as between the two X lines is the J coupling constant J_{AX} between the A nucleus and the X nucleus. The energy cubic diagram of this two-spin AX system has been shown previously in Figure 4.7b.

8.4.4 Three-spin AMX Systems

For a three-spin AMX system (Table 8.2), where all three spins are spin-1/2 and their J-coupling constants are unequal, each nucleus' resonant line is split into four equal-intensity lines, named a doublet of a doublet. The same selection rule in Eq. (8.2) applies, providing the additional constraint that only one spin is allowed to flip. Figure 8.4 shows the schematics for the peaks produced from nucleus A in a three spin-1/2 AMX system. For nucleus A, the corresponding resonant line, ω_A, is split twice, once by coupling with M and another by coupling with X. Assume the coupling with M is greater (i.e., $J_{AM} > J_{AX}$). The pattern of A lines may be predicted by considering first the A splitting by M and then a further splitting of each of the resulting pair of A lines by X. The final pattern, for nucleus A, contains four lines.

Note that there is a subtle difference between two kinds of four resonance lines of equal intensity. The term *two doublets* describes the total line pattern for a two spin-1/2 AX system (the last section, Section 8.4.3), where each doublet centers around an unperturbed resonance

126

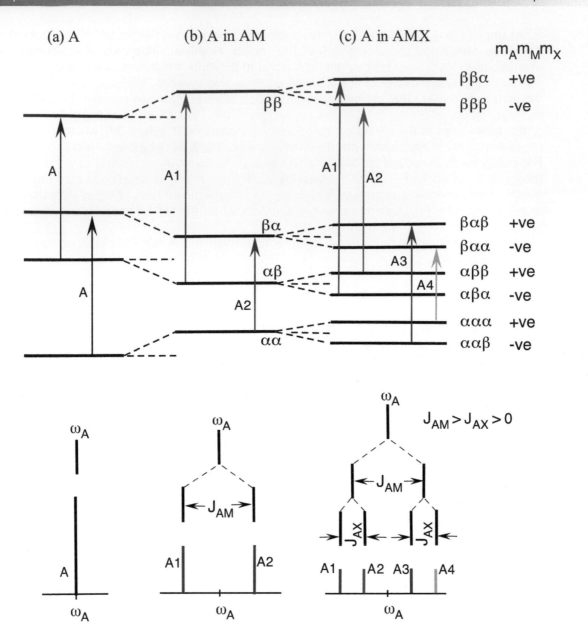

Figure 8.4 Schematic line patterns of the A resonance in an AMX system. (a) A only. (b) A in AM. (c) A in AMX, assuming $J_{AM} > J_{AX} > 0$. Note that the line intensities in (a), (b), and (c) are scaled according to the numbers of splitting, so the sum of the intensities remains constant among (a) to (c).

frequency (ω_A or ω_X). The second term, *a doublet of doublets* (or *dd* for short), describes the line pattern of a single nucleus in a three spin-1/2 AMX system (this section, Section 8.4.4), where the single nucleus (e.g., A) interacts with two different nuclei (e.g., both M and X); *a doublet of doublets* centers around *one* unperturbed resonance frequency (ω_A or ω_M or ω_X).

In the AMX system illustrated in Figure 8.4, the coupling constants associated with each of the nuclei can be measured directly from the spectrum when the frequency axis uses the unit of hertz. For nucleus A, J_{AM} is the line separation, in hertz, between lines 1 and 3 or lines 2 and 4. J_{AX} is the line separation, in hertz, between lines 1 and 2 or lines 3 and 4. The M and X nuclei's coupling constants (e.g., J_{AM} and J_{MX} for M) could be measured (not depicted in Figure 8.4) from a splitting pattern similarly but not identically from A's pattern. Altogether, the spectrum of this three-spin system would consist of 12 lines. Each part of the spectrum (a *doublet of doublets, or dd for short*) is symmetrical about its midpoint, where the midpoint also corresponds to the unperturbed Larmor frequency.

8.4.5   Coupled with Identical Spins in AX_N Systems

An AX_2 system is a system where there are two equivalent spins in a three-spin system. Looking at the schematic transition diagram in Figure 8.5b, the intensity of the middle line should be twice as high as the two adjacent lines since the middle two lines are overlapping. In fact, for an AX_N system where N is an integer, the peak intensities belonging to A can be determined from Pascal's triangle. Listed in Table 8.3 are the first six rows of Pascal's triangle. Note that Table 8.3 refers to the coupling between spin-1/2 nuclei; coupling to spin-1 nuclei will have different line patterns and different intensities.

Since the sum of the line intensities or peak areas for a molecule can be used to quantify the number of nuclei in each group of unique nuclei in the molecule (see the ethanol example in Figure 8.2), each nucleus' corresponding peaks are routinely scaled to correctly account for each nucleus in one molecule of the sample. The ratios in Table 8.3 are the nominal values, with a minimum consistently set at 1. The ratio values indicate the intensity differences among the

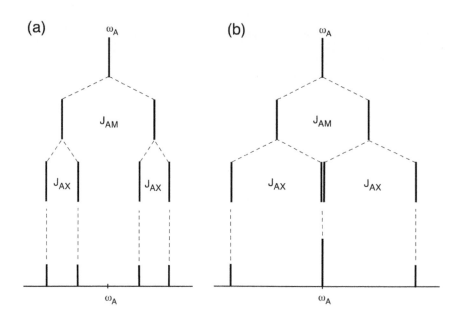

Figure 8.5   Schematic line patterns of the A resonant peaks in a three-spin system. (a) A being split by J_{AM} and J_{AX} when $J_{AM} > J_{AX} > 0$ (a coupled AMX system). (b) A being split by J_{AM} and J_{AX} when $J_{AM} = J_{AX} > 0$ (a coupled AX_2 system).

Table 8.3 Pascal's triangle for the spin-1/2 systems.

Spin systems	Peak patterns of nucleus A	Pattern name	Abbreviations
A	1	Singlet	s
AX	1 1	Doublet	d
AX_2	1 2 1	Triplet	t
AX_3	1 3 3 1	Quartet	q
AX_4	1 4 6 4 1	Quintet/pentet	p
AX_5	1 5 10 10 5 1	Sextet	Sextet

split peaks of the same spin due to the influences of its neighbor nuclei. When a molecule has more than one group of unique spins (e.g., the ethanol molecule CH_3–CH_2–OH has three groups of unique spins), a subtle normalization of the intensities among the different splitting patterns needs to be carried out as part of any scaling. Normalization is critical in identifying a nuclei count; an example will be given in Section 8.5.1.

Note that there could be two different causes when unequal intensities in peak splitting are seen. One cause is due to the overlapping of the peak split patterns as shown in Table 8.3; another cause is due to the strong coupling, which occurs when $|\omega_I - \omega_S| \sim 2\pi|J|$ or $|\omega_I - \omega_S| \ll 2\pi|J|$, resulting in unequal intensities (see Chapter 4.3). In this chapter where the first-order spectra are discussed (i.e., the weak coupling cases in Chapter 4.3), unequal line patterns due to the strong coupling are not considered.

8.5   EXAMPLES OF 1D SPECTRA

This section provides a number of practical 1D spectra from several small organic molecules that can be found commonly in a science or industry laboratory. By going through these samples, one can practice the knowledge from the earlier sections of this book, to have the joy of (re)discovering chemical structures.

8.5.1   Ethanol

A classic and easy example is the ethanol molecule (CH_3–CH_2–OH), which has been shown in Figure 8.2 (purposely acquired in low resolution). The high-resolution spectrum of the same sample is shown in Figure 8.6. It has three chemically unique groups of protons, approximately at the chemical shifts of 5.48 ppm (OH), 3.73 ppm (CH_2), and 1.30 ppm (CH_3). The intensity or peak area ratios of these three groups' corresponding peaks are 1:2:3, which quantify the number of nuclei (protons) in each chemically unique group. In applying NMR nomenclature, one may suppose this molecule is an A_3M_2X system, but the ethanol molecule is more appropriately considered as an A_3X_2 system. Because of the rapid acid-catalyzed proton exchange, the OH resonance is not usually split nor coupled with the other two groups (Figure 8.6a). An A_3X_2 system matches well in that the CH_2 resonance (Figure 8.6b) is split into a quartet (i.e., 1:3:3:1) by the CH_3 protons, and the CH_3 resonance (Figure 8.6c) is split into a triplet (i.e., 1:2:1) by the CH_2 protons.

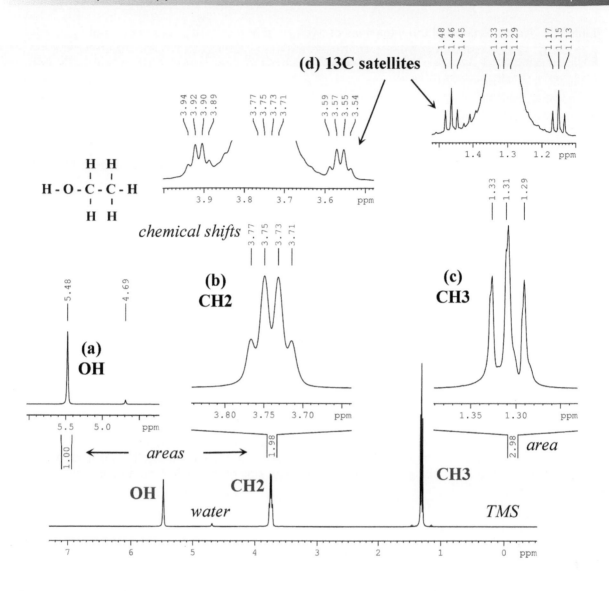

Figure 8.6 The full ^{1}H NMR spectrum of ethanol acquired without the use of any solvent (the same specimen used in Figure 8.2) is shown at the bottom on the figure. The CH_3–CH_2 form an A_3X_2 system, while the OH is a singlet. The inserts are not scaled relative to each other.

If one aims to calculate or draw a correctly scaled multi-spin spectrum by hand, one must take care to account for the differences in the number of protons in each chemically unique group of peaks. For example, plotting ethanol's eight peaks with heights of (from left to right) 1, 1,3,3,1, and 1,2,1 is incorrect. The reason is that the smallest intensity in the CH_2 splitting pattern (patterned 1:3:3:1) represents a different fraction of the total protons (two protons) than the fraction represented by the smallest peak in the CH_3 splitting pattern (1:2:1), that is, the intensity of 1 in CH_2 does not equal to the intensity of 1 in CH_3. Table 8.4 illustrates the steps in determining the height (or area) of the peaks, which accounts for height differences between different patterns (step #3), each associated with a different number of spins (step #4). These procedures start with the calculation of the smallest height

Table 8.4 Normalization of the intensities of two group of resonant peaks in ethanol.

Steps	CH$_2$	CH$_3$
(#1) The initial intensity/area patterns	1, 3, 3, 1	1, 2, 1
(#2) Sum of the initial intensities in each group	8	4
(#3) Scale the sums of the initial intensities between the groups for n = 1	1, 3, 3, 1	2, 4, 2
(#4) Multiply (#3) with the number of spins in each chemically unique group (2 in CH$_2$ and 3 in CH$_3$)	2, 6, 6, 2	6, 12, 6
(#5) Cross out the common numerical factor in (#4), which is 2	1, 3, 3, 1	3, 6, 3
(#6) Final sum of the intensities in each group	8	12

in each group (step #2). The final step (#6) serves to check if the ratios between groups of protons is preserved (in this case, CH$_2$:CH$_3$ = 8/12 = 2/3, which equals the ratio of protons between CH$_2$ and CH$_3$). The final plot should also include the OH peak, at an intensity of 4 relative to the smallest peak in the spectrum (which is scaled to 1), so that the ratio of the three separate sums of peak intensities (one sum for each chemically unique group of protons) remains at 1:2:3 (OH:CH$_2$:CH$_3$).

The ethanol spectrum displays some minor peaks shouldering the quartet and triplet (see Figure 8.6d). These minor peaks are called ^{13}C satellites, which come from the coupling of hydrogen atoms to an adjoining carbon-13 atom. Most of the carbon atoms are ^{12}C (spin zero), with ^{13}C having about 1.1% natural abundance in the carbon population. As a result, these satellites are smaller in intensity compared to Figure 8.6b and Figure 8.6c. When the main proton peak has proton-proton coupling, each satellite will be a miniaturized mirror of the main peak; for example, if the main proton peak is a doublet, then the carbon satellites will appear as a pair of miniature doublets, with one doublet on each side of the main proton peak.

8.5.2 Styrene and 2,3,4,5,6-pentafluorostyrene

Styrene or vinylbenzene is a colorless oily liquid with the chemical formula C$_6$H$_5$–CH=CH$_2$, consisting of benzene (C$_6$H$_6$) derivatized with a vinyl group (–CH=CH$_2$). (In + organic chemistry, vinyl protons refer to protons attached directly to a C=C double bond.) Figure 8.7a shows the proton spectrum of styrene, which is dissolved in deuterated chloroform (CDCl$_3$) spiked with tetramethylsilane (TMS) and measured with a 400-MHz spectrometer. On the right side of the spectrum, there are two single peaks. TMS is pre-set to define 0 ppm on the chemical shift scale. At approximately 1.5 ppm is the peak of residual water, where 1.5 ppm is an appropriate chemical shift for water with CDC13 as a solvent. [Without solvent CDC13, the chemical shift of water is 4.7 ppm (cf. Figure 8.2 and Figure 8.6).] Higher than 7 ppm on the left side of the spectrum, there is a group of complicated peaks that come from the benzene protons.

The easily resolved part of the styrene spectrum is between 5 ppm and 7 ppm, which represents the three vinyl protons. They are non-equivalent and mutually coupled, forming a three-spin AMX system that should have a total of 12 spectral lines. From the A protons, we can determine the two coupling constants (J_{AM} and J_{AX}), as J_{AM} = 17.61 Hz (e.g., 2703.00 Hz – 2685.39 Hz) and J_{AX} = 10.89 Hz (e.g., 2703.00 Hz – 2692.11 Hz). Similarly, the M protons have J_{MX} = 0.92 Hz (e.g., 2291.34 Hz – 2290.42 Hz). These calculations agree with the values in the literature [5].

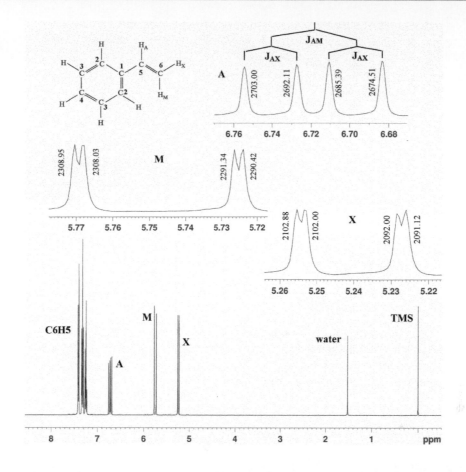

Figure 8.7a ^{1}H spectrum of styrene, where the three protons in the vinyl group (–CH=CH$_2$) form an AMX system, which does not couple with the benzene protons. The numbers 1–6 label the carbon nuclei.

The five protons in the aromatic ring (phenyl group, C_6H_5) can be replaced by five fluorine atoms, which gives the chemical a new name, 2,3,4,5,6-pentafluorostyrene (C_6F_5–CH=CH$_2$). Its proton spectrum is shown in Figure 8.7b, where the benzene proton peaks have been largely eliminated. The leftmost peak at approximately 7.2–7.3 ppm comes from residual non-deuterated chloroform (CHCl$_3$); it is common for a deuterated solvent to contain trace amounts of its non-deuterated molecule. The patterns of the three-spin AMX system remain largely unchanged, except J_{MX} (the smallest J coupling among the three) is not resolved. The M and X nuclei are called the *geminal* protons (i.e., two hydrogen atoms that are attached to the same carbon). It is possible that fluorine is causing some line broadening for the doublet of a doublet, making the doublet of a doublet appear as just a doublet. When J_{MX} is not resolved, one only has eight lines. Note also that although the sums of all individual peaks in each nucleus have equal areas (1.00, 1.01, 1.01), the individual lines have slight differences in their intensities, which might indicate the weakening of the weak scalar coupling condition (i.e., $|\omega_I - \omega_S|$ is no longer much larger than $2\pi|J|$) – cf. Figure 4.6. A detailed area analysis over the individual peaks yielded the four A peaks having the areas of 1.00, 1.08, 1.14, and 1.22, respectively; the two M peaks having the areas of 2.44 and 2.04, respectively; and two X peaks having the areas of 2.37 and 2.15, respectively (from left to right for each group of peaks; all peaks with the reference area of 1.00 at the leftmost A peak).

Figure 8.7c is the ^{13}C spectrum of styrene in deuterated chloroform, the same sample as in Figure 8.7a. Around 77.16 ppm, the three equally sized peaks come from the single carbon in

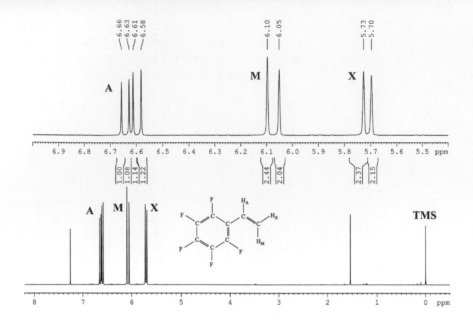

Figure 8.7b ¹H spectrum of 2,3,4,5,6-pentafluorostyrene, where the complicated peaks from the benzene protons have been eliminated. The lower spectrum is the full spectrum, while the upper spectrum is the enlarged segment for the AMX protons.

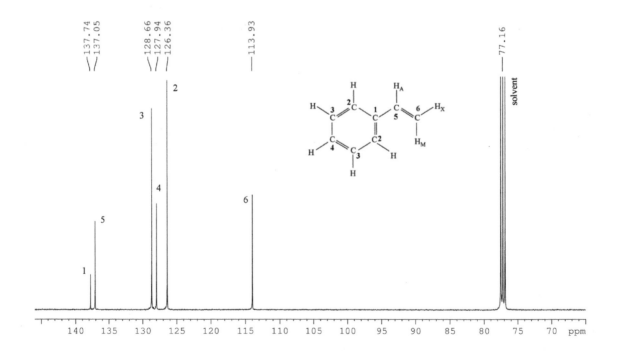

Figure 8.7c ¹³C spectrum of styrene when the coupling with the protons is eliminated.

the solvent, $CDCl_3$. Since the deuterium (^2H) has a spin of 1 (Table 2.1), it has three Zeeman energy levels (Figure 2.2b); hence spin coupling between ^{13}C and ^2H causes the carbon to have three narrow lines. The carbon spectrum is very much simplified due to the fact that ^{13}C is a dilute spin with about 1.1% natural abundance, which eliminates the spin-spin coupling between carbon atoms. In addition, the use of deuterated chloroform eliminates the proton coupling. The eight carbons in the styrene molecule are represented by the single peaks between 110 and 140 ppm, which have been well documented in the chemistry literature [5].

A few brief notes can be made about ^{13}C NMR, when comparing it to ^1H NMR. Since the gyromagnetic ratio γ of ^{13}C is about ¼ of ^1H, the resonance frequency of ^{13}C is about ¼ of ^1H at the same magnetic field (Table 2.1). Since ^{13}C constitutes only 1.1% of naturally occurring carbon, the NMR sensitivity of ^{13}C is several thousand times weaker than that of ^1H. In addition, because of the low abundance, ^{13}C NMR often lacks the homonuclear (^{13}C–^{13}C) couplings among the neighboring carbons, since a single molecule will most likely have only one ^{13}C atom. However, heteronuclear coupling between ^{13}C and ^1H can still be observed, because ^1H abundance is almost 100%. Since the interacting ^1H atoms are directly attached to the ^{13}C atom, the coupling constants in ^{13}C NMR are large, typically 50–250 Hz. The range of ^{13}C chemical shifts is 0–200 ppm with the 0 ppm being the carbon in TMS, much wider than the 0–10 ppm range in ^1H NMR. Finally, ^{13}C NMR spectra acquired with conventional parameters are not quantitative because of several specific characteristics of the carbon atoms (long T_1, decoupling, repetition time; more in Chapter 9.1).

8.5.3   Ethyl Acetate

Ethyl acetate (CH_3–COO–CH_2–CH_3) is a colorless liquid, which has an ester functional group (–COO–) connecting two alkyl groups (CH_3–, –CH_2–CH_3). It is used primarily as a solvent in chemical processes, one of which includes the decaffeination of coffee beans and tea leaves. The isolated CH_3 group produces a singlet at 2.05 ppm, while the other alkyl group (CH_3–CH_2–) has an A_3X_2 pattern. Its spectrum is shown in Figure 8.8.

8.5.4   Quinoline

Quinoline (C_9H_7N) is a heterocyclic aromatic compound, comprising a benzene ring fused to the C2 and C3 carbons of a pyridine ring. A pyridine ring is like the structure of a benzene ring, but one methine group (=CH–) is replaced by a nitrogen atom. Apparently, the direct application of this chemical is limited, but many of its derivatives have diverse applications. For example, over 200 biologically active quinoline and quinazoline alkaloids are identified; one of them is the anti-malaria drug quinine. Figure 8.9 shows the expanded ^1H spectrum of quinoline between 7 and 9 ppm. Since it is a common chemical, there are numerous previous reports on the peak assignment, for example, the Spectral Database for Organic Compounds by the National Institute of Advanced Industrial Science and Technology (AIST), Japan (https://sdbs.db.aist.go.jp), as well as journal papers [6]. The labels 2 to 8 in Figure 8.9 are for both carbons and hydrogens.

There are essentially two groups of hydrogens in quinoline. The benzene hydrogens in C_6H_6 are chemically and magnetically equivalent (hence having a single peak). All H5–H8 hydrogens become non-equivalent, after a benzene is "fused" with a pyridine. The pyridine hydrogens (H2–H4) are chemically and magnetically non-equivalent before it is fused with benzene and stay non-equivalent after fusing. The peaks for H2 and H3 hydrogens are located at both ends of the

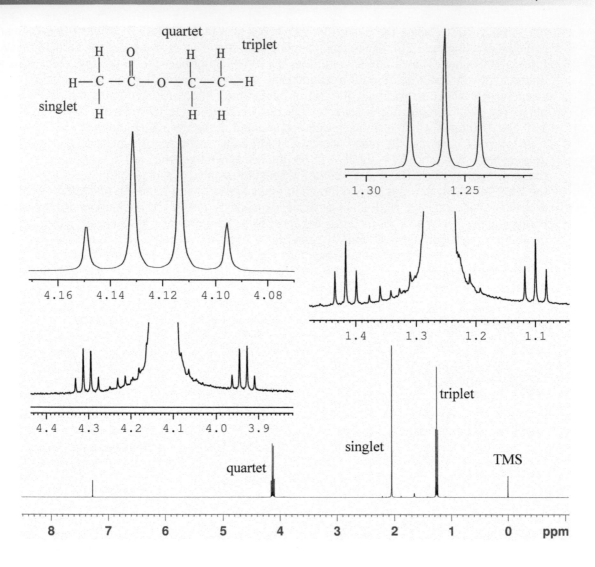

Figure 8.8 ¹H NMR spectrum of ethyl acetate, with its structure shown at the top left corner. It has an A_3X_2 pattern. The inserts are not scaled relative to each other.

quinoline spectrum (~7.35 and 8.9 ppm). Each keeps a simple quartet pattern of equal intensities for its local AMX couplings. For H2, the couplings are from $J_{2,3}$ (4.2 Hz) and $J_{2,4}$ (1.5 Hz); for H3, the couplings are from $J_{3,4}$ (8.3 Hz) and $J_{2,3}$ (4.2 Hz).

The peaks for the rest of the hydrogens (H4–H8), which are within 7.4 to 8.2 ppm, are more complicated. The enlarged inserts show that these peaks no longer have any simple patterns of equal intensity peaks (e.g., doublet, quartet) but form combinations of the simple patterns as well as coupling with the identical spins AX_N (see Section 8.4.5). As we have mentioned earlier in Section 8.5.2, the weakening of the weak scalar coupling condition can cause the split peaks to have unequal intensities (cf. Figure 4.6d), with the inner peaks of the line patterns having higher intensities than the outer peaks. Combined with some very small J-coupling constants, clear identification of these peaks is all but simple.

135

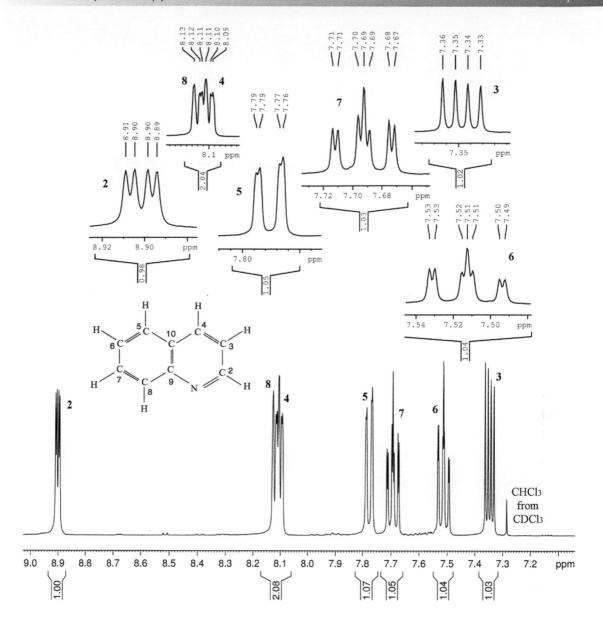

Figure 8.9 ^{1}H NMR spectrum of quinoline. See text for assignment.

References

1. Harris RK. *Nuclear Magnetic Resonance Spectroscopy – A Physicochemical View*. Essex: Longman Scientific & Technical; 1983.
2. Bovey FA. *Nuclear Magnetic Resonance Spectroscopy*. 2nd ed. San Diego, CA: Academic Press; 1988.
3. Sanders JKM, Hunter BK. *Modern NMR Spectroscopy - A Guide for Chemists*. 2nd ed. New York: Oxford University Press; 1993.
4. Gottlieb HE, Kotlyar V, Nudelman A. NMR Chemical Shift of Common Laboratory Solvents as Trace Impurities. *J Org Chem*. 1997;62:7512–5.
5. Friebolin H. *Basic One- and Two-dimensional NMR Spectroscopy*. 2nd ed. New York: VCH; 1993.
6. Seaton PJ, Williamson RT, Mitra A, Assarpour A. Synthesis of Quinolines and Their Characterization by 2-D NMR Spectroscopy. *J Chem Educ*. 2002;79(1):106–10.

9

Advanced Topics in Spectroscopy

In non-viscous (inviscid) liquids and solutions of simple chemicals at room temperature, molecules tumble rapidly, which averages out local interactions (e.g., dipolar interaction in Chapter 4) and results in uncomplicated NMR line shapes. In certain liquid-like samples, many biological tissues, and solids, these local interactions are not averaged out completely, which can cause complications in their spectra. To simplify the spectra for interpretation, special NMR techniques are needed to decouple, that is, to remove interactions among, the spins. In addition, one might actually be interested in the interactions among two nuclei that are of different kind, for example, a proton and carbon 13; such a spin system is described as *heteronuclear*. (By comparison, the word *homonuclear* in NMR describes the interactions among two nuclei that are of the same type, e.g., two protons, hence the two gyromagnetic ratios are equal.) This chapter briefly discusses a few advanced topics in NMR spectroscopy [1], which should also have relevance to MRI, since MRI images mostly non-liquid samples.

9.1 DOUBLE RESONANCE

A double-resonance experiment by definition applies two rf fields to the specimen. The main purpose of a double-resonance experiment is to decouple the coupled nuclear spins in order to simplify the spectrum and to improve its SNR. One can also use the double-resonance technique to suppress unwanted signals or to observe different features in the spectrum.

In the double-resonance experiments, the rf field that we have used in all single-resonance experiments in the previous chapters is called the observation field and still written as $B_1(t)$, which records the signal (observation). A second rf field is added in the double-resonance experiments. The second field is called the decoupling field, represented by $B_2(t)$, and is applied on a second rf channel. Both $B_1(t)$ and $B_2(t)$ are perpendicular to the polarizing magnetic field B_0. The decoupling field, B_2, commonly uses a noise signal to modulate its radio-frequency field. This noise signal is called a broad-band (BB) or noise pulse.

Three different pulse sequences for double-resonance experiments are shown in Figure 9.1. In these pulse sequences, both rf channels can be the same or different nuclear species, termed as homonuclear (e.g., two ^1H channels) or heteronuclear (e.g., one ^1H channel and one ^{13}C channel) decoupling; the decoupling field $B_2(t)$ can be a continuous-wave (CW) or gated format.

Essential Concepts in MRI: Physics, Instrumentation, Spectroscopy, and Imaging, First Edition. Yang Xia.
© 2022 John Wiley & Sons Ltd. Published 2022 by John Wiley & Sons Ltd.

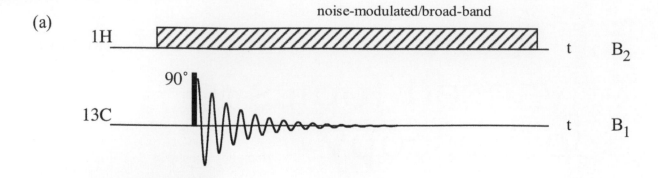

138

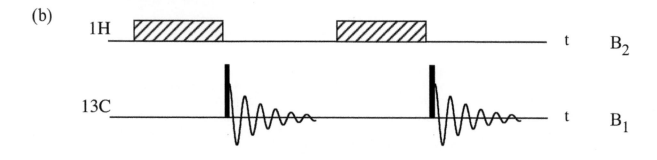

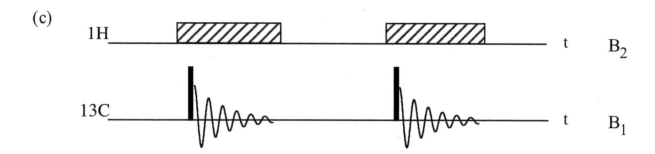

Figure 9.1 Pulse sequences for double-resonance experiments: (a) decoupling, (b) gated decoupling, and (c) inverse gated decoupling.

There are three different ways to apply this decoupling field, (a) decoupling (Figure 9.1a), (b) gated decoupling (Figure 9.1b), and (c) inverse gated decoupling (Figure 9.1c); the application of a B_2 decouples the interaction during that time period. Since coupling with ^{13}C is much weaker than coupling with ^1H (since ^{13}C is a dilute spin), the observation channel in a hetero-nuclear experiment is always on ^{13}C, not on ^1H. The effects of these pulse sequences are illustrated in the following sections.

9.1.1   Homonuclear Decoupling

A first-order three-spin-1/2 AMX system can have up to 12 lines in its spectrum due to the three J couplings (e.g., ^{1}H spectrum of styrene in Figure 8.7a). This type of line pattern provides extremely useful information about the chemical environment of the molecules. However, multiple lines lead to a reduced SNR for all peaks (since the sum of all split peak areas should equal to the area of the unsplit peak) as well as a more complicated spectrum. If a homonuclear decoupling sequence (e.g., similar to Figure 9.1a with both rf channels set on the proton) is used for the ^{1}H NMR experiment of an AMX system, setting the B_2 channel (CW irradiation) on ω_X would eliminate the coupling with the X proton. Since CW irradiation constantly flips the X spins between the two energy levels, the coupling with the X proton is averaged out completely, which reduces the number of lines in the spectrum from 12 to 4 (Figure 9.2b). This simplification can also improve the SNR of the spectrum, since each peak in Figure 9.2b would be twice as big as in Figure 9.2a. One can in principle further decouple the M nucleus at the same time with a CW irradiation set on ω_M (Figure 9.2c).

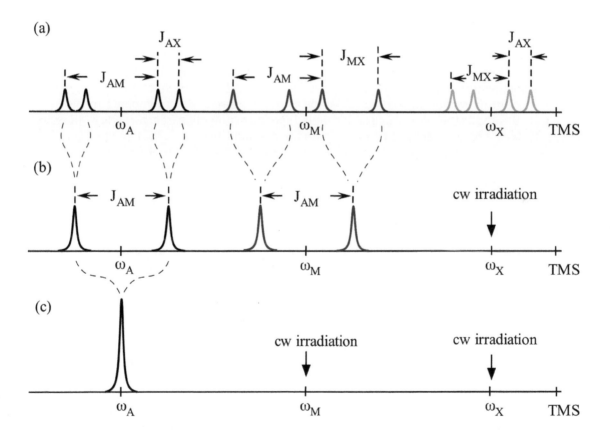

Figure 9.2   Schematic of homonuclear decoupling, where the spectrum in (a) is fully coupled. Note that the decoupling not only simplifies the spectral patterns but also improves the SNR of the spectrum.

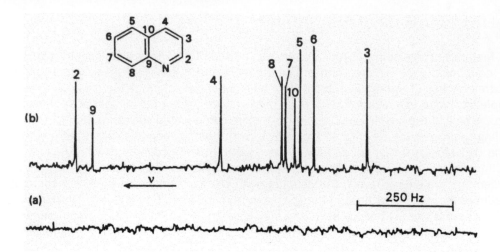

Figure 9.3 ^{13}C NMR spectra of quinoline (50% volume in CDCl$_3$) at 25 MHz: (a) coupled to protons, and (b) decoupled from protons. The intensities of (a) and (b) are on comparable scales. The line assignments are given by Pugmire et al. [2]. Source: Harris [1].

9.1.2 Heteronuclear Decoupling

The same decoupling sequence can also have the two rf channels set at different nuclei (e.g., one on ^{1}H and one on ^{13}C), which is called heteronuclear decoupling. The observation channel B_1 is commonly set on ^{13}C with the decoupling channel B_2 on ^{1}H, since the ^{13}C signal contains additional information in the molecule and is much weaker when compared to the ^{1}H signal. Figure 9.3 shows the heteronuclear decoupled spectra of quinoline from an early NMR experiment at 25-MHz ^{13}C frequency (where the instrument should have a 2.35 T magnet and a ^{1}H frequency of 100 MHz) [1, 2]. When coupled with the ^{1}H signal (Figure 9.3a), the ^{13}C signal is not visible. When decoupled from the ^{1}H signal (Figure 9.3b), the ^{13}C signal can be observed using the same amount of signal averaging. Removing the coupling between ^{1}H and ^{13}C, therefore, not only simplifies the ^{13}C spectrum but also significantly improves the SNR of the ^{13}C spectrum. This proton-decoupled ^{13}C spectrum of quinoline illustrates only one signal per unique carbon environment.

In the decoupled spectrum shown in Figure 9.3b, the coupling between the ^{1}H and ^{13}C nuclei was not present because of the use of double-resonance decoupling. At the same time, the coupling among the ^{13}C nuclei was not measurable, which was due to (a) ^{13}C is a dilute spin (at about 1.1% natural abundance), and (b) the measurement was performed using a weak magnetic field (25 MHz), so the coupling among the ^{13}C nuclei was much weaker due to the low probability of coupling between two ^{13}C nuclei in one molecule. Hence, the line patterns of ^{13}C in quinoline are much simpler and the SNR of the spectrum is much higher. Compare for yourself the ^{13}C spectrum and the ^{1}H spectrum of quinoline (Figure 9.3 vs. Figure 8.9).

The same chemical quinoline was measured in a modern 400-MHz NMR instrument (which has a 9.4 T magnet, ^{1}H frequency of 400 MHz, and ^{13}C frequency of 100 MHz). Two ^{13}C spectra are shown in Figure 9.4. Comparing Figure 9.4a and Figure 9.3a, a modern high-resolution NMR spectrometer that has a 4× stronger magnetic field and much better system engineering can resolve the couplings among the ^{13}C nuclei. Figure 9.4b shows the proton-decoupled ^{13}C spectrum from the 400-MHz NMR, where the peaks are much simpler and the spectrum has the same line pattern as that from the "ancient" instrumentation (Figure 9.3b), just as it should!

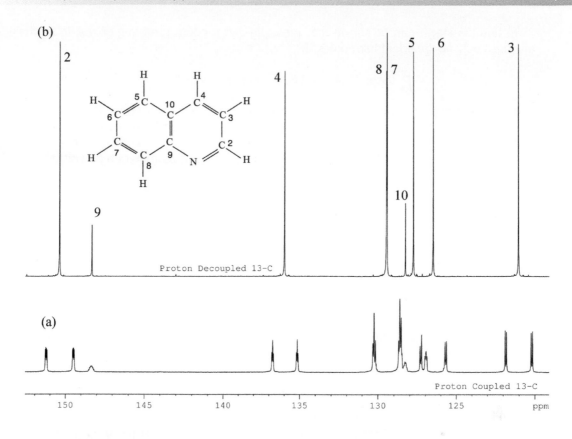

Figure 9.4 ^{13}C NMR spectra of quinoline dissolved in $CDCl_3$ at 400 MHz: (a) coupled to protons, and (b) decoupled from protons.

It should be noted that in simple (i.e., first-order) ^1H NMR (Chapter 8), the peak area/height in the proton spectrum can be used to determine quantitatively the relative number of protons at that particular chemical shift. The same determination in ^{13}C NMR is, however, not accurate. Due to more efficient relaxation (i.e., transfer of spin from carbon to hydrogen), the carbon atoms with more hydrogen atoms have stronger signals. The peak area or intensity can give a rough estimate of the number of carbons within the same type of carbons (e.g., methylene or methyl groups). In the example of quinoline in Figure 9.4b, the carbon 2 and carbon 3 are comparable; so are carbon 9 and carbon 10, as well as carbon 5, carbon 6, and carbon 7. More complicated experiments allow for recording of quantitative ^{13}C NMR spectra.

9.2   DIPOLAR INTERACTION IN A TWO-SPIN SYSTEM

The dipolar interaction for a two-spin system, $\boldsymbol{\mu}_1$ and $\boldsymbol{\mu}_2$, has been defined previously in Eq. (4.3), as

$$\mathcal{H}_D = \frac{\mu_0}{4\pi} \gamma_1 \gamma_2 \hbar^2 \left\{ \frac{I_1 \cdot I_2}{r^3} - 3 \frac{(I_1 \cdot r)(I_2 \cdot r)}{r^5} \right\}. \tag{4.3}$$

It was expanded in the format of a rank 2 tensor in spherical coordinates [Eqs. (4.4) and (4.5)], where each of the six terms from A to F contains a spin factor (the operator) and a geometrical factor (the spherical harmonics).

In liquid state, it can be shown that the rapid and random tumbling of the liquid molecules averages out the spherical harmonics [3],

$$\int_0^\pi (3\cos^2\theta - 1)\sin\theta d\theta = [-\cos^3\theta + \cos\theta]_0^\pi = 0, \tag{9.1}$$

where $\sin\theta$ is the weighting for a molecule with no preferred orientation. Therefore, the dipolar interaction does not produce line splitting in an NMR spectrum for non-viscous liquids.

In solid state, molecules are generally held rather rigidly, which results in a non-zero averaging of the dipolar interaction. For a two-spin AX system, the Hamiltonian can be written as

$$\mathcal{H} = \mathcal{H}_Z + \mathcal{H}_D = -h(f_A I_{AZ} + f_X I_{XZ}) + \mathcal{H}_D, \tag{9.2}$$

where the first part represents the Zeeman terms and the second part represents the dipolar terms as in Eq. (4.4). (Note that $\hbar\omega = hf$, since $\omega = 2\pi f$ and $\hbar = h/2\pi$).

Since the Zeeman energies are much larger than the dipolar energies, we can keep just the first term A in Eq. (4.5) and truncate Eq. (9.2) to

$$\mathcal{H} = -h(f_A I_{AZ} + f_X I_{XZ}) - hR I_{AZ} I_{XZ} (3\cos^2\theta - 1), \tag{9.3}$$

where R is the dipolar coupling constant specified in Eq. (4.4). The energies of these two-spin states are therefore

$$E = -h(f_A m_A + f_X m_X) - hR m_A m_X (3\cos^2\theta - 1), \tag{9.4}$$

where the J coupling is ignored for its smaller energies.

The patterns of the allowed transitions (and hence the spectral patterns) depend upon the molecular orientations in the specimen. There are two extreme cases. A single crystal could have only one orientation θ for r_{AX} (the distance that separates the two dipoles), which results in two A transitions and two X transitions. At the other extreme, a powder sample could have all values of θ existing randomly, which would result in a so-called powder pattern [1].

9.3   MAGIC ANGLE

The $(3\cos^2\theta - 1)$ term in Eq. (4.5) and Eq. (9.4) is a second-order Legendre polynomial in $\cos\theta$. For this term, the angle $\theta = \arctan\sqrt{2} \approx 54.7356°$ makes this term equal to zero. This means that even if the dipolar interaction is not zero, there exists a particular angle, termed the magic angle in NMR and MRI, which minimizes the influence of the dipolar interaction for the spin system. This magic angle effect has far-reaching consequences and is useful in many applications of NMR and MRI. [*Within the 0° − 360° angle space, there are four particular θs that make the* $(3\cos^2\theta - 1)$ *term equal to zero.*]

In NMR spectroscopy of solids, for example, if the specimen is loaded into a tube that is rotated along its axis at an angle of 54.7° to B_0, one would expect much narrower spectral peaks in contrast to the wider peaks in the spectrum had it been parallel to B_0. This technique is called the magic angle rotation in NMR spectroscopy. To completely eliminate the line broadening, the rate of the rotation could go as high as tens of kilohertz, which is difficult and expensive to accomplish experimentally. In MRI, the rotation of an imaging specimen (or the rotation of the imaging hardware) is theoretically beneficial and possible but unlikely with the current hardware configuration. However, the magic angle effect is well known in MRI of many types of connective tissues that have organized macromolecules such as collagen fibers [4]; in these cases, a stationary tipping of the specimen at a certain orientation in the B_0 could manipulate the measured parameters of the specimen. Chapter 15 has more discussion on the topic.

142

9.4   CHEMICAL EXCHANGE

Chemical exchange refers to a number of processes that lead to a change of the magnetic environment of nuclei, either intramolecular (within the molecule) or intermolecular (between molecules). The intramolecular exchanges could include the motion/rotation of sidechains in macromolecules such as proteins and the folding/unfolding or other conformational change of macromolecules. The intermolecular exchanges include the binding of small molecules (e.g., H_2O) to macromolecules or protonation/deprotonation equilibria. A nice example is the conformational interconversion of cyclohexane [5]. Each of these exchanges can lead to a change of chemical shift. Such changes would inevitably lead to changes of some NMR parameters, such as chemical shifts or relaxation times.

Figure 9.5 shows schematically two chemical shifts, separated by Δf, from nuclei that are in a chemical exchange governed by an exchange rate k (which has a unit of 1/s). If the exchange rate is slow on the chemical shift time scale ($k \ll \Delta f$), two distinguishable resonant peaks are observed (Figure 9.5a); at the other extreme, if the exchange rate is fast on the chemical shift time scale ($k \gg \Delta f$), only one narrow-width resonant peak is observed, centered at a weighted average frequency of the two chemical species (Figure 9.5c). At intermediate exchange rates

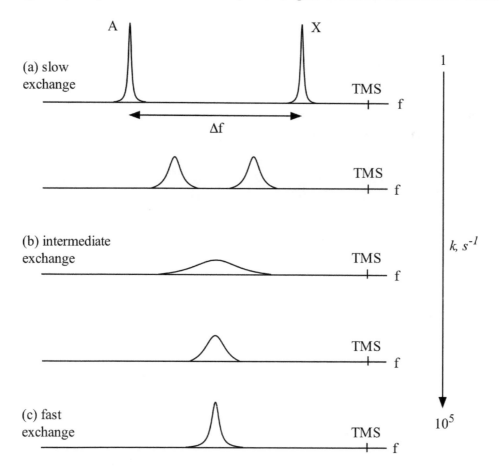

Figure 9.5   The rate of chemical exchange, k, can vary as the function of several measurement parameters, such as temperature. (a) When $k \ll \Delta f$, it is the slow exchange where two distinguishable resonant peaks can be observed. (c) When $k \gg \Delta f$, it is the fast exchange where only one narrow resonant peak can be observed, which represents two chemical species. (b) In between the slow and fast exchanges, the resonant peaks are broad and merge together.

$k \sim \Delta f$, resonant peak(s) with broad line widths can be seen (Figure 9.5b). A short form of the chemical exchange is commonly written as

$$A \underset{\rightleftharpoons}{k} X. \tag{9.5}$$

Since the line broadening of the resonant peaks can be viewed as an increase in the uncertainty of the energy levels between the transitions (Figure 3.2), T_2 relaxation would vary when there is a chemical exchange in the spin system such as this [6]. Since the exchange rate k is highly sensitive to temperature, the temperature-dependent chemical exchange is a powerful tool to study the dynamics of chemical conformation and reaction [7, 8].

9.5　MAGNETIZATION TRANSFER

Magnetization transfer (MT) refers to the dynamic process when spin polarization (i.e., magnetization) is transferred from one type of nuclear spin species to another (either the same or different nuclear species). For example, two spin species, A and B, which generate two distinguishable resonant peaks (Figure 9.6a), are exchanging magnetization via either the dipolar interaction or chemical exchange. Their magnetizations are therefore transferred from species A to species B and vice versa, which contribute to each other's resonant peaks. When a B_1 field is applied continuously at the frequency of species A for a duration of time, the population differences of species A can be eliminated, where species A is said to be saturated and its signal disappears

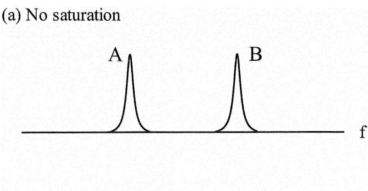

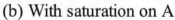

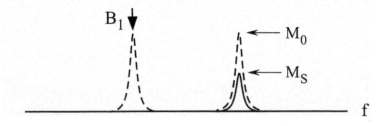

Figure 9.6　(a) Two resonant peaks represent two spin species, A and B. Due to the exchange of magnetizations, both spin species contribute to each other's resonant peaks. (b) When the population difference of species A is eliminated by a B_1 field at the frequency of species A, A does not contribute to the peak of species B, which reduces the peak of species B from M_0 to M_S.

(Figure 9.6b). The saturation of species A not only *suppresses* the peak of species A but also *reduces* the peak of species B, because a part of the species B signal was contributed by the species A signal. The reduction in the B peak depends upon the rate of magnetization transfer between the two species and on the T_1 of the species B in the absence of any exchange. An equation of MT rate can therefore be written as

$$\text{MT rate} = \frac{M_0 - M_S}{M_0}. \tag{9.6}$$

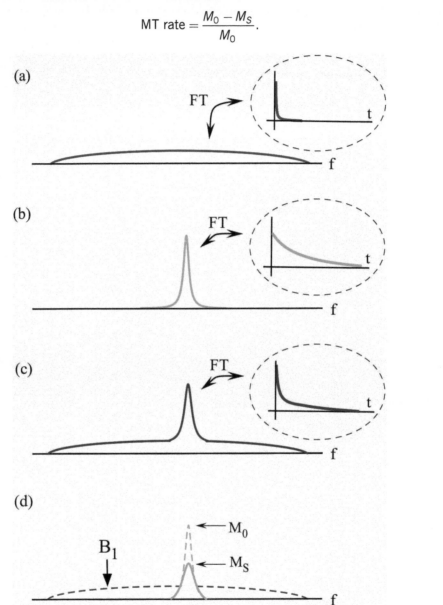

Figure 9.7 Magnetization transfer can be used to examine the exchange and interaction between two types of water molecules in tissue, where the bound water is associated with the macromolecules and the free water exists in tissue. (a) and (b) are the spectra of bound water and free water, respectively, where each insert shows schematically the signal decay in the time domain. (c) The spectrum of tissue commonly contains both bound and free water, which leads to multi-component signal decay in the time domain. (d) By an off-resonant B_1 field, it is possible to examine the exchange of magnetization between the two types of molecules in tissue from the reduction of the resonant peak.

This process is commonly used in NMR and MRI [9–11] to examine the exchange and interaction between two types of molecules; for example, the bound water molecules associated with the macromolecules and the free water molecules (Figure 9.7). Since the FIDs of macromolecules typically decay rapidly, the spectra of macromolecules have very broad peaks (Figure 9.7a). In contrast, the signals of free water decay much more slowly and its spectrum is narrow (Figure 9.7b). Since biological tissue and macromolecules in solutions contain both bound water and free water, their spectra commonly have both fast- and slow-decaying components (Figure 9.7c). A low-level saturation pulse placed at an off-resonance frequency can saturate the energy levels of the macromolecule, which suppresses the magnetization transfer between the macromolecules and the water, hence reducing the peak amplitude of the water, from M_0 to M_S. (Since all bound spins are coupled, the saturation B_1 field can be applied to any region of the broad resonance of the macromolecules.) This suppression technique can be used in biomedical MRI (e.g., MR angiography) to enhance some fine details in the images. If the application of B_1 does not reduce the peak, then there is no magnetization transfer between the two spin species. For example, since the protons in fat and water are not coupled, one cannot use the MT sequence to separate/suppress the fat signal in MRI.

9.6   SELECTIVE POLARIZATION INVERSION/TRANSFER

For a coupled two-spin system, selective polarization inversion can transfer the populations between the energy levels of the two-spin species. The transfer is accomplished by an inversion rf pulse on the transition of one of the spin species. Figure 9.8a shows the energy levels of a two-spin AX system in

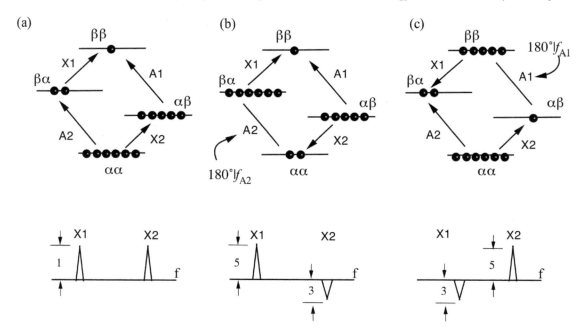

Figure 9.8   Numerical illustrations for selective population inversion. (a) The energy-level diagram for a two-spin system, where each dot represents a spin population. In the equilibrium state, both X1 and X2 transitions have a peak intensity of 1, due to the population differences between $\beta\alpha$ and $\beta\beta$ (for X1) and $\alpha\alpha$ and $\alpha\beta$ (for X2). (b) and (c) By a 180° population inversion pulse at the frequency of A2 or A1, the peak intensities of X1 and X2 can be changed.

steady state when nucleus X is being observed, where the two X lines both have the height of 1 (which add to 2). By applying a 180° inversion pulse to the system at the precise A2-line frequency f_{A2} (Figure 9.8b), the populations of $\alpha\alpha$ and $\beta\alpha$ are inverted, which results in the intensity changes of the two X lines to 5 and −3, respectively. Although the sum of the two X lines is still 2, the intensities of the two individual X lines have changed considerably; the absolute values of the intensities have both increased in this case. In addition, the difference between the two X lines has changed significantly. This means that for a coupled system, manipulating one spin state can change the transition of the other spin state. Applying the 180° rf pulse at the precise A1-line frequency f_{A1} (Figure 9.8c) can invert the populations of $\alpha\beta$ and $\beta\beta$, which results in different intensity changes of the two X lines. One can also use a low-level CW irradiation instead of the 180° pulse, which saturates the populations at the two relevant levels (this is called *selective polarization transfer*).

Note that this simple illustration is not easy to implement in practice, since one needs to apply an irradiation at a very precise frequency, which is difficult to do when several chemical shifts are present. Better pulse sequences for this type of application include Insensitive Nuclear Enhanced by Polarization Transfer (INEPT) and Distortionless Enhancement by Polarization Transfer (DEPT), among others [8].

9.7   RADIATION DAMPING

If you have been paying attention to details, you might be wondering about an odd feature in the two ethanol spectra in Figures 6.8b and 8.6. The samples in these two spectra were made from the same bottle of ethanol (CH_3–CH_2–OH). The sample in Figure 6.8 was a full tube of solvent D_2O with only a trace amount of ethanol; the spectrum was acquired on a 200-MHz NMR spectrometer, in order to illustrate the SNR issue in the signal averaging. The sample in Figure 8.6 was a full tube of ethanol without the use of any solvent; the spectrum was acquired on a 400-MHz NMR spectrometer. The same glass tube was used in both experiments. Have you noticed that the 200-MHz spectrum seems to have a narrower line width (hence better resolution) than the 400-MHz spectrum? Figure 9.9 compares the segment of the quartet in these two spectra, which is the CH_2 resonance, being split by the CH_3 protons. The conclusion is clear – the 200-MHz spectrum has higher resolution (narrower peaks). So, why does everyone want to have higher-frequency NMR spectrometers in the research labs?

What has affected the line width in the ethanol spectra is radiation damping, which was noticed in the beginning of NMR spectroscopy [12]. After the magnetization is tipped to the transverse plane by an rf pulse, the precession magnetization generates an electric current in the rf coil, where the current is acquired by the spectrometer and called the FID signal. At the time when the current is being acquired by the spectrometer, the same current ("the FID current") will interact with the spin system in such a way that it produces a torque that will try to return the magnetization to the z axis. This rotation of the magnetization by the FID current is historically called radiation damping, which is a dynamic process that can be described by [13]

$$\frac{d\theta}{dt_{rd}} = \gamma \boldsymbol{B}_{rd} = -\frac{\sin\theta}{T_{rd}}, \tag{9.7}$$

where the subscript rd stands for radiation damping, θ is the angle between the magnetization and the $\boldsymbol{B}_0$ field, and T_{rd} is the radiation-damping time that can be related to a number of experimental factors including the sample filling factor and the quality factor of the coil. Since this FID current drives the magnetization back to the thermal equilibrium, the return process of the magnetization must be *faster* than due to the relaxation alone. Hence the line width of the spectral peaks becomes wider, as shown in Figure 9.9. It has been estimated that for a 500-MHz spectrometer, T_{rd} can be around 6–10 ms and a proton signal can be as broad as 42 Hz, depending upon the tuning of the rf coil [14]. This radiation-damping effect is often overlooked, but it is common when the NMR machine is modern and the signal is plentiful, which is especially true for ^{1}H proton NMR, that is, when the NMR signal is actually too *big*.

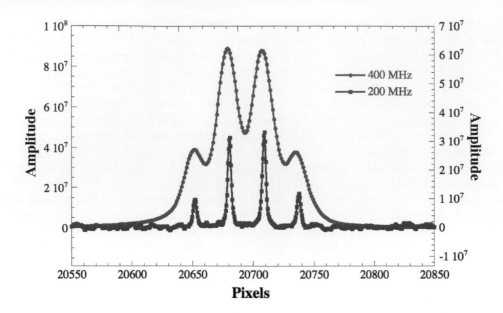

Figure 9.9 Comparison of two CH_2 resonant peaks of ethanol, which are a quartet being split by their respective neighboring CH_3 protons. One spectrum was obtained on a 200-MHz NMR spectrometer, which has a much *narrower* line width than the other spectrum from a 400-MHz NMR spectrometer. It is the radiation damping that broadens the 400-MHz spectrum. (Each dot in the plots is a data point acquired by the spectrometer. The binary versions of the spectra were read by a generic software [cf. Chapter 17]. The amplitudes between the two spectra are arbitrary units upon Fourier transform.)

References

1. Harris RK. *Nuclear Magnetic Resonance Spectroscopy – A Physicochemical View*. Essex: Longman Scientific & Technical; 1983.
2. Pugmire RJ, Grant DM, Robins MJ, Robins RK. Carbon-13 Magnetic Resonance. XIV. Aza-analogs of Polycyclic Aromatic Hydrocarbons. *J Am Chem Soc*. 1969;91(23):6381–9.
3. Hennel JW, Klinowski J. *Fundamentals of Nuclear Magnetic Resonance*. Essex: Longman Scientific & Technical; 1993.
4. Xia Y. Magic Angle Effect in MRI of Articular Cartilage – A Review. *Invest Radiol*. 2000;35(10):602–21.
5. Jensen FR, Noyce DS, Sederholm CH, Berlin AJ. The Energy Barrier for the Chair-chair Interconversion of Cyclohexane. *J Am Chem Soc*. 1960;82(5):1256–7.
6. Zheng S, Xia Y. Changes in Proton Dynamics in Articular Cartilage Caused by Phosphate Salts and Fixation Solutions. *Cartilage*. 2010;1(1):55–64.
7. McFarland EW, Neuringer LJ, Kushmerick MJ. Chemical Exchange Magnetic Resonance Imaging (CHEMI). *Magn Reson Imaging*. 1988;6:507–15.
8. Friebolin H. *Basic One- and Two-dimensional NMR Spectroscopy*. 2nd ed. New York: VCH; 1993.
9. Wolff S, Balaban R. Magnetization Transfer Contrast (MTC) and Tissue Water Proton Relaxation in Vivo. *Magn Reson Med*. 1989;10:135–44.
10. Henkelman RM, Stanisz GJ, Graham SJ. Magnetization Transfer in MRI: A Review. *NMR Biomed*. 2001;14(2):57–64.

11. Eliav U, Navon G. Multiple Quantum Filtered NMR Studies of the Interaction Between Collagen and Water in the Tendon. *J Am Chem Soc*. 2002;124(12):3125–32.

12. Bloembergen N, Pound RV. Radiation Damping in Magnetic Resonance Experiments. *Phys Rev*. 1954;95:8–12.

13. Abragam A. *The Principles of Nuclear Magnetism*. Oxford: Clarendon; 1960.

14. Mao XA, Ye CH. Understanding Radiation Damping in a Simple Way. *Concepts Magn Reson*. 1997;9:173–87.

10

2D NMR Spectroscopy

2D NMR spectroscopy is often employed when some resonance peaks in a 1D spectrum are not easily identifiable or distinguishable. The difficulty in the peak identification could be associated with complex couplings among the nuclear spins or with peaks overlapping due to nuclear spins sharing similar chemical shifts and/or coupling constants (hence overlapping each other in a 1D spectrum). A second dimension in spectroscopy may distinguish these overlapping peaks. Most 1D NMR experiments can be extended into 2D forms by varying the evolution time during the experiments. This chapter briefly describes the concept of 2D NMR spectroscopy and some basic experiments.

10.1 ESSENCE OF 2D NMR SPECTROSCOPY

In NMR spectroscopy, a time-domain signal (i.e., FID) is recorded over a certain time interval, which is called the acquisition time or sampling time. In modern spectroscopy, this time-domain signal is commonly converted by a fast Fourier transform (FFT) into a frequency-domain spectrum. This signal acquisition and subsequent conversion is the basic concept in 1D NMR spectroscopy. 2D NMR spectroscopy aims to resolve the couplings among the nuclear spins, either via homonuclear (e.g., two protons with different chemical environments) or heteronuclear (e.g., between protons and carbons) coupling.

10.1.1 The Second Dimension

In Chapter 7, the pulse sequences such as inversion recovery or spin echo are used to measure T_1 and T_2 relaxations. In these experiments, the same pulse sequence is repeated multiple times, each time with a different time delay τ between the rf pulses (Figure 10.1). Therefore, the acquired FID signal is in fact the function of two time intervals, one before the signal acquisition (called t_1) and one during the signal acquisition (called t_2).

When two time intervals are specifically examined, we name t_1 as the evolution time of the spin system under the influences of the rf pulses, where different t_1 durations provide different opportunities for the spin system to evolve, leading to different characteristics of the acquired signal. We name t_2 as the evolution time of the spin system without the influence of the rf pulses but under the influence of relaxation mechanisms and field inhomogeneity. If the experiment is

Essential Concepts in MRI: Physics, Instrumentation, Spectroscopy, and Imaging, First Edition. Yang Xia.
© 2022 John Wiley & Sons Ltd. Published 2022 by John Wiley & Sons Ltd.

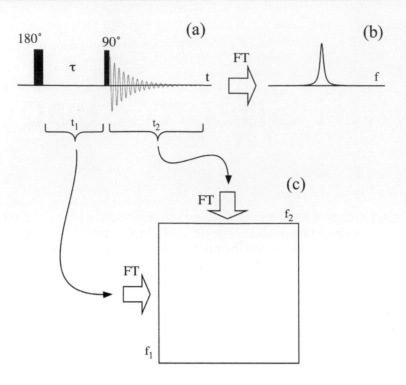

Figure 10.1 A schematic illustration of extending a 1D spectrum into a 2D NMR spectrum by adding a second dimension t_1. (a) A regular 1D pulse sequence (Figure 7.3), where the FID can be Fourier transformed into a spectrum (Hz) (b). By treating τ as a second dimension (t_1) and repeating the sequence sequentially many times, each time with a different τ, it is possible to obtain a 2D spectrum (c) with the application of 2D Fourier transform.

repeated systematically for a series of sequentially varied t_1 times, then the acquired data are two dimensional and include the information during all evolution times used, which can be extracted in the post-acquisition processes. Extraction in both dimensions is commonly done by a 2D FFT.

Compare the pulse sequence for the 2D spectroscopy experiment in Figure 10.1a and the sequence for a double-resonance experiment (Figure 9.1). The difference is that in a 2D spectroscopy experiment, the second dimension is explored explicitly with different t_1 values, which yields fine resolution in the second dimension f_1. In a double-resonance experiment, the second dimension is explored only once, which essentially resolves only one data point in the second dimension.

10.1.2 Coupled Spins

In the discussion of spin couplings, a more precise understanding and notation for the individual spin populations is often needed. For a heteronuclear AX system with ^{1}H (protons), designated A, and ^{13}C (carbon 13), designated X, we can consider multiple magnetization components. Due to the coupling between the two magnetizations, M_H and M_C, in the spin system, we should have four magnetization components. M_H has two components, $M_H^{C\alpha}$ and $M_H^{C\beta}$, and M_C has two components, $M_C^{H\alpha}$ and $M_C^{H\beta}$. In this notation, $M_H^{C\alpha}$ and $M_H^{C\beta}$ are the spin states of the protons coupled with the carbon in its α and β states, and vice versa.

Similarly for a homonuclear AX system when both A and X are protons, due to the coupling between A and X the M_A vector is now represented by $M_A{}^{X\alpha}$ and $M_A{}^{X\beta}$, which are the spin states of the A protons associated with the two components of the X protons. Similarly, the M_X vector is now represented by the $M_X{}^{A\alpha}$ and $M_X{}^{A\beta}$ vectors, which are the spin states of the X protons associated with the two components of the A protons.

With this notation, a B_1 pulse in one rf channel (e.g., proton for A) of a heteronuclear AX experiment manipulates only the nuclear spins in a specific spin species (i.e., protons). Due to the spin coupling, however, a change of one spin species will affect the population differences in the other spin species (e.g., carbon for X), via a mechanism such as selective polarization transfer (Chapter 9.6). In a homonuclear experiment, by contrast, a particular pulse will directly manipulate both spin species (i.e., both A and X populations).

10.1.3   Classification

There are several classes of 2D NMR spectroscopy experiments [1, 2]. One class aims to resolve the spin-spin or J coupling. In these experiments one frequency domain resolves the usual chemical shift δ, while a second frequency domain resolves the J coupling. Examples of this class of spectroscopy experiments include J-resolved spectroscopy [3]. A different class of 2D NMR spectroscopy experiments aims to identify the correlation among the nuclear spins, where the two frequency axes can both be the chemical shift or changes in chemical shift. Examples of this class include Correlated (or Correlation) Spectroscopy (COSY) and its variants. There are some other classes that resolve different molecular properties (e.g., the exchange among the nuclear spins or the nuclear Overhauser effect). A unique intellectual exercise in the field of 2D NMR spectroscopy is, for every new pulse sequence, to conceive of a practical acronym that is both eye-catching and memorable – COSY, DEPT, DOSY, EXSY, HETCOR, HOHAHA, HSQC, INADEQUATE, NOESY, SERPENT, TOCSY, and WATERGATE, just to name a few.

10.2   COSY – CORRELATION SPECTROSCOPY

COSY stands for *Correlation Spectroscopy*, which was among the first 2D NMR spectroscopy experiment proposed in 1971 by Jean Jeener, a professor at the Université Libre de Bruxelles, Brussels, Belgium. The COSY experiments are useful to identify spins that are coupled and have similar coupling constants. The method now has numerous variations (each variation has a new acronym) and can be done in either homonuclear or heteronuclear systems. This section provides a conceptual analysis for the simplest versions of *Correlation Spectroscopy*.

10.2.1   2D Heteronuclear COSY

Although the basic COSY is now used mostly for homonuclear systems, a heteronuclear COSY experiment can be illustrated conceptually in a vector drawing as in Figure 10.2, using a heteronuclear AX system with H (protons) as A and C (carbon 13) as X. Due to the coupling between the two magnetizations, M_H and M_C, in the spin system, we now have four magnetization components. M_H has two components – $M_H{}^{C\alpha}$ and $M_H{}^{C\beta}$ – and M_C has two components – $M_C{}^{H\alpha}$ and $M_C{}^{H\beta}$. In this notation, $M_H{}^{C\alpha}$ and $M_H{}^{C\beta}$ are the spin states of the protons coupled with a carbon's α state and β state, respectively. $M_C{}^{H\alpha}$ and $M_C{}^{H\beta}$ represent analogous spin states, but of carbons coupled with protons.

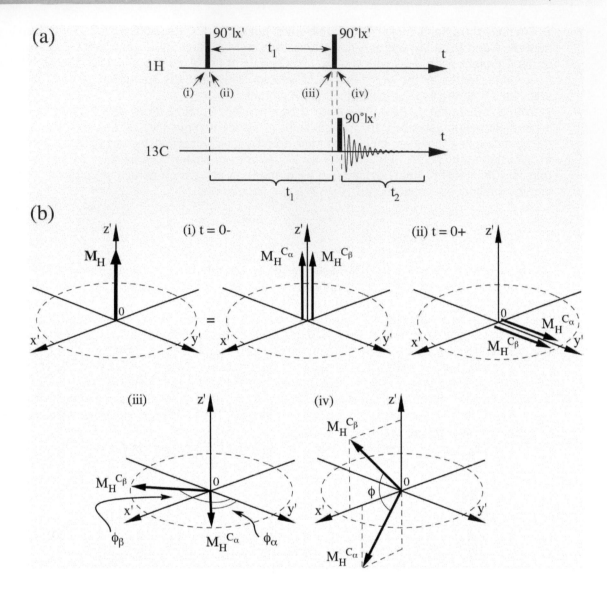

Figure 10.2   (a) 2D heteronuclear COSY sequence, where the motion of the magnetization in the four particular moments is shown in (b). The signal is acquired after the 90° pulse in the ^{13}C channel. See text for explanation.

The experiment starts with a 90° pulse in the ^1H channel, which tips both M_H components to the transverse plane of the rotating frame (Figure 10.2b ii). The different precession frequencies of the two M_H components ($f_H \pm J/2$, with J as the coupling constant between the carbons and protons) during t_1 gradually separate the two vectors $M_H^{C_\alpha}$ and $M_H^{C_\beta}$, with the following phases: $\phi_\alpha = 2\pi(f_H - J/2)t_1$ and $\phi_\beta = 2\pi(f_H + J/2)t_1$ (Figure 10.2b iii). The net difference between the orientations of the two magnetizations ($\phi = \phi_\beta - \phi_\alpha$) can be determined by $2\pi J t_1$, that is, the separation is t_1 dependent. When $t_1 = (4J)^{-1}$ or $(2J)^{-1}$, then $\phi = 90°$ or $180°$, respectively, meaning the two components are perpendicular ($\phi = 90°$) or antiparallel ($\phi = 180°$) to each other in the transverse plane. The second 90° pulse in the ^1H channel (Figure 10.2b iv) leads to the partial restoration of M_H in the z direction – with the possibility of having both $+z$ and $-z$ components.

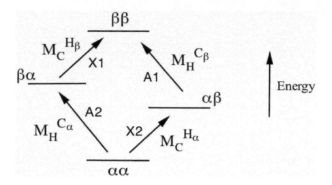

Figure 10.3 An energy-level diagram for a heteronuclear AX system in a 2D heteronuclear COSY experiment. The settings of the four energy levels are similar to Figure 4.7b, and the effect of the heteronuclear COSY pulse sequence on the spin population is similar to the selective population inversion shown in Figure 9.8.

On the topic of selective polarization transfer in Chapter 9.6, we see that the manipulation of one spin population (e.g., protons) can lead to a magnetization transfer to a different spin population (e.g., carbons) in a coupled system. Identical to the previous discussion, the magnitude of $M_H^{C\alpha}$ is proportional to the population difference between the energy levels $\alpha\alpha$ and $\beta\alpha$, and the magnitude of $M_H^{C\beta}$ is proportional to the population difference between the energy levels $\alpha\beta$ and $\beta\beta$, as shown in Figure 10.3. Similarly, the magnitude of $M_C^{H\alpha}$ is proportional to the population difference between the energy levels $\alpha\alpha$ and $\alpha\beta$, and the magnitude of $M_C^{H\beta}$ is proportional to the population difference between the energy levels $\beta\alpha$ and $\beta\beta$.

The application of the second 90° pulse in the ^{1}H channel in Figure 10.2a can make population $\beta\alpha$ larger than population $\alpha\alpha$ (shown as the negative $M_H^{C\alpha}$ vector in Figure 10.2b iv), with population $\alpha\beta$ remaining larger than population $\beta\beta$ (shown as the positive $M_H^{C\beta}$ vector in Figure 10.2b iv). Up to this moment (Figure 10.2b iv), the ^{13}C channel has not yet been utilized. However, changing the proton populations by the 90° pulses in the ^{1}H channel has already led to a change of the spin environment for the carbon populations via magnetization transfer. This is because the proton spin populations are coupled to those of carbon. Moreover, the characteristics of $M_C^{H\alpha}$ and $M_C^{H\beta}$ after Figure 10.2b iv, upon application of the 90° pulse in the ^{13}C channel, will continue to vary further from their characteristics before Figure 10.2b iii.

The 90° pulse in the ^{13}C channel turns the two ^{13}C longitudinal vectors, $M_C^{H\alpha}$ and $M_C^{H\beta}$, into the transverse plane to generate the FID for a ^{13}C detection, where $M_C^{H\alpha}$ and $M_C^{H\beta}$ have already been influenced by the longitudinal magnetizations of $M_H^{C\alpha}$ and $M_H^{C\beta}$. During t_2, $M_C^{H\alpha}$ and $M_C^{H\beta}$ precess at the frequencies corresponding to two carbon peaks, which are acquired in signal detection. This process is repeated n times, each time at a different t_1. This long series of variable t_1 causes $M_C^{H\alpha}$ and $M_C^{H\beta}$ to have different signals each time due to different amounts of magnetization transfer. The 2D FID data are Fourier transformed into a 2D spectrum, which has a diagonal symmetry.

A 2D spectrum is schematically illustrated in Figure 10.4, where the spectra on both the horizontal and vertical axes are regular 1D NMR spectra, which have the same units in the axes as the original 1D spectra. Notice the "square" in the 2D plot formed by the additional cross-peaks above and below the diagonal peaks. The cross-peaks indicate the coupling between pairs of nuclei represented by the diagonal peaks. The appearance of the cross-peaks serves a similar purpose to the use of the peak splitting in 1D spectra: to indicate the coupling between the pairs

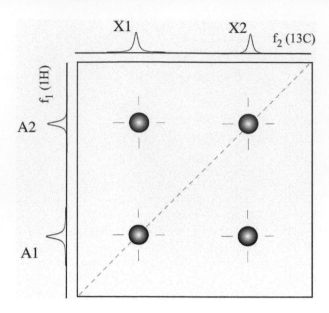

Figure 10.4 A schematic 2D heteronuclear COSY spectrum, where the peaks are symmetric about the diagonal line. The corresponding 1D spectra (known as *skylines*) are usually shown above and on the side of the 2D spectrum.

of nuclei. Consequently, the lack of the cross-peaks implies the lack of coupling between the pairs of nuclei.

10.2.2   2D Homonuclear COSY

A homonuclear COSY experiment is illustrated in Figure 10.5. The first 90° pulse turns both M_A and M_X magnetizations into the transverse plane. Due to the coupling between A and X, the M_A vector is now represented by $M_A^{X\alpha}$ and $M_A^{X\beta}$, which are the spin states of the A protons associated with the X protons. Similarly, the M_X vector is now represented by $M_X^{A\alpha}$ and $M_X^{A\beta}$ vectors. These four magnetization vectors precess simultaneously in the transverse plane of the rotating frame, similar to the two M_H vectors in the heteronuclear experiment (Figure 10.2b). These magnetization vectors fan out at four different frequencies: $f_A \pm J/2$ and $f_X \pm J/2$, with J being the coupling constant between A and X. At the end of t_1, the second 90° pulse tips the y' components of the magnetizations into the $\pm z$ axes. Magnetization transfer will occur, associated with

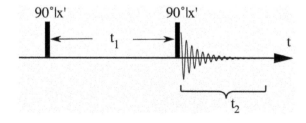

Figure 10.5   A conceptual pulse sequence for 2D homonuclear COSY experiments.

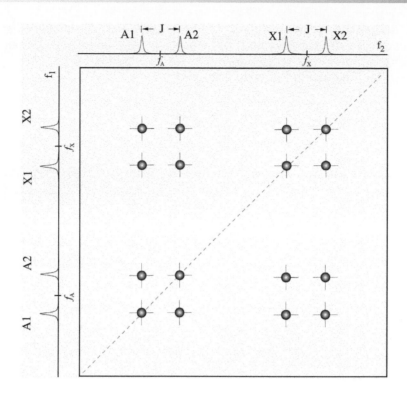

Figure 10.6 A schematic 2D homonuclear COSY spectrum, where the peaks are symmetric about the diagonal line. The coupled peaks form a rectangle in the 2D spectrum.

the changes in the M_z components (the population differences) among the coupled nuclear spins, similar to the heteronuclear discussion in Section 10.2.1. The amount of the magnetization transfer depends upon f_A, f_X, J, and t_1. The FID signals are recorded and processed with the use of a Fourier transform.

For this type of homonuclear AX system, a 1D spectrum along the t_2 direction will have four peaks (cf. Figure 8.3), at $f_A + J/2$ (for A1), $f_A - J/2$ (for A2), $f_X + J/2$ (for X1), and $f_X - J/2$ (for X2). A second FFT along the t_1 direction will produce a 2D spectrum, shown schematically in Figure 10.6, which also has a diagonal symmetry. The diagonal peaks of a pair of mutually coupled nuclei, together with their cross-peaks, form the four corners of a square. In actual spectra, the peaks are dispersion and absorption signals (in-phase and anti-phase peaks). The usefulness of 2D COSY is to discover the network of spin-spin coupling in a molecule and to determine which resonant peak corresponds to which nucleus. An example will be given in Section 10.4.

10.3   *J*-RESOLVED SPECTROSCOPY

Chemical shift and *J*-coupling constant are the two most important molecular parameters in a high-resolution NMR spectroscopy of liquids. A 2D *J*-resolved NMR spectrum resolves the coupling between the spin systems; often one axis contains the chemical shift information while the other axis contains the information about *J* coupling. The *J* coupling can be resolved by utilizing the same decoupling techniques discussed in Chapter 9.1, for example, a noise-modulated pulse (hence broad-band, BB) or an rf pulse at the precise frequency of a particular resonance (hence narrow band). These decoupling pulses saturate all nuclear transition, causing the elimination of all splitting due to this coupling.

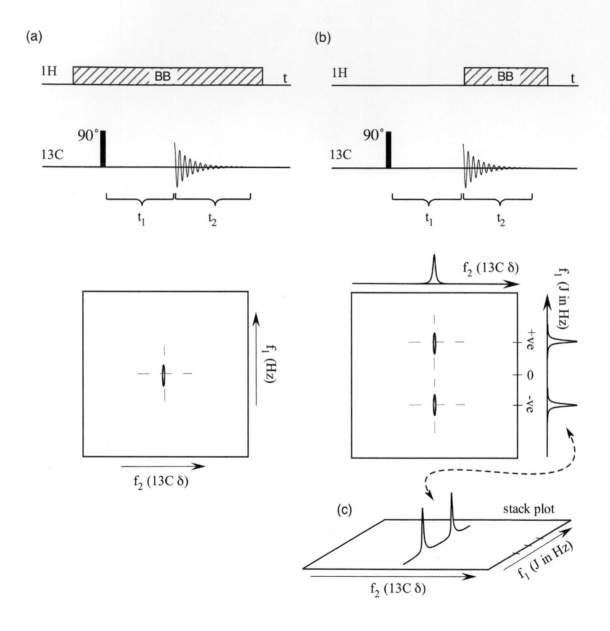

Figure 10.7   (a) A conceptual pulse sequence for 2D heteronuclear J-resolved experiments. Due to the use of decoupling pulse during t_1 in the ^1H channel, the 2D ^{13}C spectrum would have an uncoupled single peak. (b) Turning off the decoupling pulse during t_1 will resolve the J couplings in t_1. Each group of resonant peaks in the f_2 dimension would be spread out along the f_1 dimension. The J-coupling constant can be measured directly on the f_1 dimension. One can display the 2D spectrum using either (b) a 2D surface plot or (c) a pseudo-3D stack plot.

10.3.1 Basic Pulse Sequences

The simplest 2D decoupling experiment can be done by turning the BB decoupling in the ^{1}H channel before the 90° pulse for acquisition in the ^{13}C channel, as shown in Figure 10.7a. Since there is no longer any difference between $M_H^{C\alpha}$ and $M_H^{C\beta}$, there is only one M_C component in the final 2D spectrum. A slightly more complicated experiment can use the pulse sequence as in Figure 10.7b. At the start of the experiment during t_1, the coupling between two heteronuclei, H and C, is permitted. Hence two ^{13}C magnetization vectors ($M_C^{H\alpha}$ and $M_C^{H\beta}$) precess at different frequencies, as ($f_C \pm J_{CH}/2$). In the $x'y'$ rotating frame, $M_C^{H\beta}$ precesses ahead by $J_{CH}/2$, and $M_C^{H\alpha}$ precesses behind by $J_{CH}/2$, because of their association with different energy levels of M_H. The net M_C magnetization at the end of t_1 will depend upon the duration of t_1; for example, M_C magnetization can become zero at $t_1 = 1/(2J_{CH})$. At the end of t_1, the ^{1}H BB decoupling is turned on, which freezes the states of the M_H magnetization during t_2. The signals in the f_2 direction thus becomes singlets, that is, there is no difference between $M_C^{H\alpha}$ and $M_C^{H\beta}$ during t_2. A 2D FFT could present the 2D spectrum as a 2D surface plot (Figure 10.7b) or more graphically as a pseudo-3D stack plot (Figure 10.7c). More complicated couplings can be resolved on the second dimension in a similar manner. See Table 8.3 for examples with size and pattern specifications.

10.3.2 2D Heteronuclear *J*-resolved Spectroscopy

A simple heteronuclear spin system, where X is ^{13}C and A is ^{1}H, is used in Figure 10.8 to illustrate the basics of 2D heteronuclear *J*-resolved spectroscopy. Before the first 90° pulse in the ^{13}C channel, the BB field in the ^{1}H channel suppressed the spin-spin interactions. Right after the first 90° pulse at the moment *i* in Figure 10.8, both ^{13}C magnetization vectors ($M_C^{H\alpha}$ and $M_C^{H\beta}$) are tipped to the transverse plane. During the first half of t_1 (between events *i* and *ii*), two fan-out events happen at the same time. The first is the increase of separation between the two carbon magnetization vectors marked by θ, which is due to the fact that $M_C^{H\alpha}$ and $M_C^{H\beta}$ have different precession frequencies. The second is the influences of the usual field inhomogeneity and T_2 relaxation, illustrated by the dashed fans and the $+$ and $-$ signs.

The 180° pulse in the ^{13}C channel flips all magnetization vectors about the y' axis. Without the BB decoupling in the ^{1}H channel, all vectors would be refocused to the y' axis to form a spin echo at the end of t_1 and the beginning of t_2. The application of the BB decoupling pulse during the second half of t_1 removes the cause of the frequency separation θ, the *J* coupling; hence, the separation between $M_C^{H\alpha}$ and $M_C^{H\beta}$ vectors are locked during the second half of t_1, both precessing at the same frequency as the rotating frame f_C. The fan-out event due to the field inhomogeneity and T_2 relaxation remains unaffected. Right at the end of the second half of t_1, the fan-out closes up.

Therefore, after the end of t_1, the detected ^{13}C signal during t_2 is proportional to the vector sum of the two M_C magnetizations, which depends on their relative positions (the separation angle θ). θ depends upon the C–H coupling constant *J*, as well as the duration of $t_1/2$. A 2D *J*-resolved spectroscopic experiment can be carried out by repeating the sequence many times, each time with a different value of t_1. The 2D *J*-resolved spectrum would therefore have the f_1 direction corresponding to *J* coupling with units of hertz, and the f_2 direction corresponding to chemical shift with units of ppm.

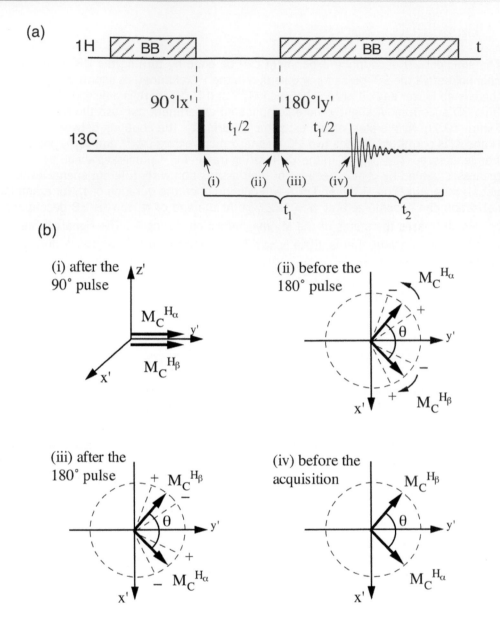

Figure 10.8 (a) The general pulse sequence for 2D heteronuclear *J*-resolved spectroscopy experiments, where the concept of spin echo is used. (b) The motion of the magnetization at four particular moments in the experiment.

10.3.3 2D Homonuclear *J*-resolved Spectroscopy

A 2D homonuclear *J*-resolved spectroscopy cannot use any BB decoupling pulse since it would affect both spins. We can consider an AX system with two different types of protons, under the manipulation of the sequence shown in Figure 10.9a.

The $90°|_{x'}$ pulse will turn both M_A and M_X proton magnetizations to the y' axis (Figure 10.9b). Each proton magnetization can be visualized as having two components, for example, M_A has $M_A{}^{X\alpha}$ and $M_A{}^{X\beta}$. Both proton magnetizations will precess during t_1. Consider M_A for now. Just as in the heteronuclear situation, the separation angle θ between the two M_A vectors will increase

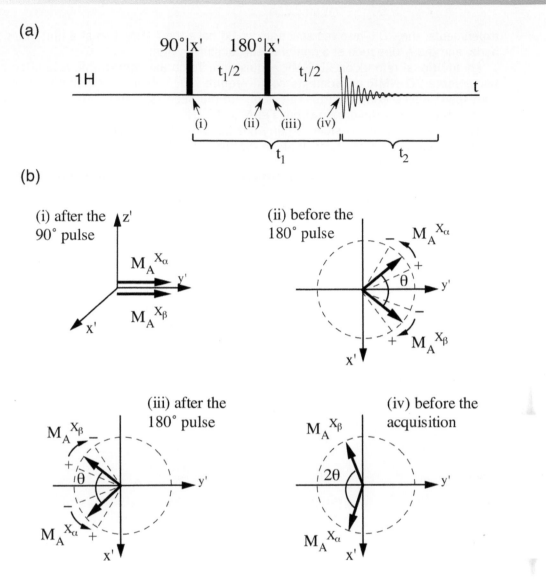

Figure 10.9 (a) The general pulse sequence for 2D homonuclear J-resolved spectroscopy experiments, where the concept of spin echo is used. (b) The motion of the magnetization at four particular moments in the experiment.

with the increase of t_1, and each M_A vector will also fan-out. M_X magnetization precesses in the same fashion. Since it is a homonuclear experiment, the $180°|_{x'}$ pulse will reflect both M_A and M_X magnetization vectors through the x' axis.

As seen in Chapter 9.6, a $180°$ pulse will invert the magnetization, changing the positive M_z to the negative M_z (i.e., exchanging α and β states) and vice versa. Hence, the $180°|_{x'}$ pulse will change not only $M_A^{X_\alpha}$ into $M_A^{X_\beta}$ and $M_A^{X_\beta}$ into $M_A^{X_\alpha}$ but also M_X magnetization in a similar manner. Both M_A and M_X protons continue to evolve during the second half of t_1, in their new spin environment. At the end of t_1, the separation angle θ between the two M_A vectors will be doubled; but the fan-out of each M_A vector due to the field inhomogeneity and T_2 relaxation will close up.

When the signal is acquired during T_2, the usual evolution will happen, where the max signal in M_A will depend upon the particular separation angle 2θ for that particular evolution time t_1, while the decay of the signal M_A will depend upon the T_2 relaxation. As in the heteronuclear experiments, the 2D J-resolved spectrum would have the f_1 direction at a unit of J coupling in hertz, and the f_2 direction at a unit of chemical shift in ppm.

An additional remark on 2D NMR acquisition: To obtain comparable resolution and sensitivity, some 2D NMR experiments could require much longer acquisition times than their corresponding 1D counterparts. This is due to the need for a fine resolution in the chemical shift in the second dimension (from TR×N to TR×N^2, where N is the number of data points in 1D). However, it is possible to carry out 2D J-resolved experiments much faster given a smaller range of J-coupling constants in the second dimension, hence needing a smaller N in the second dimension.

10.4 EXAMPLES OF 2D NMR SPECTROSCOPY

10.4.1 Homonuclear COSY of Styrene

Figure 10.10a shows a 2D homonuclear ^{1}H COSY spectrum for styrene, whose 1D NMR spectrum has been shown in Figure 8.7a. From the diagonal symmetries in Figure 10.10a, we could categorize styrene's couplings into two groups: one group contains the protons in the aromatic ring

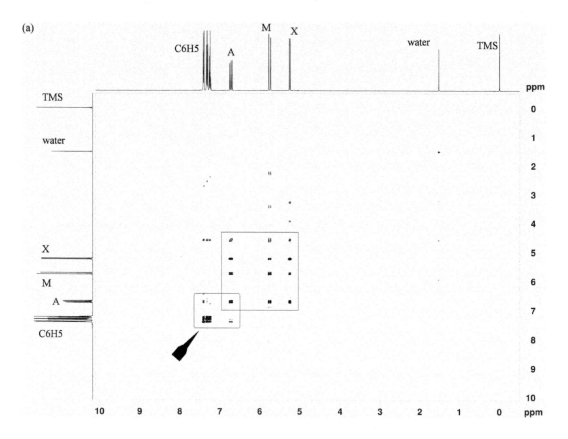

Figure 10.10a 2D homonuclear ^{1}H COSY spectrum of styrene, with $CDCl_3$ as the solvent and measured on a 400-MHz NMR spectrometer. It is the same specimen that has been measured in 1D spectroscopy, shown in Figure 8.7a and Figure 8.7c. (a) The full 2D spectrum, together with two 1D skylines on the top and left and the peak assignments from Figure 8.7a. The arrow points to the aromatic ring signals and their coupling with peak A (boxed by the small rectangle). The large square includes the AMX peaks.

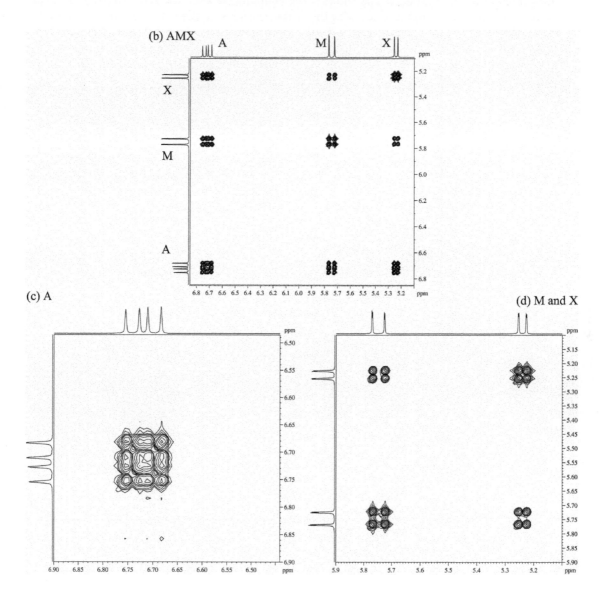

Figure 10.10b-d   The AMX peaks in the 2D homonuclear ¹H COSY spectrum of styrene shown in Figure 10.10a are enlarged in (b), then further enlarged in (c) for the nucleus A and (d) for the nuclei M and X.

(C_6H_5), showing as a densely packed small rectangle centered at ~7.1 ppm (marked by an arrow in Figure 10.10a), and the other group contains the AMX protons, which are boxed by the larger square centered at ~6.0 ppm in Figure 10.10a. When we analyzed the 1D spectra of styrene in Chapter 8, no coupling was considered between the carbon ring protons and the AMX protons; but the 2D spectrum in Figure 10.10a shows some coupling between the A hydrogen and the protons in the aromatic ring, which is evident by the cross-peaks inside the small rounded corner rectangle at ~7.1 ppm in Figure 10.10a. Figure 10.10b expands the 2D homonuclear COSY spectrum for styrene's AMX region, where the couplings among the three protons are clear by the formation of many square/rectangular shapes. Figure 10.10c and d further expand the spectral plots for the individual nuclei.

10.4.2 Heteronuclear *J*-resolved Spectroscopy of Styrene

Figure 10.11a shows a 2D heteronuclear ^{13}C *J*-resolved spectrum for styrene for the range of 110–140 ppm on the f_2 dimension. Refer to the full 1D ^{13}C spectrum, with the proton decoupling, in Figure 8.7c (this spectrum included the assignments for the carbons). At each carbon site, a set of coupling peaks (where the coupling constant is between C and the immediately

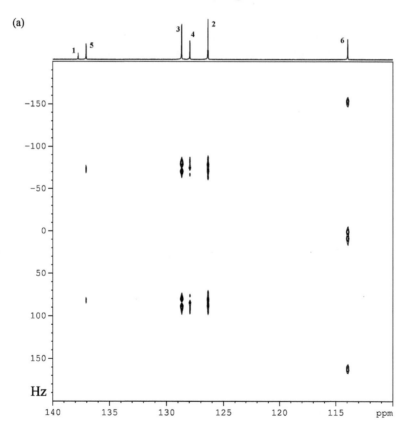

Figure 10.11a   2D heteronuclear ^{13}C *J*-resolved spectrum of styrene for the range of 110–140 ppm using a 400-MHz NMR spectrometer, with $CDCl_3$ as the solvent. It is the same specimen that has been measured using 1D spectroscopy; see Figure 8.7a and Figure 8.7c. (a) The 2D spectrum for the range of 110–140 ppm, together with the 1D skyline on the top, with peak assignments from Figure 8.7c.

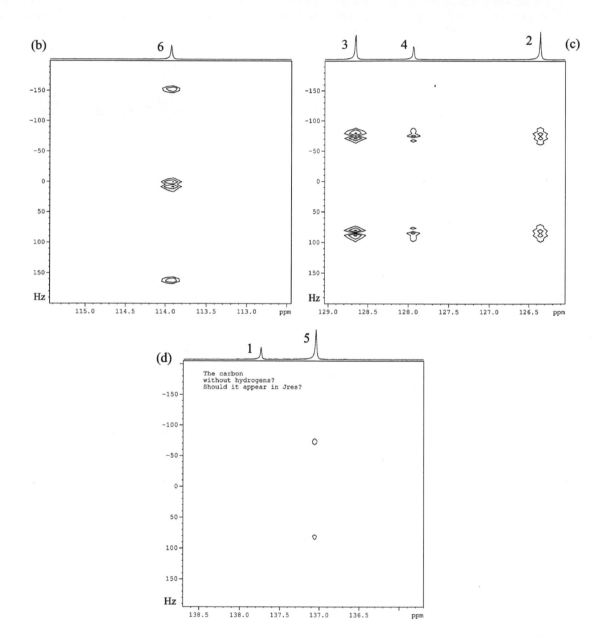

Figure 10.11b-d The enlarged spectra of the individual carbons and their *J* couplings from the 2D heteronuclear ^{13}C *J*-resolved spectrum of styrene, shown and assigned in Figure 10.11a.

attached protons) can be determined, similar to the surface and stack plots in Figure 10.7b and c. Figure 10.11b–d shows the extension plots for several local regions.

It should be noted that a straightforward analysis on the data using a sequence in Figure 10.9a will not produce a 2D J-resolved spectrum as in Figures 10.7b and 10.11. Due to the use of a spin echo in t_1, the spin evolution due to the J coupling will be resolved during t_1, while the spin evolution due to the chemical shift will be refocused at the end of t_1 (cf. Figure 7.6). During the signal detection time t_2, both effects (J coupling and chemical shift) will influence the spin evolution as usual, which gives the f_2 direction an extra f_1-dependent frequency component. Hence the multiplets of the peaks would appear at a 45° angle in the 2D spectrum if both axes are plotted in hertz [3]. An angular tilt of 2D spectra in J-resolved spectroscopy during the post-acquisition process is often carried out to generate the parallel-lined 2D plots as in Figure 10.11.

References

1. Harris RK. *Nuclear Magnetic Resonance Spectroscopy - A Physicochemical View.* Essex: Longman Scientific & Technical; 1983.
2. Sanders JKM, Hunter BK. *Modern NMR Spectroscopy - A Guide for Chemists.* 2nd ed. New York: Oxford University Press; 1993.
3. Ludwig C, Viant MR. Two-dimensional J-resolved NMR Spectroscopy: Review of a Key Methodology in the Metabolomics Toolbox. *Phytochem Anal.* 2010;21(1):22–32.

Part IV

Essential Concepts in MRI

11

Effect of the Field Gradient and *k*-space Imaging

The fundamental equation in NMR, $\omega = \gamma B$, states that the Larmor precessional frequency ω of nuclear spins is a measure of the external magnetic field B that it experiences. Hence, any variation in the external field should lead to a change in the precessional frequency. In NMR spectroscopy, specific efforts are made to improve the uniformity of the magnetic field over the sample space so that all chemically identical nuclei in the specimen resonate at the same frequency. This uniformity in B ensures that the variations of the resonance frequencies measured in a specimen bear different chemical information about the nuclear spins.

Let us now pause for a moment and think about this. If the external field B is deliberately made non-uniform, in a manner that the non-uniformity of the magnetic field over the sample space is known, can we tell, from the spreading of the Larmor frequencies, which frequency comes from which part of the sample space? The answer is *yes*! If every discrete location in a pre-defined 3D space (i.e., the sample space) can be made to have a unique value of the magnetic field, then each resonance frequency should tell us where the resonant signal comes from. In other words, the nuclear spins in the specimen become spatially encoded! Some sort of reconstruction should be able to recover this spatial information, that is, to map the spatial distribution of the spin density in the specimen. That is the principle of NMR imaging, or MRI.

Figure 11.1 shows four 1D NMR spectra acquired at 7 Tesla from water in a 4-mm inner diameter glass tube, where each spectrum in b-d was acquired with a deliberate change of the electric current being sent to the x-shim coil for the superconducting magnet of the spectrometer. As the shim current deviates away from the optimal value, the deterioration of the Lorentzian line shape is clear. The final spectrum seems to adopt the shape of the glass tube! This was in fact an example of 1D MRI.

Essential Concepts in MRI: Physics, Instrumentation, Spectroscopy, and Imaging, First Edition. Yang Xia.
© 2022 John Wiley & Sons Ltd. Published 2022 by John Wiley & Sons Ltd.

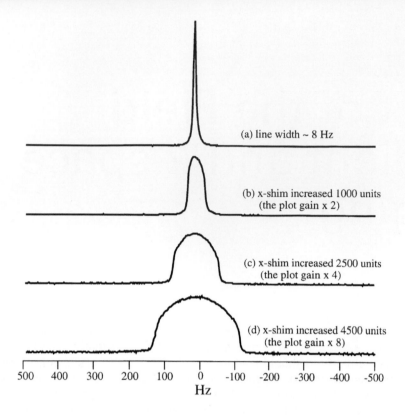

Figure 11.1   1D profiles of a tube of water. (a) Shimmed to a line width of ~8 Hz. (b) x-shim was increased about 1000 units (the plot's gain was doubled). (c) x-shim was increased about 2500 units (the plot's gain was increased about four times). (d) x-shim was increased about 4500 units (the plot's gain was increased about eight times). Source: Xia [5].

11.1   SPATIALLY ENCODING NUCLEAR SPIN MAGNETIZATION

By linearly varying the magnetic field across the sample space (Figure 11.2b), all of the nuclei in a 1D stripe that are perpendicular to the direction of the field gradient will experience the same field strength and contribute to the signal amplitude at the same frequency. In other words, spatial displacements along that direction are turned into frequency displacements. In addition, the more spins in any narrow range of frequencies, the higher the signal at this particular frequency location. The final profile takes a form that represents the 1D shape of the sample, known as the *1D image* or *1D projection profile* (cf. Figure 11.1). The first demonstration of the MRI concept was done by Paul Lauterbur (State University of New York at Stony Brook) in a paper [1] published in 1973, where the signals were acquired from a test object (two tubes of water) in the electromagnet of a Varian A-60 spectrometer and a first-order shim was

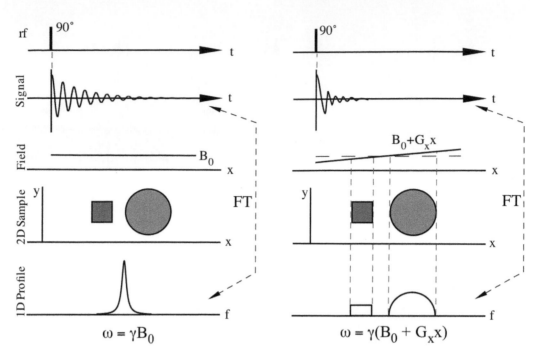

(a) Without the field gradient   (b) With the field gradient

Figure 11.2   Spatially encoding the nuclear spins across the sample space using a field gradient. (a) 1D projection of the specimen without any field gradient, that is, the magnetic field is uniform. (b) 1D projection of the specimen with an applied field gradient.

purposefully made non-uniform. By rotating the tubes manually, the 1D profiles were back-projected to form the first 2D image by any NMR method (Figure 1.5).

In practice, a constant field gradient can be accomplished by the superposition of a uniform polarizing field B_0 (produced by a magnet) and a field gradient G (produced by an additional set of gradient coils). To distinguish any spatial location in a 3D space, three orthogonal gradients in the component of G parallel to B_0 are needed, namely G_x, G_y, G_z. Since the direction of the main polarizing field B_0 is always defined as the z axis, the field gradients are denoted as the variations of B_0 along three orthogonal axes (Figure 11.3), that is,

$$G = \nabla B_0 \tag{11.1}$$

or

$$G_x = \frac{\partial B_z}{\partial x} \text{ , } G_y = \frac{\partial B_z}{\partial y} \text{ and } G_z = \frac{\partial B_z}{\partial z} \tag{11.2}$$

Note that only the gradient along the three orthogonal axes matters; other gradients (e.g., $\partial B_y / \partial x$, ...) do not alter the resonance frequencies significantly, which can be seen in an example in Figure 11.4. While the component parallel to B_0 ($\partial B_z / \partial x$) does increase the total field linearly, the component in a different orthogonal direction (e.g., $\partial B_y / \partial x$) only causes an insignificant increase of the total field.

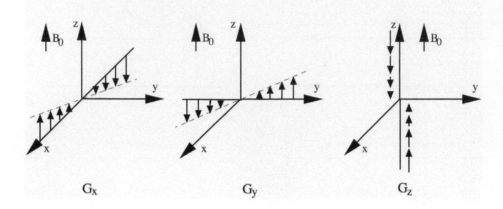

Magnetic field gradients (along the axes)

G_x G_y G_z

Figure 11.3 Components of the 3D field gradient, which are used to encode a 3D physical space where the specimen is located.

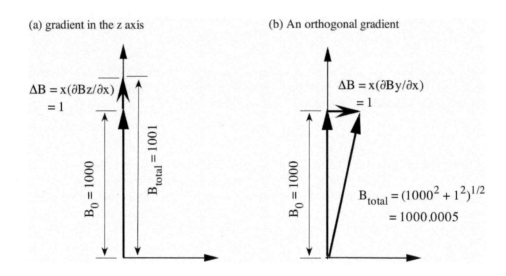

(a) gradient in the z axis

$\Delta B = x(\partial Bz/\partial x)$
$= 1$

$B_0 = 1000$

$B_{total} = 1001$

(b) An orthogonal gradient

$\Delta B = x(\partial By/\partial x)$
$= 1$

$B_0 = 1000$

$B_{total} = (1000^2 + 1^2)^{1/2}$
$= 1000.0005$

Figure 11.4 Insignificant effect of the field gradient in other orthogonal directions.

The expression of the precession frequency in an MRI experiment can be written as

$$\omega(\boldsymbol{r}) = \gamma(B_0 + \boldsymbol{G} \cdot \boldsymbol{r})\boldsymbol{k}, \tag{11.3}$$

where $\omega(\boldsymbol{r})$ is written purposefully to indicate ω is now a function of the spatial location $\boldsymbol{r}$, $\boldsymbol{k}$ is the unit vector along the z axis, and the effect of chemical shift has been ignored for the moment.

11.2 *k* SPACE IN MRI

The nuclear spin density $\rho(r)$ is defined as the contribution of all nuclear spins in each image voxel (volume element) towards the detected signal $S(t)$. For an ideal system, $\rho(r)$ is directly proportional to the transverse magnetization $M_{y'}$, and consequently, equal to M_0 if a perfect 90° pulse has been applied. In practice there is another proportionality constant between $\rho(r)$ and the transverse magnetization $M_{\perp'}$, where the constant will depend upon many practical factors such as the sensitivity and efficiency of the receiver.

The acquisition of the signal in MRI occurs in the absence of the rf field but in the presence of the applied imaging gradients and is given by

$$S(t) = \int \rho(r) \exp(-i\omega(r)t)\, dr, \tag{11.4}$$

where $\omega(r)$ was given previously by Eq. (11.3) and the integral is over the defined 3D space. In the heterodyne detection frame where the reference frequency is γB_0, the above equation becomes

$$S(t) = \int \rho(r) \exp(-i\gamma \boldsymbol{G} \cdot \boldsymbol{r}t)\, dr. \tag{11.5}$$

This spatial mapping procedure would be clearer if a reciprocal vector space, called *k* space [2], is introduced by defining

$$\boldsymbol{k} = \frac{1}{2\pi} \int_0^t \gamma \boldsymbol{G}(t')dt', \tag{11.6}$$

where the spatial frequency vector *k* has units of m^{-1}. When the gradient pulse is constant over the pulse time duration t, the above expression becomes

$$\boldsymbol{k} = \frac{1}{2\pi} \gamma \boldsymbol{G} t, \tag{11.7}$$

where *G* is the imaging gradient. The MRI signal expressed in the formalism of *k* space is therefore

$$S(\boldsymbol{k}) = \int \rho(\boldsymbol{r}) \exp(-i2\pi \boldsymbol{k} \cdot \boldsymbol{r})\, dr. \tag{11.8}$$

Equation (11.8) states that the spin density $\rho(r)$ can be obtained by a Fourier transform of the detected signal, as

$$\rho(\boldsymbol{r}) = \int S(\boldsymbol{k}) \exp(i2\pi \boldsymbol{k} \cdot \boldsymbol{r})\, dk. \tag{11.9}$$

Equations (11.8) and (11.9) are a pair of equations related by Fourier transformation (Appendix A1.2), which states that the signal can be acquired in *k* space during an experiment and the image of the nuclear spin density can be reconstructed in the conjugate *r* space via a Fourier transform.

Since Fourier transform employs the sinusoidal functions (sine and cosine functions), there are symmetry relationships that can be exploited [3]. For example, if $\rho(r)$ is a real function, then the image reconstruction procedure can utilize the following symmetry relationship:

$$S^*(\boldsymbol{k}) = S(-\boldsymbol{k}) \tag{11.10}$$

where S^* is the complex conjugate of S. This relationship comes from the fact that the real spectrum is identical in the opposite *k* direction and the imaginary spectrum has the opposite sign in the opposite k direction. The recognition of this symmetry implies that to reconstruct a 2D image, the signal only needs to be sampled in two quadrants (e.g., from 0° to 180°) of k space. This type of relationship has been widely used in MRI. In practice, however, it has been found that the employment of this symmetry relationship can introduce some image artifacts when the instrument is not perfect, since the superimposed artifact signal does not have the same symmetry as the image signal [4].

11.3   MAPPING OF k SPACE

The origin of k space is the location where all imaging gradients equal to zero, that is, at the Larmor frequency. As one can see from Eq. (11.7), the direction of the k-space vector is determined by the direction of the field gradient G, while the magnitude of the k-space vector is determined by the product of G and the time t, or the area under a gradient pulse. The position in k space is therefore determined by the time effect of the imaging gradients. There are numerous ways in which the entire k space can be sampled, where a larger number of k-space lines results in better resolution for the image. It is the method of traversing k space that results in different MRI techniques.

Figure 11.5 shows four essential steps to traverse k space in the format of a 2D Cartesian grid. In Figure 11.5a, keeping the magnitude of the G_x constant and sampling n points sequentially at a fixed rate results in the acquisition of n data points along the k_x direction (at $k_y = 0$). In Figure 11.5b, applying a G_y pulse for a short duration moves the k-space vector to a desired location along the k_y axis. In Figure 11.5c, repeating the sequence n times with the linear increment of the G_y gradient in n steps results in the movement of the k-space vector n times, each time moving to a different location along the k_y direction. Having a negative gradient changes the direction of traverse in k space, as shown in Figure 11.5d.

The combination of these approaches can be used to form various 2D k-space sampling patterns, as shown in Figure 11.6. To sample the first quadrant of k space in the Cartesian grid (Figure 11.6a), one needs to repeat the sequence n times, each time acquiring one line of n data points at a specific k_y location (determined by the fixed G_y pulse); stepping up the G_y pulse linearly for n times in the repetition results in the sampling of the first quadrant of k space. To sample the first quadrant of k space using the radial grid (Figure 11.6b, c), one needs to repeat the sequence n times, each time varying the magnitude of G_x pulse according to $\cos\theta$ and the magnitude of G_y pulse according to $\sin\theta$, where θ is from 0° to 90° in n steps. When $\theta = 45°$, the diagonal line is sampled (Figure 11.6b). Several fancier trajectories of traversing k space are shown in Figure 11.7, each having its own strengths and weaknesses.

11.4   GRADIENT ECHO

In Chapter 7.3, we show that the use of two rf pulses such as in Figure 7.6 can form a spin echo, in which the dephasing of the magnetization in the transverse plane due to the inhomogeneous effects (such as the non-uniformity of the magnetic fields and chemical shift) can be refocused. A pair of gradient pulses can also be used to form an echo, which is called the gradient echo and is shown in Figure 11.8a, which uses a bipolar pair of gradient pulses in which the magnitude of the second pulse is inverted and the areas of the two gradient pulses are matched. The 90° rf pulse tips the magnetization to the transverse plane as usual. The application of the first gradient pulse causes the signal to dephase in a similar manner as the field inhomogeneity that creates the FID. The use of a gradient just makes the dephasing effect immensely stronger, which effectively eliminates the signal (which is the average of all individual spin packs). If, however, a second gradient pulse, with the identical shape/area but negative magnitude, is applied at a later time, all of the

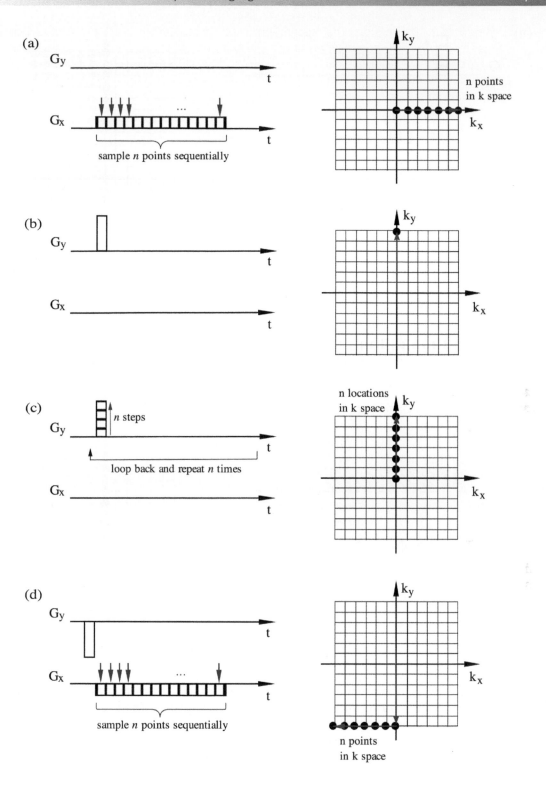

Figure 11.5 Moving around in 2D *k* space, where the center of *k* space is defined by the Larmor frequency (i.e., when the two mapping gradients are both zero). (a) Sampling a k_x direction, where the row of dots indicates a series of sampling points. (b) Moving to a particular k_y position. (c) Increasing k_y in *n* discrete steps, each increment nested in a loop. (d) Sampling a negative k_x direction not from the *k*-space origin.

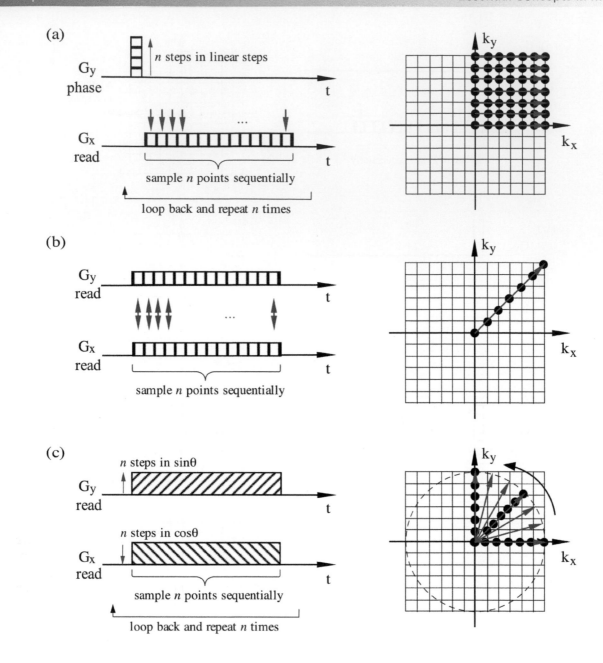

Figure 11.6 (a) Covering the first quadrant in **k** space in a Cartesian grid – the sequence is repeated *n* times, each time with an increment of G_y gradient (hence moving the start of the sampling upwards). (b) By increasing the G_x and G_y gradients simultaneously with the same amplitude, one can sample a diagonal line in **k** space. (c) By increasing the G_x and G_y gradients simultaneously, with the amplitudes following $\cos\theta$ and $\sin\theta$ where θ is from 0° to 90°, one can sample the first quadrant in **k** space in a radial grid.

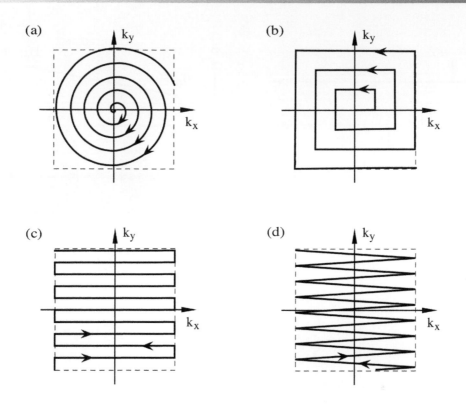

Figure 11.7   Four non-standard 2D mapping trajectories, each formed by tailoring the G_x and G_y gradients in some specific ways. The goal is to cover the entire **k** space.

location-dependent winding of the nuclear spins can be unwound (Figure 11.8c iii). Hence, an echo signal can be recovered after the second gradient pulse. The gradient echo has numerous uses for image formation in MRI and also in the measurement of flow and diffusion by NMR and MRI, which we will discuss more extensively in the later chapters.

Note that since the gradient echo only refocuses the spins that have been dephased by the gradient itself, the dephasing of the magnetization due to the inhomogeneous effects is not refocused by the gradient echo. Note also that it is possible to combine the gradient echo with the spin echo, which forms a powerful method to quantify the molecular motion (see Chapter 15.1).

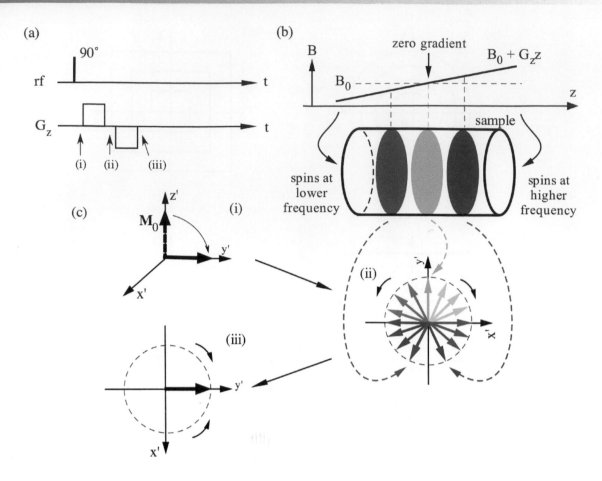

Figure 11.8 The dephasing and re-phasing by a gradient echo sequence in (a), where the magnetization in three time events is drawn schematically in (c). (b) A sample is subject to the pulses in (a). At three different locations of the sample along the direction of the gradient, the nuclear spins precess differently. The green pack of spins locates at the center of the gradient, hence precesses at the same frequency of the rotating frame (the Larmor frequency). The red and blue packs of spins experience the least and most of the field gradient, hence precess the slowest and fastest in the rotating frame, respectively. A second gradient pulse with equal magnitude and opposite direction can re-phase the different packs of nuclear spins, as shown in (c) from the moment *ii* to the moment *iii*. The drawing does not include the reduction in the final magnetization vector due to T_2 relaxation.

References

1. Lauterbur PC. Imaging Formation by Induced Local Interactions: Examples Employing Nuclear Magnetic Resonance. *Nature*. 1973;242:190–1.
2. Mansfield P, Grannell PK. NMR "Diffraction" in Solids? *J Phys C: Solid State Phys*. 1973;6:L422.
3. Bracewell R. *The Fourier Transform and its Applications*. New York: McGraw-Hill Book Company; 1965.
4. Callaghan PT, Eccles CD, Xia Y. NMR Microscopy of Dynamic Displacements: K-space and Q-space Imaging. *J Phys E: Sci Instrum*. 1988;21:820–2.
5. Xia Y. Introduction to Magnetic Resonance. In: Blümler P, Blümich B, Botto R, Fukushima E, editors. *Spatially Resolved Magnetic Resonance: Methods, Materials, Medicine, Biology, Rheology, Geology, Ecology, Hardware*. Weinheim, Germany: Wiley-VCH; 1998. pp. 713–39.

12

Spatial Mapping in MRI

Spatial mapping in MRI is inherently three dimensional over the usual 3D physical space. Figure 12.1 shows a number of ways that a 3D space can be sampled. One could even go beyond the 3D physical space to carry out 4D or 5D imaging with additional dimensions in molecular motion or chemical environment, which will be discussed in Chapter 15. In this chapter, we discuss the ways to encode the 3D physical space. In fact, it is the manner that a 3D k space is sampled that determines the advantages and disadvantages of each imaging protocol.

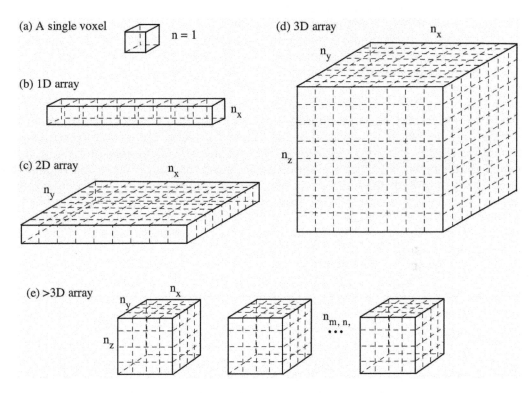

Figure 12.1 Imaging matrices, assuming each dimension has $n = 256$ voxels and each data voxel is 32 bits, or 4 bytes, in depth. (a) It is possible to select one single voxel in imaging, which only takes 4 bytes of storage space. (b-c) One could select a 1D linear array, which has a size of 1 kB (= 256 × 4), or a 2D slice, which has a size of 256 kB (= 256^2 × 4). (d) A proper 3D data array will have a size of 64 MB (= 256^3 × 4). (e) It is also possible to extend the imaging dimension to be beyond the 3D physical space, which is used in quantitative imaging (Chapter 15).

Essential Concepts in MRI: Physics, Instrumentation, Spectroscopy, and Imaging, First Edition. Yang Xia.
© 2022 John Wiley & Sons Ltd. Published 2022 by John Wiley & Sons Ltd.

12.1   SLICE SELECTION IN 2D MRI

Even with a 3D physical sample, MRI commonly acquires one or several 2D images, which reduces the size of each data array from N^3 to N^2. The resultant 2D images are consistent with the 2D visual display on computer monitors. The means selectively and non-invasively exciting the nuclear spins within a 2D slice is achieved by using simultaneously a narrow-band, shaped rf pulse and a gradient field in the direction orthogonal to the slice plane (imaging plane), which can be described graphically and mathematically.

12.1.1   Graphical Description of 2D Slice Selection

Without any field gradient, a 3D object of any shape would have a typical Lorentzian line shape with narrow width, as long as the main field $\boldsymbol{B}_0$ is uniform (Figure 12.2a). The application of a 1D field gradient along the z direction significantly broadens the width of the spectrum, from a narrow Lorentzian peak to a broadly spread profile in the frequency domain (Figure 12.2b). In this example of a cylinder with its long axis parallel to the z axis, any location along its length would have a different resonance frequency since the gradient is also along the z axis. The left end of the cylinder would resonate at a frequency lower than ω_0 and the right end of the cylinder higher than ω_0; at the z center of the 3D cylinder, which is defined by the center of the z gradient (cf. Figure 11.3), the resonance frequency should be ω_0. In other words, the 3D cylinder has been frequency encoded along the z direction by this 1D gradient.

At this moment if a narrow-band, shaped rf pulse is applied at a particular frequency ω_{rf}, only the nuclear spins at that particular frequency would be excited (i.e., selected), shown as the 2D x-y disc (an imaging slice) in Figure 12.2b. Since the nuclear spins in any particular cross section of the 3D cylinder would resonate at the same frequency, a circular disc would be selected, which is called the imaging slice and localized along the z axis. When the rf pulse turns the magnetization towards the transverse plane, the MRI signal would only come from that selected slice. Since any 2D slice along the z axis would be of the circular shape and have the same number of nuclear spins in the slice, the intensity profile of the spectrum for this cylinder should be similar to a rectangle (except for some edge effects), identical to the situation in 1D imaging (Figure 11.1).

Note that two descriptive terms are used to describe the rf pulse in the slice selection (cf. Chapter 2.10). The first term, "narrow-band," refers to the fact that the frequency bandwidth of the excitation rf pulse needs to be small in comparison with the spectral width that is broadened by the gradient. A narrow-band rf pulse, which can also be termed as a *soft* rf pulse, would have a long duration in time, hence a narrow range in frequency (Figure 12.3a). The second term, "shaped," refers to the fact that in order to select a rectangular slice in the frequency domain, the waveform of the rf pulse in the time domain must have an envelope of sinc waveform (sinc $= \sin\theta/\theta$, shown in Figure 12.3a), since the sinc function and the rectangular function are a Fourier transform pair (Table 2.2). In contrast, a *hard* rf pulse would have a short duration in time and hence a wide range in frequencies (Figure 12.3b). By changing the frequency of the excitation $\boldsymbol{B}_1$ pulse (ω_{rf} in Figure 12.2b), the location of the slice can be moved along the length of the cylinder. By adjusting the strength of the gradient, the thickness of the 2D slice can be adjusted, since a larger gradient would cause a larger frequency spreading, resulting in a thinner slice.

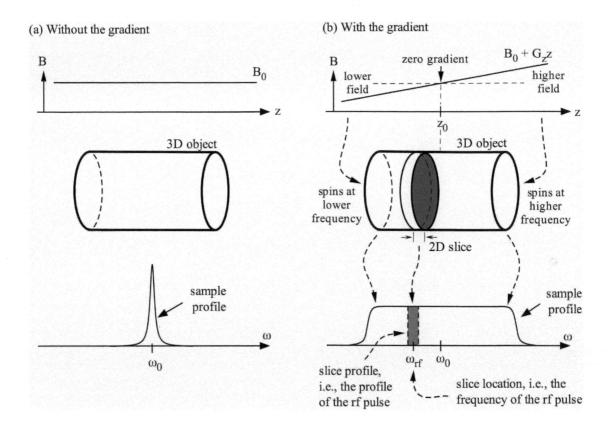

(a) Without the gradient

(b) With the gradient

Figure 12.2 Slice selection in MRI. (a) Without any gradient in the sample space, the nuclei inside a 3D object that has any physical size and shape will resonate at the same frequency, which results in a Lorentzian shape when the object contains simple liquids. (b) With a gradient set along the length of the 3D object (z), the left end of the object will experience a lower field, hence resonate at frequencies lower than ω_0; similarly, the right end of the object will resonate at frequencies higher than ω_0. Hence, the length of the physical object is encoded into a frequency spectrum. Since any 2D slice along the direction of z in this object is a circular disc, which contains the same number of nuclei, the frequency profile of this object remains constant for the majority of the object, except perhaps some edge effects on both ends.

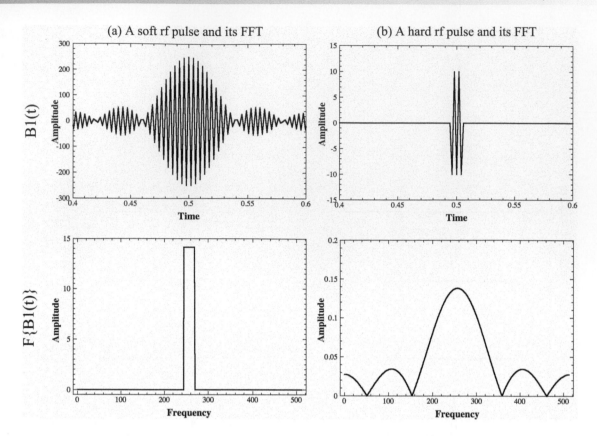

Figure 12.3 (a) A soft rf pulse means long in the time duration. When the envelope of the shaped pulse is sinc, the frequency profile of the pulse will be a narrow rectangle. (b) A hard rf pulse means short in the time duration. When the shape of the pulse is constant in its envelope, the frequency profile of the pulse will be a wide sinc. The data points in each of the four time and frequency data are consistent at 512. The shape of a pulse is the envelope of the pulse, not the high-frequency rf oscillation of the pulse.

12.1.2   Mathematic Description of 2D Slice Selection

Figure 12.4a shows a pulse sequence for slice selection, which consists of a soft sinc pulse together with a slice selection gradient (the first gradient pulse in the drawing). A mathematic description [1] for the evolution of $\boldsymbol{M}$ in the presence of an rf field in slice selection would benefit from the use of two rotating frames $x'y'z'$ and $x''y''z''$, as shown in Figure 12.4b. The usual $x'y'z'$ frame, in which $\boldsymbol{B}_1(t)$ is stationary and lies along the x' axis, is incorporated into the analysis because it simplifies the motion of $\boldsymbol{M}$ by eliminating the effect of $\boldsymbol{B}_0$. The second $x''y''z''$ rotating frame, in which $\boldsymbol{B}_1(t)$ is no longer stationary, is introduced into the analysis because it makes the evolution equations of M_x and M_y more symmetrical, from which they can be treated as the real and imaginary parts of a transverse magnetization, $M_{\perp''}$, as

$$\frac{d\boldsymbol{M}_{\perp''}}{dt} = \gamma M_0 \boldsymbol{B}_1(t) \exp(-i\omega'(t+\tau)), \tag{12.1}$$

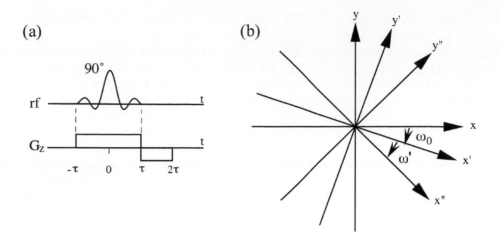

Figure 12.4   (a) A slice selection sequence in MRI, where a sinc pulse is applied together with an imaging gradient. A negative refocusing gradient is used in this sequence; this negative gradient should have one half of the area of the positive gradient for the purpose of phase compensation. (The gradient area is used schematically to indicate the product of the gradient strength and the gradient duration.) (b) The double rotating frames in the analysis of 2D slice selection.

where τ is the half duration of the slice selection rf pulse. The solution of Eq. (12.1) contains an integral that corresponds to a Fourier transform of $\boldsymbol{B}_1(t)$ from $-\tau$ to τ, namely, $\mathcal{F}\{\boldsymbol{B}_1(t)\}$, as

$$\boldsymbol{M}_{\perp''} = \gamma M_0 \, \mathcal{F}\{\boldsymbol{B}_1(t)\} \exp(-i\omega'\tau), \tag{12.2}$$

where $\mathcal{F}\{\boldsymbol{B}_1(t)\}$ is the shape of the slice. Since the $x'' y'' z''$ frame has rotated $\omega'2t$ clockwise with respect to the $x'y'z'$ frame during the rf pulse time, one obtains

$$\boldsymbol{M}_{\perp'} = \boldsymbol{M}_{\perp''} \exp(i\omega'2\tau) = \gamma M_0 \, \mathcal{F}\{\boldsymbol{B}_1(t)\} \exp(i\omega'\tau). \tag{12.3}$$

The extra phase shift term, $\exp(i\omega'\tau)$, is a function of position in the sample space and so leads to a dephasing of the signal. It can be removed by using a re-phasing gradient that is applied in the opposite direction, that is, the negative gradient in the sequence shown in Figure 12.4a. Therefore we have

$$\boldsymbol{M}_{\perp'} = \gamma M_0 \, \mathcal{F}\{\boldsymbol{B}_1(t)\} \tag{12.4}$$

or

$$M_{y'} = \gamma M_0 \, \mathrm{Re}\left[\mathcal{F}\{\boldsymbol{B}_1(t)\}\right]. \tag{12.5}$$

The variation of $M_{y'}$ with respect to ω' is termed the slice profile. At 2τ, $M_{y'}$ is proportional to the real part of the Fourier transform of the applied rf pulse, which means that in order to obtain a rectangular slice of spins, a sinc-modulated rf pulse is required in the time domain since the rectangular function and the sinc function are an FT pair.

12.1.3   Sequences and Parameters in 2D Slice Selection

Figure 12.5 shows four versions of the slice selection sequence, with different combinations of hard and soft rf pulses. In Figure 12.5a, the negative gradient, which rephrases the extra term in Eq. (12.3), can be considered as a gradient echo. When a unipolar gradient is used, one can combine the concept of spin echo into the slice selection, as in Figure 12.5b. In addition, one can also use a hard 90° pulse for the initial excitation, as in Figure 12.5c. Finally, one can use two soft rf pulses for the slice selection, as in Figure 12.5d.

Although these four sequences serve the same purpose in 2D slice selection, there are three subtle differences among them. First, for the sequences that use the soft pulses only (Figure 12.5a,12.5d), the nuclear spins in other parts of the sample are not excited by the slice selection and hence can be tagged immediately for subsequent additional imaging measurement; in contrast, when the sequence contains one hard pulse (Figure 12.5b, 12.5c), the nuclear spins that are not inside the imaging slice have also been disturbed (excited) by the hard pulse. Second, since a soft pulse has a longer time duration than a hard pulse (e.g., 1 ms vs 10 µs), the use of soft pulse increases the echo time of the imaging sequence, which is undesirable in MRI if some parts of the specimen have short T_2 (see Section 12.8). Finally, the use of spin-echo format (Figure 12.5b–d) in the slice selection generally yields better quality images than those produced using the gradient-echo format (Figure 12.5a).

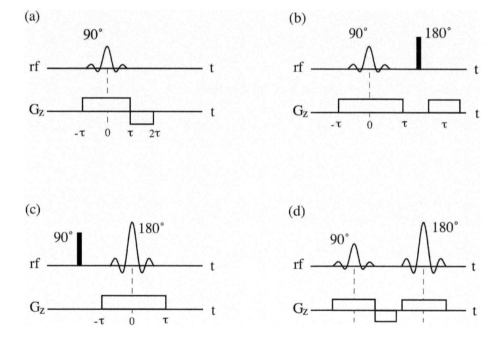

Figure 12.5   Pulse sequences in slice selection. (a) A slice selection sequence with a negative refocusing gradient. (b) The spin-echo form of a slice selection sequence with a hard 180° pulse, where a positive refocusing gradient is used. (c) A soft 180° slice selection sequence. (d) A slice selection sequence with two soft rf pulses. The second gradient in (a), (b), and (d) should have one half of the area of the first gradient, for the purpose of phase compensation.

When the precession frequency of $B_1(t)$, ω_{rf}, is identical to ω_0 (i.e., B_1 is stationary in the $x'y'z'$ frame), the slice is selected at the center of the field. By changing the precession frequency of $B_1(t)$, the position of the slice plane where M and $B_1(t)$ precess at the same frequency can be adjusted. This means the position of the slice can be shifted along the gradient axis, as

$$z = (\omega_{rf} - \gamma B_0)/\gamma G_z. \tag{12.6}$$

The width of the selected slice is determined by the bandwidth of B_1 field, Δf, and the magnitude of the gradient, as

$$\Delta z = \Delta \omega / \gamma G_z, \tag{12.7}$$

where $\Delta \omega$ is equal to $2\pi \Delta f$. It is clear that a larger gradient would result in a thinner slice (unless you adjusted Δf at the same time). The relationships between ω, z, and Δz in the slice selection are shown graphically in Figure 12.6.

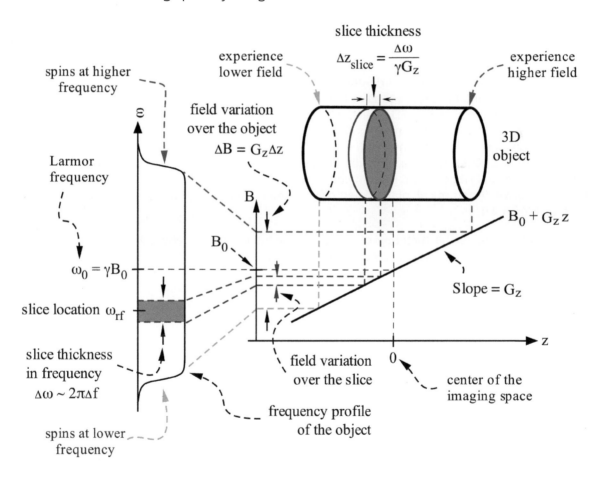

Figure 12.6   Relationships among the slice selection parameters. The slice thickness is determined by the magnitude of the slice selection gradient. A smaller gradient (i.e., a smaller slope graphically) causes the object to have a smaller range of frequency spread, which would result in a thicker slice being selected. The center of the object, which sits at the zero gradient, resonates at ω_0. By moving the frequency of the B_1 pulse ω_{rf}, one can move the location of the slice along the z axis, that is, along the length of the object. The thickness of the slice is determined by the frequency profile of the B_1 pulse.

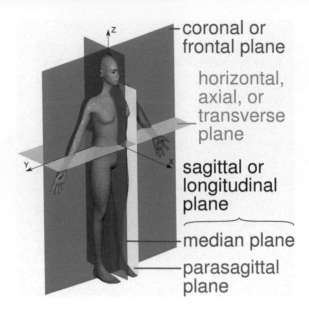

Figure 12.7   Three common slice orientations in clinical MRI of human subjects are axial or transverse or horizontal plane, sagittal or longitudinal plane, and coronal or frontal plane. One can also select an oblique slice in a 3D object, when the slice selection gradient is set obliquely. In addition, there are other common terms that describe the orientations in a human object (or animals), such as superior and inferior (i.e., head and foot directions), anterior and posterior (i.e., towards front and back), lateral and medial (i.e., outside and inside directions), etc. Source: David Richfield and Mikael Häggström, Human anatomy planes, Wikipedia. Available at https://en. wikipedia.org/wiki/Anatomical_plane#/media/File:Human_anatomy_planes,_labeled.svg. Licensed under CC BY-SA 4.0.

By changing the direction of the slice selection gradient, any 2D slice in a plane that is orthogonal to the gradient vector can be selected. In clinical MRI of humans, there are three specific terms that define the 2D slice selected in each of the three orthogonal directions in a human body, shown in Figure 12.7. Of course, there are numerous other ways that a slice can be selected obliquely over a 3D object, when more than one gradient is turned on simultaneously as the slice selection gradient.

12.2   READING A GRAPHICAL IMAGING SEQUENCE

This section describes conceptually the ways to read and understand a graphical (i.e., schematic) version of imaging pulse sequence. Figure 12.8 is an example: a 2D imaging pulse sequence that employs a radial or polar sampling raster (Figure 11.6c), which can be used to form an image using the back-projection reconstruction (cf. Section 12.3).

In Chapter 6.3, we introduced the concept of pulse sequence in NMR spectroscopy, which commonly has one or two rf channels – each rf channel manipulates one type of nuclear spins (e.g., ^{1}H) and commonly has a few rf pulses along the axis of time (the horizontal axis in sequence drawings). In advanced NMR spectroscopy, some multidimensional sequences can become very complicated; they are designed with a deep understanding of NMR in quantum

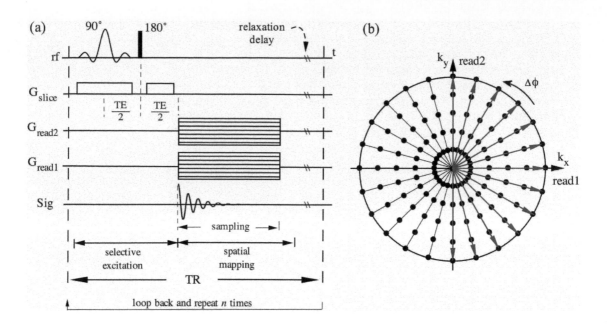

Figure 12.8 2D back-projection imaging sequence (a) that uses a radial sampling raster (b). The center of the echo arrives at the beginning of the sampling period, which is called half-echo sampling. The imaging experiment is repeated n times, each time two read gradients have a different set of amplitudes, one following $\sin\theta$ and one following $\cos\theta$, hence the sampling is along a radial line from the origin in $\boldsymbol{k}$ space. The second G_{slice} gradient should have one half of the area of the first G_{slice} gradient. The double backslashes near the end of each time axis indicate an extra time delay is needed in order to restore the magnetization to the longitudinal z axis – the actions of the imaging sequence take about 20–30 ms in total, while the extra time delay can be as long as $5 \times T_1$ in order to restore the magnetization to its thermal equilibrium. When the nuclear spin density $\rho(r)$ is a real function, the symmetry relationship specified by Eq. (11.10) can be used; in this case, only one half of $\boldsymbol{k}$ space needs to be sampled [indicated by the red arrows in (b)], where G_{read1} only needs to vary positively.

mechanical theory to probe a particular spin property. In comparison, the imaging sequences in MRI always have multi-channels, commonly having one rf channel and three gradients channels.

All imaging sequences also use time as the horizontal axis, which defines the direction of the operation – any pulse earlier on the time axis gets executed first, followed by the pulses appearing later on the time axis. The rf channel in the imaging sequence can be identical to the rf channel in simple sequences in spectroscopy, which applies the $\boldsymbol{B}_1(t)$ pulses (both hard and soft) to the specimen. The three gradient channels in the imaging sequences define the 3D imaging space. In 2D imaging, one of the gradient channels is used together with the rf channel to provide the slice selection. This gradient is termed as the slice gradient (G_{slice} or G_s). The direction of this slice selection gradient defines the orientation of the 2D slice through a specimen (cf. Figure 12.2). The other two orthogonal gradients map out positions in the plane of the 2D slice. They are called the mapping gradients, whose trajectories can be visualized by the aid of a $\boldsymbol{k}$-space map, shown in Figure 12.8b.

In 2D imaging using a Cartesian grid (Figure 11.6a), the mapping gradient that turns on during the signal acquisition is called the read gradient (G_{read} or G_r); while the mapping gradient that turns on before the signal acquisition is called the phase gradient (G_{phase} or G_p), since it introduces a phase twist to the transverse magnetization. In 2D imaging using a radial grid (Figure 11.6c), both mapping gradients are turned on during the signal acquisition; hence, both of them are read gradients and can be labeled as G_{read1} and G_{read2}. In the imaging instrument, the three gradients are commonly labeled as G_x, G_y, and G_z (cf. Figure 11.3 and Chapter 13), which are defined by the physical direction of the $\boldsymbol{B}_0$ field (cf. Figure 1.2). A total of six different combinations can be found between the three physical gradients G_x, G_y, and G_z and three imaging directions G_s, G_p, G_r in 2D and 3D MRI. In the MRI literature, the most common combination is $G_s = G_z$, $G_r = G_x$, and $G_p = G_y$.

The 2D imaging sequence shown in Figure 12.8a can be divided along the time axis into two segments, the selective excitation segment and the spatial mapping segment. Starting from the left, the first action of the sequence is to execute the selective excitation segment, which is used to select a 2D imaging slice from a 3D object (cf. Figure 12.2). The selective excitation segment can use any of the four slice selection sequences shown in Figure 12.5. If the slice selection gradient is set at the z direction, a 2D axial slice will be selected from the 3D object (cf. Figure 12.7). Due to the use of a spin-echo sequence in the rf channel, the center of the echo signal will be at the end of the selective excitation segment, which can be mapped out immediately. The 2D mapping can use any of the 2D $\boldsymbol{k}$-space sampling grids shown previously in Figure 11.6 and Figure 11.7. This particular sequence has a radial sampling raster, where the two read gradients are applied simultaneously following the cosine and sine relationships, respectively. Recall Eq. (11.7), which relates the gradient amplitude G and its duration t with the traverse in $\boldsymbol{k}$ space. For each line of signal acquisition, both G_{read1} and G_{read2} have some fixed values. When $\phi = 0°$, G_{read1} is at the max and G_{read2} is zero; the sampling will start from the $\boldsymbol{k}$-space origin horizontally toward the maximum positive value of the k_x line. When the sequence is looped back for the second time, G_{read1} follows $\cos(\Delta\phi)$ and G_{read2} follows $\sin(\Delta\phi)$, which samples the first tilted line in the first quadrant of $\boldsymbol{k}$ space. When $\phi = 90°$, a $\boldsymbol{k}$-space sampling line will go from the origin to the maximum positive value of the k_y line (vertically up in Figure 12.8b). Different combinations of positive and negative values for G_{read1} and G_{read2} can sample the other quadrants in $\boldsymbol{k}$ space.

Note a graphic version of an imaging sequence is commonly drawn as a schematic, that is, not strictly to scale (i.e., the elements are often not in proportion). For example, the total time needed for both selective excitation and spatial mapping in Figure 12.8a would typically be 15–25 ms. Before looping back for a second $\boldsymbol{k}$-space acquisition, however, one needs to wait for the magnetization to return to the thermal equilibrium (which can be as long as several seconds, up to $5 \times T_1$). If the time axis was drawn in scale in Figure 12.8a, all actions would be compressed to a very short duration of time, which leaves most of the drawing containing nothing (i.e., waiting). In the current drawing, the delay to allow for spin relaxation is simply represented by a double backslash on the time axis. The repetition time, TR, is the total time between the repetitive acquisitions (typically in several hundreds of milliseconds to several seconds). The echo time, TE, is the time between the spin excitation to the echo center of the signal (typically in 5–20 ms).

The relative sizes among gradient pulses are also not drawn to scale in a schematic pulse sequence drawing. In an imaging sequence, only when all three physical gradients have the same calibration values (in units of mT/m) will the areas of the gradient pulses correspond in proportional size to the effects of the gradients on the magnetization. In reality, different gradients have vastly different strengths (due to the coil geometry, the number of turns in the coil, the size of the coil, the amount of electric current to a coil, the effect of active shielding, etc.). So the sizes of the gradient pulses in these imaging sequences are just illustrative, with a few simple rules in their design. For example, the second G_{slice} gradient in Figure 12.8a should be ½ of the area of the first G_{slice} gradient (for phase compensation). To sample a "square" Cartesian raster (Figure 11.6a), the areas of G_{phase} and G_{read} should match; to sample a "circular" radial raster (Figure 11.6c, Figure 12.8b), the two read gradients should match their areas (i.e., strengths).

12.3   2D FILTERED BACK-PROJECTION RECONSTRUCTION

The filtered back-projection (FBP) reconstruction [2] was the method used by Paul C. Lauterbur in 1973 for the first ever image reconstruction in MRI (Figure 1.5). The FBP pulse sequence, described in the previous section (Figure 12.8), uses the radial grid for k-space sampling where the two mapping gradients are applied simultaneously following the cosine and sine relationships, respectively (cf. Figure 11.6c). As shown in Figure 12.8b, the sampling in FBP always starts from the k-space origin, going outwards on a radial trajectory. The sampling point along a radial line in k space is given by

$$\phi = \tan^{-1}(G_{read2} / G_{read1}) \tag{12.8a}$$

and

$$|\boldsymbol{G}| = \left(G_{read1}^2 + G_{read2}^2\right)^{1/2}. \tag{12.8b}$$

Based on the orientation of k space in Figure 12.8b, the two read gradients can be written down from Eq. (11.7), as

$$\Delta k_{read1} = \frac{1}{2\pi}\gamma G_x(\Delta t_x) \tag{12.9a}$$

and

$$\Delta k_{read2} = \frac{1}{2\pi}\gamma G_y(\Delta t_y). \tag{12.9b}$$

For each signal acquisition (along a single k-space line), G_x and G_y have constant amplitudes. The traverse in k space during signal acquisition is governed by the increment of time t; the sampling occurs at a fixed rate during this time. The signal and the spin density given previously by Eq. (11.8) and Eq. (11.9) can now be rewritten in the radial coordinate as

$$S(k,\phi) = \int_0^\pi \int_{-\infty}^{+\infty} \rho(x, y) \exp(-i2\pi \boldsymbol{k} \cdot \boldsymbol{r}) r dr d\phi \tag{12.10}$$

and

$$\rho(x, y) = \int_0^\pi \left\{ \int_{-\infty}^{+\infty} S(k, \phi) \exp(i2\pi \boldsymbol{k} \cdot \boldsymbol{r}) |k| dk \right\} d\phi, \tag{12.11}$$

where the third dimension, z, is ignored in the equations due to the application of slice selection.

It is worth noting that the inner integral in Eq. (12.11) represents a 1D FT where the signal has been multiplied by a ramp $|k|$, given by $k = (k_x^2 + k_y^2)^{1/2}$ [2]. This step is termed as "filtering," which removes the "star artifacts" or "blurring" produced by the simple back-projection reconstruction process, which is shown graphically in Figure 12.9. Without the ramp filtering, a straightforward back-projection using the projection spectra would result in "star artifacts" around the objects, due to the positive intensity of the profiles over the entire image at all projections (Figure 12.9a). The filtering changes the profiles to $P^*(r)$, which graphically have the negative edges/spikes around the objects. These negative edges cancel out the positive star artifacts immediately outside the objects, reaching a near-zero background (Figure 12.9b).

The result of this one-dimensional transform is termed the "filtered back-projection," as

$$P^*(r, \phi) = \int_{-\infty}^{+\infty} S(k, \phi) \exp(i2\pi \boldsymbol{k} \cdot \boldsymbol{r}) |k| dk. \tag{12.12}$$

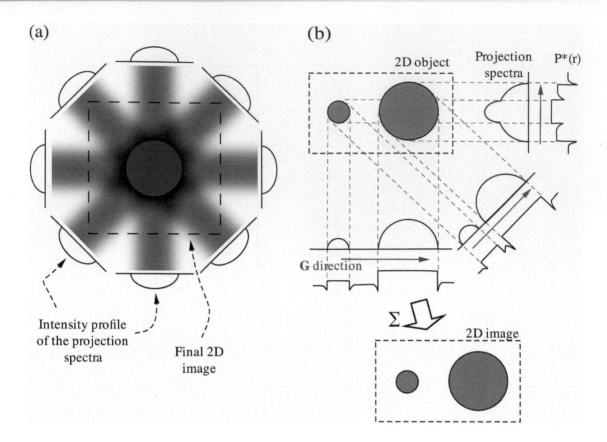

(a)

Intensity profile
of the projection
spectra

Final 2D
image

(b)

2D object

Projection
spectra

P*(r)

G direction

Σ

2D image

Figure 12.9 The image reconstruction in the filtered back-projection method. (a) A straightforward projection reconstruction using the unfiltered projection spectra will cause the image to have a star artifact outside the sample object (the blue disc), due to the positive intensity of the profiles projected across the image. (b) The filtering introduces the negative spikes in the P* profiles immediately outside the sample object, which cancel out the positive intensity outside the sample object when the profiles are back-projected.

$P^*(r, \phi)$ contains information of all points, along the gradient direction, satisfying

$$r = x \cos\phi + y \sin\phi. \tag{12.13}$$

Eq. (12.12) and Eq. (12.13) state that the result of a 1D FT of the projection data at a given angle ϕ represents the value of the 2D FT of the object function $\rho(x, y)$ along the radial direction in $\boldsymbol{k}$ space, with the given angle ϕ.

This filtered profile is then back-projected into the image plane by summing through all angles,

$$\rho(x, y) = \sum_{j=1}^{n} P^*(x_j, y_j, \phi_j) \Delta\phi. \tag{12.14}$$

where n is the number of directions used when imaging projection. When the symmetry relationship given by Eq. (11.10) is employed in imaging, the step increment size in the angular space is therefore $\Delta\phi = \pi/n$ (i.e., only ½ of $\boldsymbol{k}$ space is sampled). The final image in the back projection needs to be transformed into a Cartesian matrix for the 2D display on computer monitors, which have the Cartesian grid (cf. Section 12.5). This back-projection method was developed long before the invention of MRI, for x-ray CT image reconstruction [2].

12.4 2D FOURIER IMAGING RECONSTRUCTION

The most widely used imaging reconstruction in MRI is called Fourier imaging or the "spin-warp" method [3–5], which employs a Cartesian sampling grid (Figure 11.6a). Fourier imaging can sample the signal in either one half of $\boldsymbol{k}$ space (Figure 12.10) or the full $\boldsymbol{k}$ space (Figure 12.11). In the beginning of the three imaging sequences in Figures 12.10 and 12.11, the operations always start with the selective excitation. After the slice selection, the G_x and G_y gradients in the spatial mapping segment are applied separately during times t_x and t_y, which gives the $\boldsymbol{k}$-space vector as

$$k_x = \frac{1}{2\pi}\gamma G_x t_x \tag{12.15a}$$

and

$$k_y = \frac{1}{2\pi}\gamma G_y t_y. \tag{12.15b}$$

For the sequence that samples one half of $\boldsymbol{k}$ space (Figure 12.10a), the magnetization within the slice is prepared at the end of the slice selection. The read gradient is turned on at this moment, which records the center of the echo (similar to the situation in the filtered back-projection pulse shown in Figure 12.8a). To shorten the echo time TE, the phase gradient can occur at the same time as the slice re-phase gradient (the negative gradient in G_{slice}). After acquiring one line of signal, the magnitude of the G_{phase} changes its value by $G_y/(n/2)$ where n is the total number of steps in the k_y direction, so the next $\boldsymbol{k}$-space sampling is at a different k_y location. (Note that G_{phase} needs to be positive and negative.) All samplings in this sequence start from the $k_x = 0$ position, for only ½ of $\boldsymbol{k}$ space. It does not matter which half of $\boldsymbol{k}$ space one samples. The drawing in Figure 12.10a implies the sampling of the first and fourth quadrants. If one wants to sample the first and second quadrants, k_y will have only the positive values and k_x will have both positive and negative values. When sampling only half of $\boldsymbol{k}$ space (Figure 12.10a), one only gets one half of the total signal; however, the image should still be good due to the symmetry relationship of the MRI signal [Eq. (11.10)]. A unique advantage of the half-echo sampling is the fact that the sampling starts at the maximum of the signal, which could be beneficial if the sample has very short T_2 (in a full echo sampling, the signal could be significantly decayed before one reaches the echo center if T_2 is on the order of the sampling time, typically in milliseconds).

For the two sequences that sample the full $\boldsymbol{k}$ space (Figure 12.11), Figure 12.11a is the spin-echo version. In this sequence, the placement of the 180° rf pulse delays the formation of the

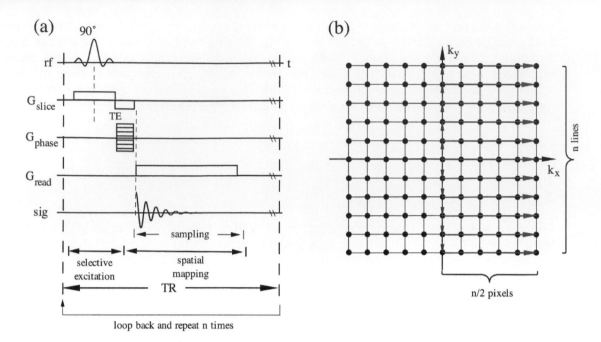

Figure 12.10 2D FT sequence (a) that uses a Cartesian sampling raster (b). This sequence only samples half of **k** space (the first and fourth quadrants, the right half of the space indicated by the red arrows), since the read gradient stays positive. The area of the maximum and minimum G_{phase} gradients should be the same as the G_{read} gradient. Similar to the back-projection imaging in Figure 12.8, the sampling starts at the center of the echo. The imaging experiment is repeated n times, each time having a different level of the phase gradient, hence moving to a different k_y position in **k** space (indicated by the blue arrows). The double backslashes near the end of each time axis indicate an extra time delay.

echo. The simultaneous application of the G_{phase} pulse and the first G_{read} pulse moves the starting point of the **k**-space sampling to the left edge of the **k**-space grid (shown with the blue dashed arrows in Figure 12.11c). When G_{phase} is at its most positive value, the start of the sampling will be from the top left corner of **k** space; when G_{phase} is zero, the start of the sampling will be at $k_y = 0$; when G_{phase} is at its most negative value, the start of the sampling will be at the bottom left corner of **k** space. If the first G_{read} pulse is one half in area/strength of the second G_{read} pulse, the max signal would occur in the middle of the second read gradient, which also coincides with the center of the spin echo.

The second full **k**-space sequence (Figure 12.11b) is the gradient-echo version. Without the 180° rf pulse, the first G_{read} gradient needs to be negative in magnitude. After the end of the slice selection, the negative G_{read} pulse and the G_{phase} pulse still move the starting point of the sampling to the left edge of **k** space. As long as the first G_{read} pulse is one half of the second G_{read} pulse, the signal will reach its maximum at the center of the signal acquisition, as a gradient echo. Between the two full **k**-space samplings, the spin-echo version (Figure 12.11a) yields better quality images than the gradient-echo version (Figure 12.11b), since the spin echo can refocus the field inhomogeneities. Among the three 2D FT imaging sequences, the half-echo version (Figure 12.10) has the shortest TE, while the spin-echo version (Figure 12.11a) has the longest TE.

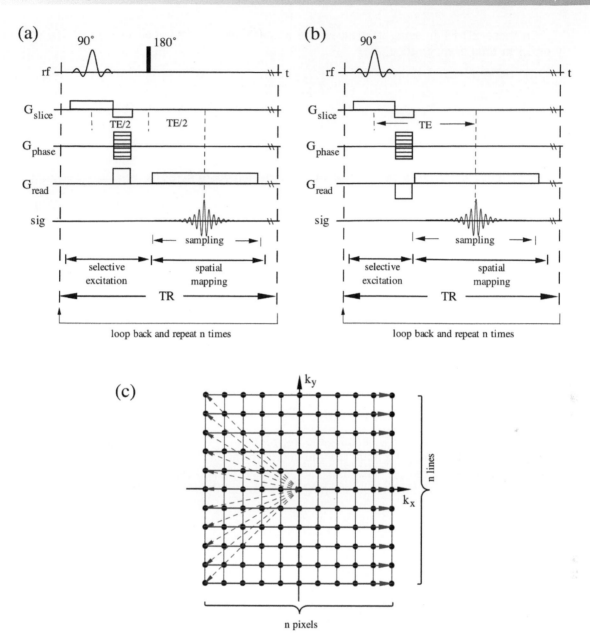

Figure 12.11 The spin-echo version (a) and the gradient-echo version (b) of a 2D FT sequence that use a Cartesian sampling raster (c) when the entire **k** space is sampled. In (a) the use of a 180° rf pulse serves two purposes: to form a spin echo to improve the image quality and to place the spin-echo center at the center of the read gradient (hence the first G_{read} pulse should be one half of the second G_{read} pulse in magnitude). The imaging experiment is repeated n times, each time having a different level of the phase gradient, hence moving the start of the sampling to a different k_y position in **k** space (indicated by the blue arrows). The double backslashes near the end of each time axis indicate an extra time delay for spin relaxation.

In these 2D FT imaging sequences, the **k**-space vectors described in Eq (12.15) can vary either in time or magnitude, as

$$\Delta k_x = \frac{1}{2\pi}\gamma G_x(\Delta t_x) \tag{12.16a}$$

and

$$\Delta k_y = \frac{1}{2\pi}\gamma(\Delta G_y)t_y. \tag{12.16b}$$

Equation (11.8) can now be rewritten in the Cartesian coordinates as

$$S(k_x,k_y) = \int_{-\infty}^{+\infty}\int_{-\infty}^{+\infty}\rho(x,y)\exp\left(-i2\pi(k_x x + k_y y)\right)dxdy, \tag{12.17}$$

where the third direction, the slice selection direction, has been ignored since it represents a simple averaging over the slice direction. It is clear that the signal, $S(k_x,k_y)$, is just the 2D Fourier transform of the spin density $\rho(x,y)$, as

$$\rho(x,y) = \int_{-\infty}^{+\infty}\int_{-\infty}^{+\infty}S(k_x,k_y)\exp\left(i2\pi(k_x x + k_y y)\right)dk_x dk_y. \tag{12.18}$$

12.5 SAMPLING PATTERNS BETWEEN THE CARTESIAN AND RADIAL GRIDS

Figure 11.6 shows two fundamental sampling grids in 2D **k** space; one is a Cartesian grid in the FT imaging method and the other is a radial grid in the back-projection imaging method. Many fast imaging sequences also use other non-Cartesian grids, as shown in Figure 11.7. Several issues in the pattern of individual imaging grids can subtly influence the image quality in MRI.

12.5.1 Sampling Density

Between the Cartesian grid (Figures 11.6a, 12.10, 12.11) and a radial grid (Figures 11.6c and 12.8), it is clear that the sampling density over **k** space is uniform in the Cartesian grid. In contrast, a radial grid would have a non-uniform sampling density over **k** space. For the classical radial grid (Figure 11.6c, Figure 12.8b), its inner part is more densely sampled than the outer part. Since the inner part of **k** space corresponds to the low-frequency components of the signal and consequently the coarse/broad features in the image, while the outer region of **k** space corresponds to high-frequency components and hence the fine details in the image, different sampling patterns in **k** space in many non-Cartesian grids could subtly emphasize different features in the images, resulting in some types of non-uniform representation.

12.5.2 Echo Time

The echo time (TE) in MRI is defined as the time between the center of the first 90° rf pulse to the center of the maximum signal in acquisition, as in Figures 12.8, 12.10, and 12.11. In general, a Cartesian grid needs slightly more time to complete the slice selection operation, since the phase gradient needs to be executed before the onset of the read gradient. This leads to a slightly longer TE for the imaging experiments. Some portion of the echo time is used as the waiting time so that a shaped pulse can reach its stable value (cf. Figure 5.8); while another portion of the echo time is needed for the application of the pulses (a soft pulse takes longer to apply). The echo time for a typical 2D MRI experiment is on the order of several to tens of milliseconds. For specimens that have a short T_2 relaxation, a long echo time will cause additional signal loss and unwanted T_2 contrast in imaging. In comparison to the full-echo sampling using a Cartesian grid, the half-echo radial sampling could be designed to start the signal acquisition at the leading edge of the gradient, which offers in principle the potential to catch the full signal.

12.5.3 Re-gridding

Any sampling pattern that is not in a regular Cartesian grid would have to be re-gridded into a Cartesian grid in order to be processed (e.g., in Fourier transform) and displayed (e.g., on computer monitors). Any re-gridding process is essentially an interpolation process, to determine the value of a data point at a location where the data has not been sampled. Several algorithms can be used in the creation of a new voxel at a new location, all of which consider the values of its neighboring voxels, the trends of the neighboring voxels, and the distances between the new voxel and the neighbors. A 1D example is shown in Figure 12.12a, where the simplest estimation comes from the insertion of multiple points on a straight line that fits between any two acquired points. A better estimation is to consider more neighbors, more than the immediate two, and fit

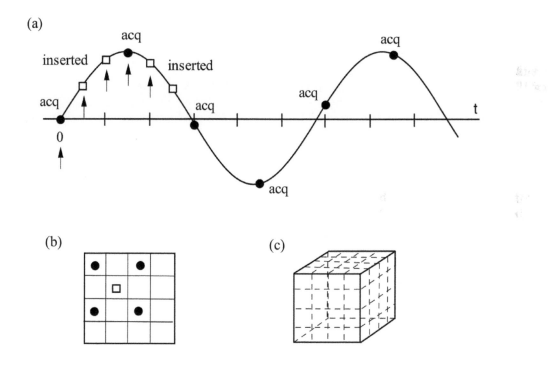

Figure 12.12 Re-gridding or interpolation process based on the acquired data points (solid circular dots). (a) Any new data point (open square) in a 1D space has at least two nearest neighbors along a line. The line could be straight or curved between the two acquired points, based on the best estimate. (b) Any new data point in a 2D space has at least four nearest neighbors among a surface. (c) Any new data point in a 3D space has many nearest neighbors in a cubic structure.

these acquired voxels with some curves, for example, a polynomial or spline. This re-gridding principle can be extended into 2D and 3D space (Figure 12.12b and c), where the neighboring points come from a 2D or 3D space.

12.6 3D IMAGING

There are two different approaches to image a 3D object: a true 3D sampling over k space or the acquisition of a series of 2D images one after the other.

Figure 12.13 shows two true 3D imaging sequences, both starting with a hard rf pulse to excite the nuclear spins in the entire sample. Figure 12.13a is the 3D back-projection reconstruction imaging that uses a 3D radial sampling grid (Figure 12.13c). As in the 2D situation, the sampling starts at the maximum of the signal from the origin of k space. Figure 12.13b is the 3D Fourier imaging sequence that uses a 3D stack of 2D Cartesian sampling grids (Figure 12.13d). One can replace the stack of 2D Cartesian grids in Figure 12.13d using any of the 2D spiral grids or zigzag grids shown in Figure 11.7. Since the acquisition is going to repeat many times (TR $\times n_2 \times n_3$), one can shorten the experimental time by using an rf pulse that tips the magnetization for less than 90°, represented by the $\alpha°$ excitation pulse.

The mathematical expressions for 2D imaging can be easily extended to the corresponding 3D imaging. For example, the 3D back-projection imaging can use a spherical radial coordinate to express the spin density in 3D back-projection imaging as

$$\rho(r) = \int_0^\pi \int_0^\pi \left\{ \int_{-\infty}^{+\infty} S(k,\theta,\phi) \exp(i2\pi k \cdot r) k^2 dk \right\} \sin\theta d\theta d\phi. \tag{12.19}$$

The equivalence of Eq. (12.12), the "filtered profile," in the 3D case is given by

$$P^*(r) = \int_{-\infty}^{+\infty} S(k,\theta,\phi) \exp(i2\pi k \cdot r) k^2 dk, \tag{12.20}$$

which contains information of all points, along the gradient (radial) direction r, satisfying

$$r = x \sin\theta \cos\phi + y \sin\theta \sin\phi + z \cos\theta. \tag{12.21}$$

A different strategy in 3D MRI is to acquire a series of 2D images, each at a different slice location along the length (height) of the object, to cover the entire 3D object. This is often done by repeating the single-slice or multi-slice 2D imaging sequence. The multi-slice imaging sequence utilizes the fact that if the slice selection only uses soft pulses (Figure 12.5a and 12.5d), the nuclear spins outside the current slice are, in principle, not disturbed hence available for immediate excitation subsequently. Consequently, the relaxation waiting time in the imaging experiment can be utilized to obtain additional images at different slice positions (with an adjustment of ω_{rf}, as in Figure 12.2) with no increase of the experimental time. In practice, since the nuclear spins located just outside of any 2D slice can be disturbed by the imperfection of the imaging slice (due to the use of a truncated sinc pulse; see Chapter 13.1), it is common to leave a gap between any two neighboring images in the first round (e.g., first time to sample all odd numbered images) and then come back to sample the gaps (e.g., to sample all even numbered images in a second round).

There are several pros and cons between the multiple 2D imaging approach and the true 3D imaging approach. First, the multiple 2D approach is simpler in image acquisition and reconstruction, since each 2D image can be examined individually and the requirement for computer memory in 2D image reconstruction is much less than for true 3D imaging. Second, the multiple

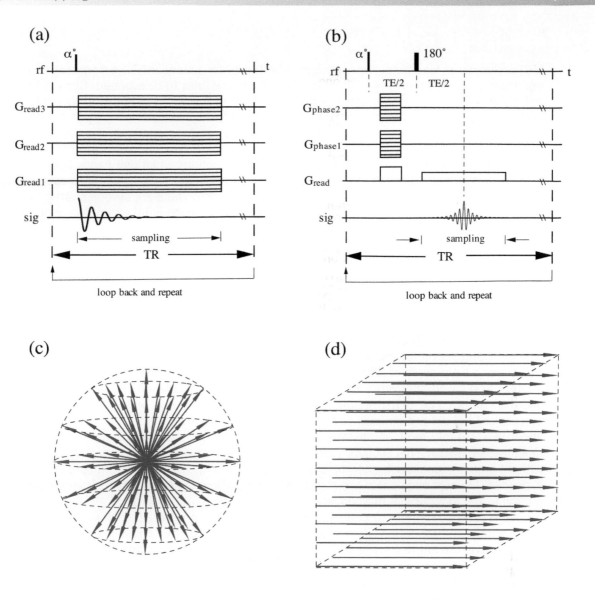

Figure 12.13 3D MRI sequences. (a) 3D back-projection imaging using a 3D radial raster (c), which can be compared with the 2D version in Figure 12.8. (b) 3D Fourier imaging using a 3D Cartesian raster (d), which can be compared with the 2D version in Figure 12.11. The $\alpha°$ in the rf pulses indicate a tipping angle smaller than 90°.

2D imaging sequence would have a longer echo time (TE) than true 3D imaging, since any 2D imaging sequence needs to use slice selection (a longer TE could cause extra signal loss as well as extra T_2 weighting). Third, a true 3D imaging sequence can achieve an isotropic (cubic) image voxel, while multiple 2D imaging sequences often have an elongated voxel size, with the slice dimension being larger than the transverse resolution. This elongated voxel size would have different voxel averaging in different directions [6]. Finally, if a long imaging experiment is interrupted in the middle of data acquisition, it is possible to recover the already acquired multiple 2D raw data but not any partially acquired 3D raw data.

12.7   FAST IMAGING IN MRI

MRI experiments take time. A straightforward estimate for a standard 2D experiment time is the repetition time (TR) multiplied by the number of *k*-space lines to be scanned. Since a full magnetization in the transverse plane maximizes the SNR of the image, one can ensure a full magnetization by using a 90° rf pulse and by setting TR to $5 \times T_1$, where the $5 \times T_1$ parameter would allow the full return of the magnetization to the *z* axis after it is tipped into the transverse plane by a 90° pulse. Therefore, a standard 2D MRI experiment typically takes several minutes or longer for the data acquisition. There are, however, situations when one needs faster imaging (e.g., to minimize the body motion of the human and animal during imaging) or to capture a fast occurring or changing event (e.g., to image a beating heart). There are several strategies for the reduction of the imaging time, divided between whether *k* space is fully sampled or not.

12.7.1   Fast MRI with Fully Sampled *k* Space

If we do not need the full magnetization for our signal, we can speed up the imaging experiments considerably, by using small tipping angles and shortening TR (as represented by an $\alpha°$ excitation pulse in Figure 12.13). A small tipping angle can preserve the longitudinal magnetization – for example, if we tip the magnetization by, say, 15°, which would preserve about 96.6% (cos15°) of the magnetization for the subsequent tipping. However, since the intensity of the transverse signal using a tip angle of 15° is only about 26% (sin15°) of that excited by a 90° pulse, this strategy works well only when the SNR is not a critical consideration (i.e., the sample is large and the resolution is not high). Both FLASH (fast low angle shot) [7] and SSFP (steady-state free precession) [8] are examples of fast imaging sequences utilizing this strategy, which can offer about 50–100-fold reduction in imaging time. Figure 12.14 shows one example of this type of imaging sequences. This sequence looks very much like an ordinary 2D imaging sequence but without a double backslash after the signal acquisition. This implies no waiting for T_1 relaxation is needed; the next acquisition can start immediately where the signal comes from the preserved portion of the longitudinal magnetization.

The second strategy is to cover more "area" in *k* space during the "transverse coherence time" within one repetition. Pulse sequences such as echo planar imaging (EPI) [9, 10] can actually cover the entire *k* space within one scan! The key to the technique is to use fast gradient switching to traverse the entire *k* space using a sampling pattern such as those in Figure 11.7. One example of this type of imaging sequences when the Cartesian grid is used is shown in Figure 12.15. In this sequence, after the regular slice selection, the first negative G_{phase} pulse and the first negative G_{read} pulse position the sampling point to the lower left corner of *k* space. Immediately after, each pair of G_{read} and G_{phase} pulses samples one *k*-space line horizontally (during G_{read}) and then moves up *k* space by $1/n$ position (by G_{phase}). G_{read} is positive when *k* space is sampled from $-k_x$ to $+k_x$ and negative when *k* space is sampled from $+k_x$ to $-k_x$. n number of zigzags will traverse the entire *k* space.

This fast imaging strategy of course requires the sample to have a long T_2, otherwise there would be little signal left for the later *k*-space sampling points. In addition, fast sampling in EPI-type imaging sequences requires a broad-band receiver (e.g., to sample 128 points within 1 ms, the bandwidth of the receiver channel needs to be 128/0.001 = 128 kHz). Fast gradient switching in EPI requires firstly smaller inductance in the coils, which is contrary to the consequence of bigger or more powerful gradient coils. Fast gradient switching also induces bigger eddy currents in the surrounding metals that need to be compensated – otherwise the eddy current would severely degrade the linearity of the imaging gradient.

Both these strategies trade shortened imaging time for reduced SNR, which is feasible only when SNR is not an issue; for example, clinical MRI, where the main concern is the experimental

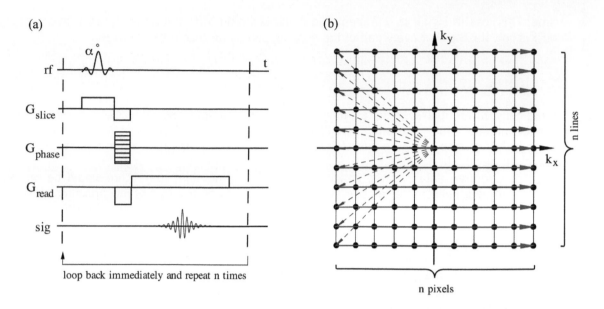

Figure 12.14 A 2D fast imaging sequence when a small tip angle rf pulse is used. Compared with a conventional 2D imaging sequence (Figure 12.10), the tipping angle of the rf pulse in this fast sequence is small, which preserves the majority of the longitudinal magnetization for subsequent excitation without the need for an extra relaxation waiting (i.e., this sequence does not have the double backslashes near the end of each time axis, which indicates an extra time delay). The blue dashed lines/arrows indicate the positioning of the starting point of sampling by the G_{phase} and the negative G_{read} gradient.

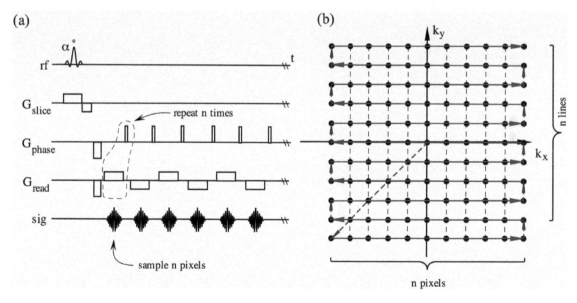

Figure 12.15 A 2D fast imaging sequence when the magnetization is tipped once and fast gradient switching is used to sample the entire $\boldsymbol{k}$ space. The first G_{read} pulse together with the first G_{phase} pulse position the starting point of sampling in $\boldsymbol{k}$ space (shown by the long blue-line arrow). Each pair of two pulses in the red dashed circle samples one line in $\boldsymbol{k}$ space (during G_{read}, indicated by the red arrows) then moves up to a new k_y location (due to G_{phase}, indicated by the short blue-line arrows). Strong T_2 weighting can happen in this type of imaging sequence since every subsequent echo signal is being weighted more and more by the transverse relaxation.

time. This loss in signal is detrimental in high-resolution MRI since the signal in microscopic MRI is so valuable that every part of the imaging procedure has to be optimized.

12.7.2 Compressed Sensing Technique

A radically different strategy in fast imaging is to under-sample *k* space using compressed sensing techniques, which shortens the experimental time by sampling fewer *k*-space trajectories [11]. At first glance, one wonders where this "*free lunch*" comes from, since the Nyquist criterion must be satisfied.

If the acquired data and the image are in the same domain, then there would be *no* free lunch – all data need to be sampled. It is like how an image from a low megapixel camera can never be processed or interpolated into a high-quality fine-resolution image. In MRI, however, the acquired FID data and the images are in the reciprocal domains, related to each other by Fourier transformation. There are inherent sparsity and compressibility in the MRI data, which allow the images to be recovered from randomly under-sampled *k*-space data using some non-linear reconstruction algorithms.

A simple example of signal sparsity in the reciprocal domain can be illustrated with the use of a sinusoidal function (Figure 12.16). If the desired information is presented in the time domain, then one must sample this sinusoidal function faithfully with at least the Nyquist criterion (cf. Figure 6.3). If, however, we want to know the frequency of this sinusoidal function, then Fourier transformation of this sinusoidal function contains only two possible values, at $+f$ and $-f$. (The signal in Figure 12.16a only has one channel of data, which could have two different frequencies upon FT due to the periodicity of the sinusoidal function. If both real and imaginary channels of the signal are given as in MRI, then the FT of this signal should contain only one frequency.) In this example, a sinusoidal function is said to be strongly sparse in the

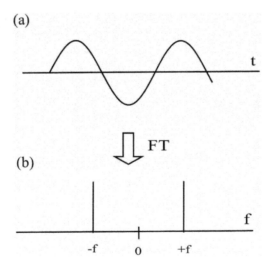

Figure 12.16 An illustration for the concept of compressed sensing in MRI, when the sampling and result are in the reciprocal domains. The useful information in this sine wave is only its frequency.

reciprocal domain since very few coefficients contain all of the information. Similar approaches are widely used in image compression, for example, to save digital images into different qualities of JPEG files.

The use of compress-sensing in MRI starts with a probability density function to generate and optimize a varied density *k*-space sampling pattern. Different sampling patterns put different emphasis on the information that is kept and ignored in data acquisition. It is common to fully sample the center of *k* space, and usually the sampling becomes gradually sparser toward the high-frequency area in *k* space, as shown in Figure 12.17 [12]. In single-slice 2D FT MRI, only the phase encodes can be under-sampled (Figure 12.17a), which offers modest accelerations. In the acquisition methods that use multi-slice 2D MRI, any 3D MRI, and quantitative MRI (Chapter 15), higher accelerations are possible. The detailed approaches in compressed sensing can be found in the references [11, 12].

It should be noted that different types of images in MRI have different sparsities (e.g., angiography images vs cardiac cine images). In addition, even for the same type of tissue, images at different resolutions could have different sparsities, depending upon the details of the image features, in particular, imaging resolution. Two recent studies [12, 13] demonstrate that both clinical MRI and μMRI can benefit from the use of compressed sensing in image acquisition, and μMRI benefits more from the use of compressed sensing by acquiring much less data, without losing significant accuracy in the quantification of T_2 maps in osteoarthritic cartilage.

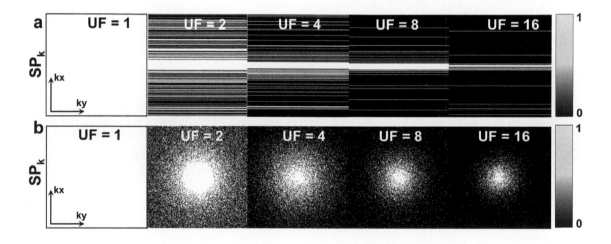

Figure 12.17   Example of the sampling patterns in MRI compressed sensing, with the optimized *k*-space sampling patterns at different under-sampling factors (UF). The white color indicates the sampled locations and the black color indicates the unsampled space. UF = 1, 2, 4, 8, 16 represent a fully sampled *k* space (100% sampled, hence all white color), a half-sampled *k* space (50% sampled), a quarter-sampled *k* space (25% sampled), a *k* space sampled 1/8th (12.5% sampled), and a *k* space sampled 1/16th (6.25% sampled), respectively. (a) 1D under-sampling patterns, (b) 2D under-sampling patterns. Depending upon the particular imaging experiments and the fineness of the image features, it is possible to sample only 1/4 or 1/8 without introducing significant errors. Source: Wang et al. [12].

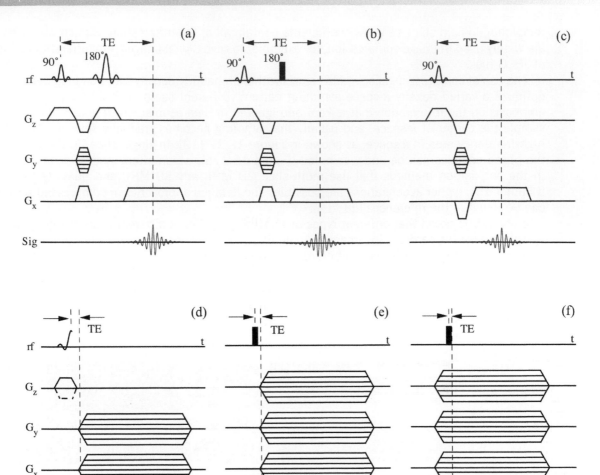

Figure 12.18 TE in imaging sequences. 2D imaging sequences using the Cartesian trajectory, (a) spin-echo version with two soft rf pulses, (b) spin-echo version with one soft rf pulse and one hard rf pulse, and (c) gradient-echo version. The gradient-echo sequence can have a shorter TE than the two spin-echo sequences. To further shorten the echo time, several center-out trajectories (radial, spiral, cones, or twisted projections) can be used in the UTE and ZTE sequences, as (d) 2D UTE, (e) 3D UTE, and (f) 3D ZTE. Sequence (a) has the longest TE, richest image contrast, best overall-quality image, and is most robust; sequence (f) has the shortest TE, lowest image contrast, and potential image artifacts.

12.8 ULTRA-SHORT ECHO AND ZTE MRI

According to Eq. (2.6), the macroscopic magnetization M is a spatial density of magnetic moments; so more nuclear spins per unit volume (i.e., per voxel) would result in a bigger signal. However, once M is tipped to the transverse plane, it is modulated by the spin-spin or T_2 relaxation time [Eq. (2.17)]. In MRI experiments, the echo time TE is typically about 5–15 ms in a spin-echo-based MRI sequence. During TE, the transverse magnetization is relaxing according to Eq. (2.17), but the maximum signal is only formed at TE. For samples with T_2 much longer than the echo time, this short delay simply implies a slightly reduced signal. In many biological specimens such as connective tissues (e.g., tendon, cartilage), however, T_2 and T_2^* can be very short in comparison with TE (cf. Chapters 2.6 and 7.3 for T_2 vs T_2^*). For example, the deep regions of articular cartilage can have a T_2^* of 0.3–2 ms, and tendon and ligaments can have T_2^* at much less than 1 ms. For an MRI sequence with TE of 5 ms, a specimen with $T_2 = 1$ ms would have an intensity of less than 0.7% of the intensity of a long T_2 specimen, hence appearing black in the image. This illustrates a simple conclusion in MRI that *a black color in an image in MRI does not necessarily mean nothing.*

Figure 12.18 compares six MRI sequences, each with an increasingly shortened TE. Figure 12.18a is a typical 2D spin-echo imaging sequence sampled on the Cartesian grid; the use of a second soft rf pulse produces a spin echo at the center of the acquisition; this sequence likely offers the best quality images among the six sequences and is also the most robust sequence. When the second soft rf pulse is replaced by a hard 180° rf pulse (Figure 12.18b), the TE of the sequence can be shortened. Figure 12.18c replaces the spin echo with a gradient echo in the slice selection to further shorten the TE of the imaging experiment, which also brings with it some small deterioration of image quality. All three sequences sample the full-echo signal hence yield the best SNR.

To further shorten the echo time, a general strategy is to start the sampling as soon as it becomes available, which goes into the half-echo sampling and the sequence designs in Figure 12.18d–f that use several 2D or 3D center-out trajectories (radial, spiral, cones, or twisted projections). Figure 12.18d is a 2D ultra-short TE (UTE) sequence, which uses ½ of the soft pulse for slice selection. Figure 12.18e is a 3D UTE sequence, which uses a hard pulse for excitation. Figure 12.18f is a 3D zero TE (ZTE) sequence, where the imaging gradients are turned on before the application of the rf excitation pulse (so the echo time literally can be as short as ½ of the hard pulse duration, in µs). For the six sequences from (a) to (f) in Figure 12.18, the image contrast decreases, the image quality (or the robustness of the imaging experiment) deteriorates, but the TE shortens. Figure 12.19 shows the two images from the same specimen (a glass phantom with water in the inner tube and the outer tube), by the spin-echo imaging sequence and by the UTE sequence. Although both images have the same tube structure, the non-uniformity of the UTE image intensity is noticeable.

12.9 MRI IN OTHER DIMENSIONS (4D, 1D, AND ONE VOXEL)

The MRI signal is three dimensional in nature. The use of slice selection enables the acquisition of 2D images. As shown in Figure 12.1, an MRI signal can also be acquired from a single voxel, a single stripe of voxels, and 3+ dimensions. The 3+ dimension imaging comes from the new physical dimension in image contrast, such as chemical shift or fluid flow, which are in addition to the 3D physical dimensions. One can obtain the physical parameters in each and all voxels in a 3+ dimensional imaging experiment. This will be discussed in detail in Chapter 15.

A 1D image in MRI consists of a single stripe of voxels, which can be accomplished by a sequence as shown in Figure 12.20. The use of a second soft rf pulse with the slice selection gradient at an orthogonal direction carves out a strip of voxels from the 2D slice selected by the first slice selection segment. This type of sequence can be used in any specimen with a known symmetry, in which one only wants to know the information along one direction. For example, in articular cartilage (cf. Chapter 15.2), if one can position the specimen in a particular way in the magnet, then the only information of interest could be the depth-dependent parameters.

203

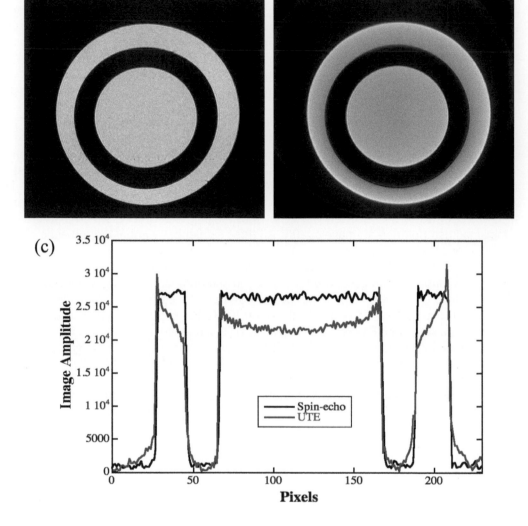

Figure 12.19   A double-tube phantom filled with water, imaged by a spin-echo sequence (a) and a UTE sequence (b), both with 256 × 256 pixels. Although the images look similar, the UTE image has some intensity artifacts, which are shown in (c) as the profiles across the center of the phantom.

Single-voxel imaging can be accomplished in principle by the simultaneous application of a selective –90° pulse and a non-selective 90° pulse in the presence of a linear field gradient, as shown in Figure 12.21. The combination is able to rotate the magnetization outside of the selected slice into the transverse plane, where it decays by $T_2{}^*$ processes, while the magnetization in the selected slice remains unperturbed. After three pairs of such pulses in the presence of three orthogonal gradients, only the magnetization in a cube that is intercepted by the three slice planes remains. A subsequent non-selective pulse can excite the spins in this remaining cube for the detection. This type of volume-selective imaging sequences or the localized spectroscopy approach has been used in the clinic to obtain localized chemical information non-invasively from a particular anatomic location in a human brain; for example, to study ^{1}H and ^{31}P metabolism in the brain (cf. Chapter 15.3).

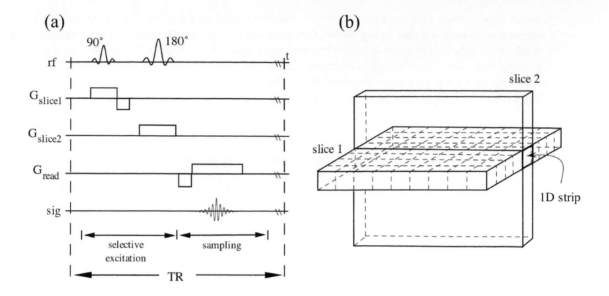

Figure 12.20   A 1D imaging sequence, which utilizes two slice selections in two orthogonal planes.

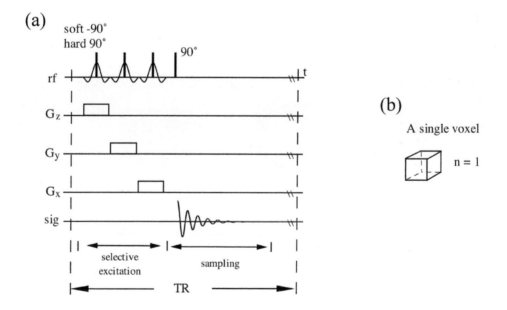

Figure 12.21   A single-voxel imaging sequence, where the signal comes from the surviving voxel, intersected by the three orthogonal slice selection planes.

Several points should be considered in the technique. First, the location of the selected cube needs to be defined accurately, which often involves a 3D imaging of the object prior to the localized spectroscopy. Second, the shape of the selected cube is usually not very precise, which may be an important issue when one wants to quantify the signal using a volume-based unit such as millimoles/mm^3. Third, the sensitivity of the method could be low since a big receiver coil is required to accommodate the entire sample while the signal only comes from a small cube inside a large sample, hence a low filling factor. One method to improve the sensitivity in this situation is to use a surface coil (Figure 5.5d), if the region of interest is close to the surface of the object (e.g., near the skin). Surface coils improve SNR in two ways by getting the tissue closer to the coil and by receiving less sample-generated noise. An optimal SNR can be achieved when the radius of the coil equals the depth of the cube interested.

12.10   RESOLUTION IN MRI

Since the publication of the first NMR image in 1973 [14], its potential for medical diagnosis was noticed immediately, and whole-body MRI scanners now have become an indispensable diagnostic tool in the hospital. In addition to its widespread use in clinical medicine, MRI is increasingly being used in laboratories to measure many physical and biophysical phenomena. With the scaling down of the receiver coil and the fine-tuning of the instrument, the resolution of MRI can be as fine as 10 μm^3. When the voxel resolution of the image is less than $(100 \ \mu m)^3$, NMR imaging is called NMR microscopy [15, 16] since a human eye cannot resolve a volume element smaller than $(100 \ \mu m)^3$. Note that in imaging science, the Rayleigh criterion is often used to define the term "resolution," based on the distinction of two points of light from each other. In MRI literature (and this book), the term "resolution" is used interchangeably with the pixel/voxel size of an image.

Figure 12.22 shows a resolution scale in MRI. The resolution in medical MRI scanners is typically in the order of 1 mm^3, limited largely by practical factors such as electronics, computer capability, and patient movement/scan time; whereas NMR microscopy (µMRI) covers the

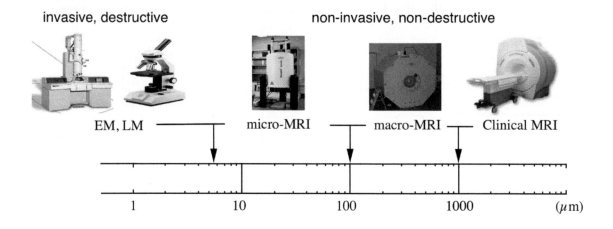

Figure 12.22   A resolution scale in MRI and beyond. There are three different sizes of magnets in MRI, which accommodate humans in clinical MRI, animals in macro-MRI, and the smallest animals and tissue blocks in micro-MRI (µMRI). Beyond the resolution of µMRI, there are light microscopy (LM: resolution limited by the wavelength of the light to a fraction of 1 µm) and electron microscopy (EM; capable of reaching a nanoscale resolution). µMRI occupies a unique position in this resolution scale, providing a translational bridge between invasive imaging tools (LM and EM) and clinical MRI. Source: Xia [20].

resolution range from 100 μm down to about a few μm. For a given instrument, the SNR obtainable from a volume element of the sample is, in principle, proportional to the number of spins in this volume. As the voxel size decreases, the number of spins contributing to the signal in a voxel also decreases, and so does the SNR. A fundamental limitation in microscopic MRI is the self-diffusion coefficient of the molecules during the echo time of the imaging sequence [17].

A final question to this short section on MRI resolution is, *how high a resolution do we need for a particular experiment?* The answer should not be "as high as possible," since a high-resolution imaging experiment takes longer time and needs a more expensive instrument. The answer to this question depends on the size of the structural features (the details) that one tries to discover in an imaging experiment, among other trade-offs. There is a well-known expression, *"Le bon Dieu est dans le detail,"* which is often attributed to the French novelist Gustave Flaubert (1821–1880). It literally means, *"The good God is in the detail,"* which has a more popular translation, *"The devil is in the details."*

Most readers of this book likely want to use MRI as an imaging tool in laboratories and clinics. Take the detection of osteoarthritis (OA) and other related joint diseases as an example, which are the number one cause of disability in the US population [18]. Osteoarthritis is a progressively degenerative joint disease and thought to be initiated when an imbalance occurs between chondrocyte-controlled anabolic and catabolic processes. This imbalance is characterized at different stages by different imbalances in various metabolic and enzymatic activities and by different morphological degradation of its extracellular matrices, tissue loss, joint space narrowing, subchondral bone sclerosis, and osteophyte formation. Only during the later stage of degradations can a patient feel joint pain, deformity, restricted motion, and dysfunction. Any successful diagnostics of OA that has the ability to slow down or pause the degradative progression towards a clinical disease should be the detection of its meaningful changes at an early stage, at the molecular and morphological levels. These early tissue degradations are subtle and occur at localized sites, which can be masked out easily in any volume-averaging process in imaging (i.e., when the imaging solution is low and the voxel size is large). Hence any early detection of OA, and many other diseases, needs high sensitivity in detection protocols (cf. Chapter 15) as well as high resolution in MRI [19, 20], which is challenging in clinical MRI scanners.

At an American Physical Society meeting on December 29, 1959, Richard Feynman delivered an inspirational presentation, titled *"There's Plenty of Room at the Bottom."* In his talk, Dr. Feynman said *"It is very easy to answer many of these fundamental biological questions; you just look at the thing! ... Unfortunately, the present microscope sees at a scale which is just a bit too crude. Make the microscope one hundred times more powerful, and many problems of biology would be made very much easier. I exaggerate, of course, but the biologists would surely be very thankful to you ..."* (emphasis added). If we replace the word "microscope" with "MRI," the same quote could become equally inspirational for the readers of this book, MRI students and researchers. Do your best to improve the spatial resolution in your imaging work; you will be rewarded with satisfaction and success.

References

1. Balies DR, Bryant DJ. NMR Imaging. *Contemp Phys*. 1984;25(5):441–75.
2. Brooks RA, Di Chiro G. Principles of Computer Assisted Tomography (CAT) in Radiographic and Radioisotopic Imaging. *Phys Med Biol*. 1976;21(5):689–732.
3. Kumar A, Welti D, Ernst RR. NMR Fourier Zeugmatogaphy. *J Magn Reson*. 1975;18:69–83.
4. Edelstein WA, Hutchison JMS, Johnson G, Redpath TW. Spin Warp Imaging and Applications to Human Whole-body Imaging. *Phys Med Biol*. 1980;25:751–6.
5. Johnson G, Hutchison JMS, Redpath TW, Eastwood LM. Improvements in Performance Time for Simultaneous Three-dimensional NMR Imaging. *J Magn Reson*. 1983;54:374–84.

6. Xia Y. MRI of Articular Cartilage at Microscopic Resolution. *Bone and Joint Res.* 2013;2(1):9–17.

7. Haase A, Frahm J, Matthaei D, Hänicke W, Merboldt K-D. FLASH Imaging: Rapid NMR Imaging Using Low Flip-Angle Pulses. *J Magn Reson.* 1986;67:258–66.

8. Gyngell ML. The Application of Steady-State Free Precession in Rapid 2DFT NMR Imaging: FAST and CE-FAST Sequences. *Magn Reson Imaging.* 1988;6:415–9.

9. Mansfield P. Multi-planar Image Formation Using NMR Spin Echoes. *J Phys C: Solid State Phys.* 1977;10:L55–L.

10. Mansfield P, Pykett IL. Biological and Medical Imaging by NMR. *J Magn Reson.* 1978;29:355–73.

11. Lustig M, Dohoho D, Pauly JM. Sparse MRI: The Application of Compressed Sensing for Rapid MR Imaging. *Magn Reson Med.* 2007;58:1182–95.

12. Wang N, Badar F, Xia Y. Compressed Sensing in Quantitative Determination of GAG Concentration in Cartilage by Microscopic MRI. *Magn Reson Med.* 2018;79(6):3163–71.

13. Wang N, Badar F, Xia Y. Resolution-dependent Influences of Compressed Sensing in Quantitative T2 Mapping of Articular Cartilage. *NMR in Biomedicine.* 2020;33(12):e4260.

14. Lauterbur PC. Imaging Formation by Induced Local Interactions: Examples Employing Nuclear Magnetic Resonance. *Nature.* 1973;242:190–1.

15. Aguayo JB, Blackband SJ, Schoeniger J, Mattingly M, Hinterman M. Nuclear Magnetic Resonance Imaging in a Single Cell. *Nature.* 1986;322(10):190–1.

16. Eccles CD, Callaghan PT. High-Resolution Imaging. The NMR Microscope. *J Magn Reson.* 1986;68:393–8.

17. Callaghan PT, Eccles CD. Diffusion-Limited Resolution in Nuclear Magnetic Resonance Microscopy. *J Magn Reson.* 1988;78:1–8.

18. CDC. Prevalence and Most Common Causes of Disability Among Adults – United States, 2005. *Morbidity and Mortality Weekly Report (MMWR).* 2009;58(16):421–6.

19. Xia Y. Resolution "Scaling Law" in MRI of Articular Cartilage. *Osteoarthritis Cartilage.* 2007;15(4):363–5.

20. Xia Y. The Critical Role of High Imaging Resolution in MRI of Cartilage – The MRI Microscope. In Xia Y, Momot KI, editors. *Biophysics and Biochemistry of Cartilage by NMR and MRI.* London: Royal Society of Chemistry; 2017. 455–70.

13

Imaging Instrumentation and Experiments

An instrument for MRI builds upon the instrumentation architecture and experimental considerations for NMR spectroscopy (Chapters 5 and 6). The major additional requirements for MRI are the hardware and software for spatial mapping and specific software for image analysis. Now let us look at them one by one.

13.1 SHAPED PULSES

In NMR spectroscopy, the shapes of the rf pulses in the time domain are usually rectangular (i.e., having a constant amplitude over the duration), which can be generated easily by chopping the constant-level oscillation at radio frequency into the fixed-duration pulses of rectangular shape (cf. Figure 5.7). In MRI, when slice selection is used, the shape of the rf pulse in the time domain needs to be modulated into a sinc form (sinc = $\sin\theta/\theta$), since the sinc and the rectangle are a Fourier transform pair. The quality of its frequency profile (i.e., how close the slice profile is to a rectangular shape) is determined by the number of side lobes in the sinc pulse – a perfect rectangle in the frequency domain requires the summation of an *infinite* number of oscillations in the time domain. Figure 13.1a shows four sinc pulses with the same central lobe but with different side lobes (sinc1 has one side lobe, sinc2 has two side lobes, ...). As one can see from Figure 13.1b, the frequency profile of sinc8, although not perfect, approximately resembles a rectangle. One can see the deterioration in the frequency profile of sinc4; however, both of the edges are still nearly vertical. The deterioration in the frequency profile of sinc2 becomes more visible, where the two sides are not as vertical and the function is non-zero outside the main slice. The frequency profile of sinc1 has additional visible deterioration, where the top surface of the slice is no longer flat.

Why don't we just use sinc8 or a sinc with more than eight side lobes for slice selection in MRI? According to Eq. (2.27), the tipping angle of the rf pulse in slice selection by a sinc pulse is largely determined by the area of its central lobe. So the four sinc functions (sinc1, sinc2, sinc4, sinc8) shown in Figure 13.1a essentially have the same tipping angle (e.g., 90°) in the

Essential Concepts in MRI: Physics, Instrumentation, Spectroscopy, and Imaging, First Edition. Yang Xia.
© 2022 John Wiley & Sons Ltd. Published 2022 by John Wiley & Sons Ltd.

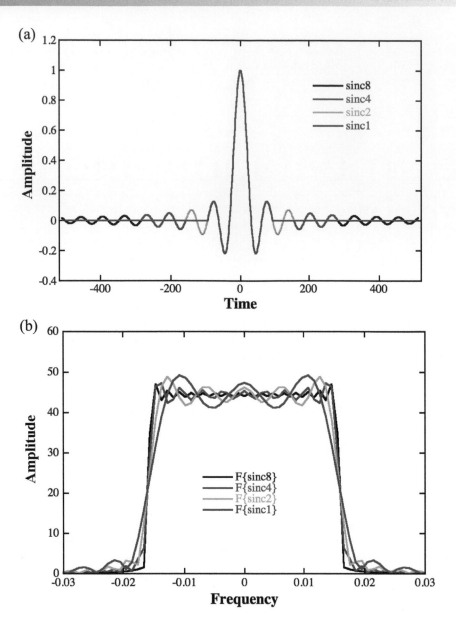

Figure 13.1 (a) Various sinc pulses, each having the same central lobe but a different number of side lobes (sinc1 [red] has one side lobe, sinc2 [green] has two side lobes, sinc4 [blue] has four side lobes, sinc8 [black] has eight side lobes). (b) The Fourier transform of the sinc pulses in (a). It is clear that as the number of side lobes decreases, the rectangular profiles begin to deteriorate, especially along the vertical edges, which are no longer vertical, and there is intensity outside of the vertical edges. All profiles have the same 1024 data points in fast Fourier transform.

slice selection. However, a sinc2 pulse is about 1.67 times longer than a sinc1 pulse, a sinc4 pulse is about 3 times longer than a sinc1 pulse, and a sinc8 pulse is about 5.42 times longer than a sinc1 pulse. Making the sinc pulse longer will lengthen the echo time (TE), and a longer TE is often undesirable in MRI (cf. Chapter 12.8). In practice, therefore, the selection of a sinc pulse in MRI is a compromise between having a short TE and having a good-quality slice.

Occasionally, to further shorten the TE (i.e., if the T_2 of the sample is short), one can even use a sinc pulse without any side lobe (sinc0) or just a Gaussian envelope. This means that the slice profile in MRI is never truly rectangular.

As one can see from Figure 13.1b, the limited side lobes in the slice selection pulse also lead to numerous ripples beyond the slice thickness, which is the reason that a multi-slice imaging experiment (Chapter 12.6) always leaves a gap between the neighboring slices – for example, acquiring the odd numbered slices first and the even numbered slices subsequently so that the effect of ripples on the magnetization can settle down during the interval.

These waveforms for the shaped pulses in the instrument are pre-generated and stored as the binary files in the computer memory. At the time of execution, a waveform is sent out one line (level) at a time in a sequential manner at a rate determined by the time resolution of the pulse sequencer. The output is used to modulate a constant-level rf oscillation into a shaped rf pulse before it is sent to the rf transmitter for further amplification. A well-resolved waveform requires hundreds of digital levels. The more discrete levels, the better the waveform is defined, the more memory it uses, and the faster one has to clock it out within a given time duration (e.g., 1 ms).

In addition to the shaped rf pulses, the gradient pulses in MRI systems with large magnets are often shaped as well, for a different reason. In principle, a gradient pulse in imaging should look like a rectangle. In reality, an ideal rectangular pulse never exists, as shown in Figure 5.8 in the example of rectangular pulses, where the leading and falling edges are never vertical due to the limited bandwidth of the amplifiers. Vertical rising and falling edges of gradient pulses imply an infinite dB/dt, or dI/dt where I is the electric current, which requires an infinite bandwidth for the amplifiers. Any sudden change in the gradient current also induces large eddy currents (see the next section). In addition, the larger the gradient coil, the more inductance the coil has; hence, the leading and falling edges of the gradient pulses in clinical MRI are purposely designed to be the shape of a trapezoid (cf. Figure 12.18, Section 13.2.3).

13.2   THE GRADIENT UNITS

The gradient unit in an MRI system includes a set of gradient coils and gradient power supplies, together with additional digital controls from the pulse programmer. The quality of the gradient unit determines the image quality in MRI. The linear field gradient vector in MRI is commonly produced by the superposition of a uniform polarizing field $\boldsymbol{B}_0$ and an additional gradient field $\boldsymbol{G}$. To distinguish any spatial location in a 3D space, three orthogonal gradients in the component of $\boldsymbol{G}$ parallel to $\boldsymbol{B}_0$ are sufficient, namely G_x, G_y, G_z. Any gradient not parallel to one of the three principal axes can be generated by using more than one gradient simultaneously. Since the direction of the main polarizing field $\boldsymbol{B}_0$ is customarily defined as the z axis, the gradient field in MRI is denoted as the variation of the polarizing field along three orthogonal axes in Eq. (11.2), which have been shown schematically in Figure 11.3.

13.2.1   Gradient Power Supplies

Good-quality audio amplifiers can be used as the power supply to drive the gradient coils. For imaging, the power of each gradient amplifier can go easily over 1000 watts. A few sensitive overheat protection switches are a must for such applications. There are several important considerations for the gradient amplifiers in an MRI system.

First, the linearity of the field gradient is critically important since it determines the distortion and artifact of the images. For example, a non-uniformity of 0.4% or less in the gradient is needed if the image distortions are to be avoided in any 256 × 256 image (1/256 = 0.4%). One can easily use the Biot–Savart equation [Eq. (5.1)] in electricity and magnetism to calculate the non-uniformity of any field gradient.

Second, the magnitude of the gradient has several important consequences in MRI experiments, influencing the spatial resolution and the ability to measure dynamic motion. A stronger gradient is needed in order to achieve a finer spatial resolution. For example, to image an object with a sub-millimeter resolution, a gradient on the order of 0.5 T/m is needed. A strong gradient would enable the detection of delicate vascular flow and slow self-diffusion in viscous polymer fluids by quantitative MRI (cf. Chapter 15). For example, using a pulsed-gradient spin-echo (PGSE) sequence and at $\Delta = 5$ ms and $\delta = 1$ ms, a gradient of 1.4 T/m would give a signal attenuation of 73% if the self-diffusion coefficient of the sample is 2×10^{-9} m^{2}s^{-1} (as it is for water at room temperature), but the same gradient strength and PGSE parameters would only cause about 0.01% attenuation if the diffusion coefficient drops to 2×10^{-13} m^{2}s^{-1} (which it is for some polymer fluids at room temperature). A gradient as strong as 7 T/m was needed for the detection of motion [1, 2].

Finally, the noise/ripple level of the electric output of the gradient power supply is a critical parameter. For example, some commercial power supplies could have a ripple on the order of 0.05% of its maximum output. Although this ripple is satisfied for spatial mapping in MRI, it could become a serious problem for the accurate measure of flow and diffusion in MRI. This is because the flow and diffusion measurements rely on subtracting the net phase shifts sustained by spins produced by two identical and intense gradient pulses at two different instances of time (Chapter 15.1). The presence of noise/ripple will therefore set a lower limit for the measurement. One project has showed the source of this ripple in a commercial power supply came mainly from the AC transformers and fans in the power supply. By special modifications to the power supply, the noise level of a power supply can be reduced from 0.05% to as low as 0.005% [3].

13.2.2   Gradient Coils

The choice of gradient coil configuration is mostly dictated by the geometry of the magnet. There are several basic configurations of coils (Figure 13.2) that can generate a field gradient: a Maxwell pair (which is also called an opposed Helmholtz coil) (Figure 13.2a) [4], a planar coil (Figure 13.2b) [5], and the double saddle coil (Figure 13.2c) [6]. A Helmholtz coil is a two-loop configuration with the directions of the electric current in both loops running in the *same* direction, which can generate a uniform magnetic field similar to the solenoid (Figure 5.5a). When the directions of the electric current in the two loops of a Helmholtz coil are opposite to each other, the coil is called an opposed Helmholtz coil, or a Maxwell pair, which could be circular or square in shape. Using the right-hand rule, it can be seen that the field at the center of the coils is zero, which forms a *z* gradient along the *z* axis. Two pairs of saddle coils (Figure 5.5b), which is also called the double saddle coil or Golay coil (Figure 13.2c), can be used to generate the gradients in *x* and *y* directions for a cylindrical magnet. Note that the wiring patterns in a modern instrument are far more complicated than these basic configurations. Using multi-step optimization in computer simulation, one can add more wires, each with a different curve and arc, into the basic coil to improve the homogeneity of the field gradients. These modern designs are called the fingerprint coils [7].

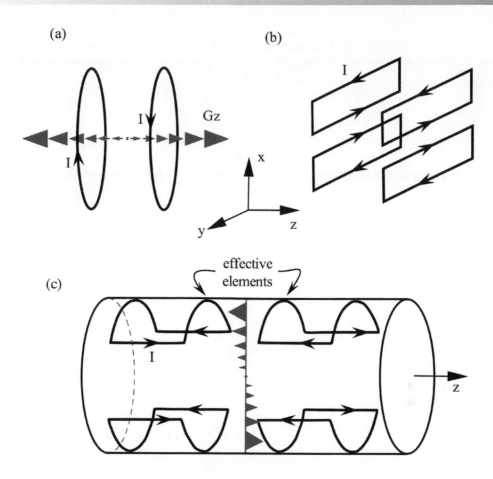

Figure 13.2 Basic configurations for the gradient coils. (a) A Maxwell pair (which is also called an opposed Helmholtz coil. Source: Based on Tanner [4]. (b) A planar coil. Source: Based on Anderson [5]. (c) A double saddle coil (Golay coil). Source: Modified from Hoult and Richards [6]. The effective elements in (b) and (c) are the four central wires. Both the planar coil and the double saddle coil can be rotated about the z axis by 90° to generate a gradient in a different orthogonal direction.

For a modern superconducting magnet with a cylindrical shape, a Maxwell pair (G_z) and two sets of saddle coils (G_x and G_y) are sufficient to form a complete 3D gradient for MRI, which would give easy access to the center of the magnet. For a magnet with a double-donut shape (electromagnet, permanent magnet, open superconducting magnet), G_z can be produced by a Maxwell pair and G_x and G_y can be produced by two pairs of planar coils (Figure 13.2b). However, experimental circumstances may arise when the gradient coils have to be re-designed for each individual project. Figure 13.3 shows two examples of the imaging probes [2], one aimed to maximize the sample space in a vascular flow experiment [8] and the other aimed to maximize one of the gradients to 7.7 T/m in a diffusion imaging experiment [1].

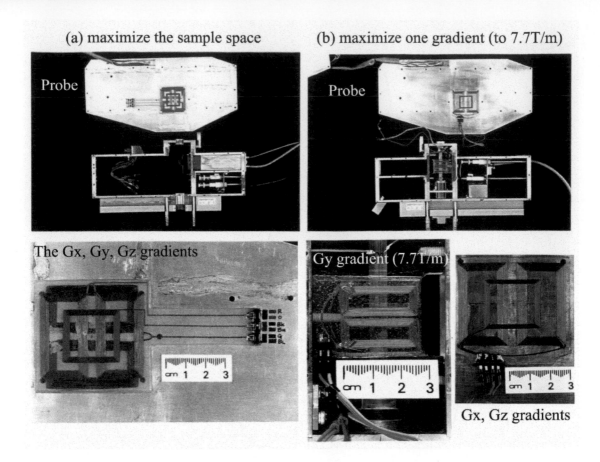

(a) maximize the sample space   **(b) maximize one gradient (to 7.7T/m)**

Probe

The Gx, Gy, Gz gradients

Probe

Gy gradient (7.7T/m)

Gx, Gz gradients

Figure 13.3 Two examples of imaging probe design for the magnet orientation as in Figure 1.2b and 1.2c, one to maximize the sample space (a) and the other to maximize one of the imaging gradients for diffusion imaging (b). Source: Reproduced with permission from Xia et al. [2].

13.2.3 Eddy Current Reduction

Eddy currents are electrical currents induced in metallic conductors by a changing magnetic field (dB/dt or dI/dt) experienced in the conductor. According to Faraday's law of induction, electrical currents can be induced within nearby metallic conductors by a time-varying magnetic field. The magnetic field generated by the gradient pulses in MRI is a particularly troublesome source of eddy currents. The induced eddy currents destroy the desired linearity of the gradient fields.

Modern design of gradient coils uses active shielding [7], which places another set of gradient coils, also with complicated patterns, between the main gradient coils and the magnet. By running the electric currents in the opposite direction, it is possible to cancel most of the magnetic field experienced by the metal in the magnet bore, hence reducing the eddy currents significantly. In addition, a pre-emphasis unit is commonly used in the MRI instrument to purposefully distort the input electric current for the gradient pulses (Figure 13.4a–c), in order to receive a corrected gradient pulse. It is also common in clinical MRI to use a trapezoidal pulse as the input waveform for the gradient current (Figure 13.4d) instead of a rectangular pulse, thereby suppressing eddy currents.

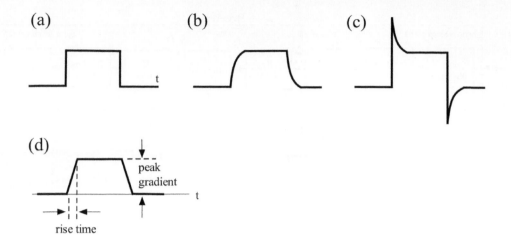

Figure 13.4 (a) The desired shape of the gradient pulses is rectangular. (b) The obtained shape of the rectangular gradient pulses at the output, whose sharp leading and falling edges are lost. (c) A purposely distorted shape of the gradient pulses' input, which aims to correct the distorted gradient pulse shown in (b), to recover the shape in (a) at the output. (d) The practical shape of many gradient pulses in clinical MRI scanners, which is a trapezoid, and the associated parameters to characterize the trapezoid.

The performance of the gradient unit in MRI scanners can be estimated by a parameter called the slew rate (Figure 13.4d),

$$\text{Slew Rate} = \text{Peak Gradient} / \text{Rise Time}. \tag{13.1}$$

Since the gradient field commonly uses units of milli-Tesla per meter (mT m^{-1}), the slew rate has units of Tesla per meter per second (T m^{-1} s^{-1}) when the rise time is in milliseconds. For example, if a gradient that has a peak amplitude of 100 mT m^{-1} can ramp from 0 to the maximum in 1 ms, the slew rate of the gradient is 100 T/m/s. The slew rate is the determining factor that sets the lower limit of the minimum echo time (TE) in imaging sequences.

From the point of view of instrumentation and experiment, large slew rates are better. However, in addition to the induction of the eddy currents in the magnet, rapidly changing gradients can also induce eddy currents in the human body, which can stimulate nerve axons. This neuronal stimulation limits how fast you can change strong field gradients in clinical MRI, where the use of the trapezoidal pulses is beneficial and the use of the fastest available slew rate requires caution.

13.3 INSTRUMENTATION CONFIGURATIONS FOR MRI

Figure 13.5 is a schematic instrumentation diagram for an MRI system. Compared with the block diagram for an NMR spectrometer (Figure 5.7), the additional items for imaging include the imaging hardware (i.e., the gradient units, the waveform generator) and image acquisition software that controls the imaging hardware. Based on their physical size and imaging usages (Figure 12.22), there are three types of MRI systems.

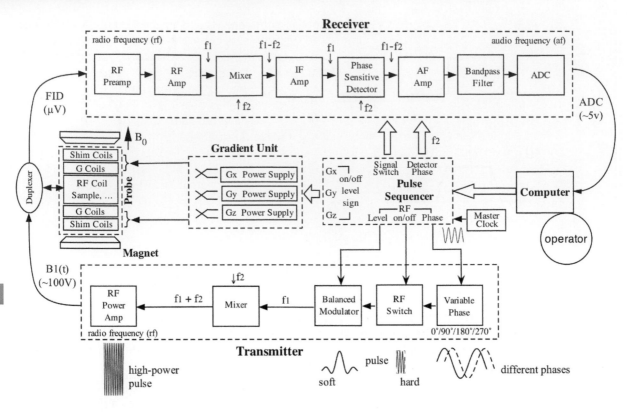

Figure 13.5 A schematic configuration for an MRI system, which is based on the system for NMR spectroscopy (shown in Figure 5.7).

The whole-body (or clinical) MRI scanners in the clinic have the largest cylindrical-shaped magnets (Figure 1.2d) and cost the most; their voxel resolution is on the order of 0.5–1 mm. These horizontal superconducting magnets commonly have 60–90 cm in clear bore diameter, accommodating an adult. The emphasis in these clinical MRI scanners is user friendliness and versatility. For the operator, this means it is easy to operate and maintain; for patients, this means it is easy to access and comfortable; for clinicians, this means it has good-quality images and provides meaningful information. At the same time, these clinical scanners often mandate a shorter experiment time, since a long imaging time would make the images more susceptible to patient movement and cost more. In recent years, open MRI designs have been developed, whose magnets have a double-donut shape (Figure 1.2b). This type of open-magnet system helps to reduce claustrophobia for some patients. One can also rotate the double-donut magnet by 90°, as in Figure 1.2c, which is useful for imaging the musculoskeletal system of humans in the weight-bearing condition (e.g., to examine any bulging spinal discs under loading). These types of open-magnet systems currently have all lower B_0 field than the typical cylindrical-shaped magnet, hence have a worse SNR and a slower imaging speed.

On the other end of the spatial scale shown in Figure 12.22, NMR microscopy or μMRI systems can have a resolution less than 100 μm, which is the smallest size that can be visualized by a human eye [9–11]. This type of system commonly uses a vertical bore superconducting magnet (typically with a bore size of 9–12 cm), similar to the magnets used in high-resolution NMR spectroscopy (Figure 1.2a). Between the clinical MRI and μMRI, there exist a number of instrument configurations for macro-imaging, which use horizontal magnets with a bore size of

20–40 cm. This type of macro-MRI system is commonly found in academic and pre-clinical research labs, where a range of animals and other objects can be imaged.

In recent years, there are numerous new developments in two different types of small MRI systems. The first type is called the extremity MRI systems, which aim to image the extremities of a human, that is, arms and legs (or even a human head). By getting the center of the magnet field to be close to one end of the magnet lengthwise, it is possible to image the elbows and knees of a person when the main body is outside the magnet. Some of these systems can even have a rotatable magnet [12], which helps in different situations such as the compensation of the magic angle effect in tissues with oriented macromolecules. The second type of small MRI system is the near-desktop or near-mobile MRI system [13], mainly based on some permanent magnets such as the Halbach magnet. With further improvements in the hardware and software, the potential uses for these types of dedicated small MRI systems would be significant.

13.4   IMAGING PARAMETERS IN MRI

When planning an MRI experiment, the previous considerations in planning a spectroscopy experiment (Chapter 6) become the background issues that go without saying. For example, the rf circuit is assumed to be optimized, calibrated, and tuned. A major exception is the much-reduced desire for an ultra-uniform B_0 field. In high-resolution spectroscopy, extensive shimming and high-speed sample spinning are used together to make B_0 as uniform as possible. In imaging, specimen rotation is currently not possible, and the application of imaging gradients destroys the non-uniformity of the main magnetic field. Hence a reasonable (i.e., quick) shimming of the B_0 field is often sufficient.

If you need to set up an imaging protocol from scratch or do a major modification, there are numerous imaging-related parameters for a successful MRI experiment. The complexity in setting up an MRI experiment is due to the fact that there are simply many acquisition controls to be designed and set, including at least three gradient channels in addition to the usual rf channels; and all of these parameters have specific vendor-dependent names. Here, we discuss briefly several top-level parameters in imaging, based on the assumption that the system hardware and imaging protocols have been maintained and set up for you.

13.4.1   Imaging Resolution and Data Size

In Chapter 12, we have discussed the three-dimensional nature of MRI (Figure 12.1) and some factors that influence the choice of imaging resolution (Chapter 12.6 and 12.10). Given a specimen enclosed in a cubic space, signal needs to be acquired from all resolvable locations in three dimensions. Each discrete point in an image is called a voxel (volume element). Although one would always prefer a high resolution in any imaging, high resolution naturally leads to a large data size. There are two parameters that determine the data size of an image: the voxel depth (dynamic range of the voxel) and cubic dimensions (digital resolution).

In Chapter 6, we have discussed the dynamic range for NMR spectroscopic data. To properly represent the value of an image intensity, at least 16 bits is commonly required for each image voxel so that the dynamic range of a pixel is $\pm32,768$ (2^{16}). In modern spectrometers, the dynamic range of a voxel typically has 32 bits in depth, which allows $2^{32} = 4,294,967,296$ possible values. The reason for a high dynamic range is that one could have a closer look at some weak-signal regions when a strong background signal is present. With this in mind, when each dimension of a sample cube is resolved by 256 voxels, the size of a 3D image file would be 64 MB (256 x 256 x 256 x 4), where 4 bytes equals 32 bits. A high resolution and/or a large dimension leads to a large file size, which would be slow to acquire, access, and analyze.

13.4.2   Parameters that Relate to the Experimental Time

The matrix size is never an isolated parameter in imaging. Other essential imaging parameters include the number of accumulations (*N*), the repetition time (TR), the echo time (TE), and the total experimental time.

The number of accumulations *N* is the number of repeated acquisitions for the same *k*-space sampling line. Selecting $N > 1$ improves the SNR and satisfies the requirement of phase cycling in data acquisition (cf. Chapter 6.6), at the expense of a longer imaging time. Since the signal voltage adds linearly with the number of accumulations, whereas the noise power adds as the square root of the number of accumulations (due to the random nature of the noise), one needs to repeat the signal acquisition four times in order to double the SNR (cf. Chapter 6.7).

The repetition time TR allows the relaxation of the magnetization, which has been disturbed by the rf pulses in the current acquisition, to the longitudinal axis before the next acquisition. As shown in Eq. (2.16), the return of the magnetization depends on the T_1 relaxation time. The longer the T_1, the longer it takes for the magnetization to return to thermal equilibrium. Although a full restoration of the magnetization to the *z* axis needs at least $5 \times T_1$, one in practice never waits for $5 \times T_1$. For a given experimental time, it can be shown that the best SNR for *a fixed total experimental time* can be accomplished by repeating TR at about $1.26 \times T_1$.

The echo time TE specifies the time delay between the spin excitation and the center of the signal acquisition (as shown in Figures 12.8, 12.10, and 12.11). TE is typically 5–20 ms in a spin-echo-based MRI experiment. For specimens that contain short T_2 components, TE must be shortened (as discussed in Chapter 12.8). One approach to shortening TE is to use the sinc pulse with fewer side lobes (Figure 13.1), which leads to a deterioration of the image quality.

For classical spin-echo imaging with a Cartesian grid of 256 × 256, if all phase steps are sampled and TR is 2 seconds, the total experimental time would be 256 × 2 = 512 seconds or 8.5 minutes, for an accumulation $N = 1$. This imaging time is already too long for human MRI, where patient movement cannot be avoided completely.

13.4.3   SNR in MRI

For imaging, the SNR of an image is, in principle, proportional to the number of nuclear spins in an imaging voxel. With an improvement of imaging resolution, the voxel size decreases, which leads to a quick decrease in the number of spins contributing to the signal. For example, when one halves the linear dimension of an image voxel (say, from 200 × 200 × 200 μm to 100 × 100 × 100 μm), the volume of the imaging voxel reduces to 1/8 of the original value. Hence, when everything else is equal, a better resolution imaging experiment naturally means a smaller signal. So the judgment becomes: *Is this image too noisy to be useful?*

In Figure 6.14b, we showed a practical way of estimating SNR in MRI. We also discussed in Chapter 6.11 the essential parameters in the SNR determination for NMR spectroscopy, which considers the signal due to the magnetization and the thermal noise in the instrument, as in Eq. (6.9). A more intuitive equation for SNR in NMR and MRI can be derived from a scaling law [14], as Eq. (6.10). This equation contains three noise terms, coming from the sample, the receiver coil, and the preamplifier. The sample noise accounts for the dielectric and magnetic losses in biological samples that contain conductive tissues, where a bigger-size sample leads to a larger noise. The noise from the receiver coil accounts for the physics of NMR and MRI, in terms of the B_0-dependent magnetization and the efficiency of the rf coils. The noise from the preamplifier, which is the first amplifier in the receiver channel, has the most important impact on the performance of the receiver, since the noise of the preamplifier is further amplified by all later amplifiers.

218

For clinical MRI, the sample size dominates the SNR consideration. For smaller samples and the use of higher field, the quality of the receiver coil and the preamplifier can become most influential. It is therefore beneficial in microscopic MRI to find super-low-noise preamplifiers or even use liquid nitrogen to cool the amplifier (e.g., a cryoprobe). Of course, there are other effects that can influence the SNR of an image. For example, a short T_2 relaxation or faster diffusion can lead to a faster decay of the MRI signal, further decreasing the SNR of an image.

13.4.4   Fundamental Constraints in MRI

Given sufficient money (e.g., best instrument) and skillful researchers (i.e., best technicians and graduate students), *three* fundamental parameters are the ultimate constraints in MRI: image resolution, SNR, and experimental time. It is simply not possible, using a current state-of-the-art instrument, to maximize all three of these parameters at the same time, that is, to obtain a high-resolution image with highest SNR in a super short time. The art of setting up a successful MRI experiment is, therefore, *to compromise*. That is, to maximize the useful information that is truly critical for the particular project; in doing so, one often sacrifices or relaxes some other parameters to gain the otherwise-unavailable knowledge. In this optimization process, any symmetry of the sample's structure could become very beneficial [15]. The direction of the highest resolution should be set along the direction in the specimen that has the most-complicated or fastest-changing structures.

13.5   IMAGE PROCESSING SOFTWARE

Additional software is required for post-acquisition image processing, to reconstruct both 2D and 3D images in MRI. Each commercial manufacturer of imaging hardware offers a package of software. There is also third-party software for image analysis. One notable general purpose software is called *ImageJ*, which is an open source image processing program designed for scientific multidimensional images (developed by Wayne Rasband at NIH, https://imagej.net). This software could be an excellent choice for any imaging researcher, for several reasons. First, it is available at no cost and runs on several computational platforms (Mac, PC, Linux). Second, its source code is available openly. Third, the built-in functions and macros are extensive and sufficient for most of the routine analysis. Fourth, there is a large community of users who deposit their custom plugins/scripts for others to use. Chances are, what you want to do has already been done before by others. Finally, if you are truly creative, you might one day want to do the things that have not been done before. In these situations, you can use *ImageJ* to read the raw data (see Chapter 17) and write your own plugins/scripts on the *ImageJ* platform.

13.6   BEST TEST SAMPLES FOR MRI

A best sample to test your new experimental setup in MRI, called *a phantom*, should not be anything from your real experimental study, or anything that you can find from your kitchen or supermarket. Sure, the internal structures of many vegetables and fruits are complex, which makes their images beautiful. Sure, vegetables and fruits do not move around – so that you can spend as much time as you need to tweak (i.e., optimize) the imaging parameters and to sort out any experimental issue. However, how do you know the delicate features in your cute images of vegetables and fruits are faithful, or not? Are these features real, or not? How do you know, for sure?

The best test sample for MRI is actually just a tube of water (or a sphere of water if you are testing a 3D sequence). You can dope the water with a tiny amount of copper sulfate ($CuSO_4$), which is a common chemical found in many laboratories that can reduce the T_1 of the water, so that you can repeat the data acquisition quicker. You can also make a water phantom by putting a small plastic cube with some known size (e.g., a Lego block from your childhood collection) into the water. These simple phantoms will be the best test samples for your MRI experiment, since you know what to expect from the image. For a tube of water, the image must have a circular shape (unless your glass tube is not circular), the tube must have a uniform intensity (if you select a slice in the middle of the water, away from the meniscus surface), the edges of the tube must be sharp (unless you tilt the tube in the probe), and the background must have no ghost trace of the tube. Figure 13.6a shows the axial image of a concentric double-tube phantom of doped water together with a horizontal line profile, and Figure 13.6b shows the sagittal image of a glass sphere of doped water together with a horizontal line profile. These types of simple tubes and spheres should always be your first test samples in setting up an imaging protocol, or whenever you suspect something is not quite right with your system.

A second-best test sample for MRI can be made from agarose gel powder, which is a linear polymer purified from a plant source. Or you can go to a supermarket to buy the agar powder, or

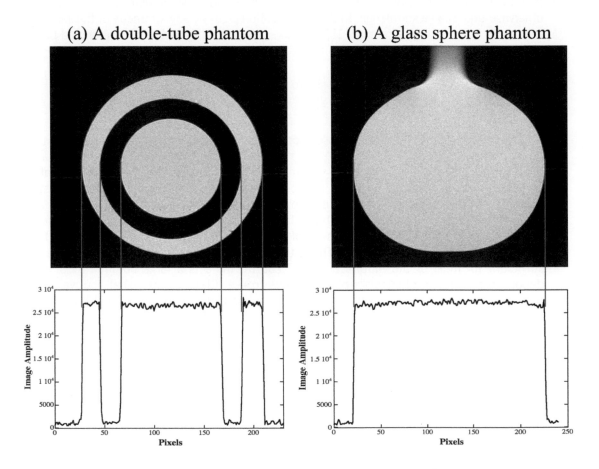

Figure 13.6 Two images of water phantoms. From these images, one can examine the shape and uniformity of the phantoms (e.g., the cross-sectional profiles must have a uniform intensity when the slice is in the middle of the tube, away from the meniscus surface), the sharpness of the edges (unless the tube is tilted on purpose), and the cleanness of the background (i.e., no ghost trace of the tube).

even gelatin powder, which is an irreversibly hydrolyzed form of collagen. You mix them in boiling water, pour them into a small plastic container, and wait for it to solidify during the cooling. If you do not want to be bothered by making them yourself, you can go to a supermarket to buy a ready-to-eat Jell-O dessert, which is made of gelatin – just remember to buy a plain one, not the one with pieces of fruit inside. Home-made agarose gel is convenient, since you can add a bit of copper sulfate to the water, so that your test experiments go faster and you can also test the quantitative imaging of T_1 and T_2 relaxation times and self-diffusion coefficient. Once the gel cools down and solidifies, you can cut the gel block into whatever shape you want, orient it in the magnet any way you want, then image it. Figure 13.7 is an example of a piece of agarose gel superglued to a piece of Teflon plastic, to represent articular cartilage that is attached to the

221

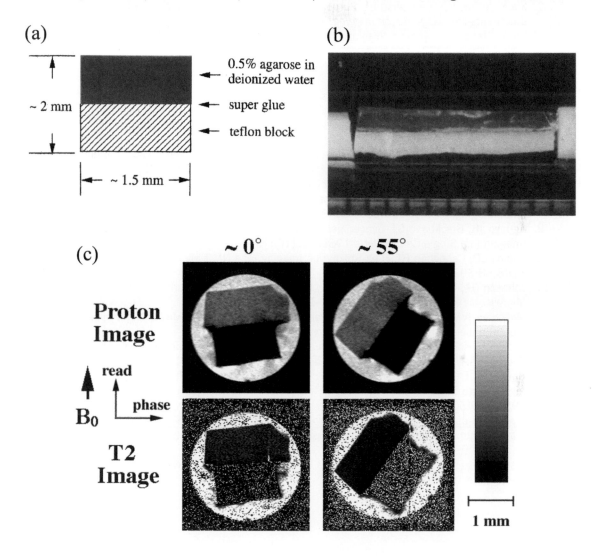

Figure 13.7 A phantom of agarose gel glued to a piece of Teflon. (a) A schematic drawing, (b) an optical photograph of the phantom in a glass tube (the gel is on the top of the white-color Teflon), (c) the orientation-dependent imaging experiments to examine any magic angle effect on the T_2 relaxation (there was none). Source: Xia et al. [16].

underlying bone [16]. Only by doing this kind of phantom testing can you be sure that any unexpected feature in your real experiments comes from the specimen, not an artifact in your experimental protocol. In this case, the T_2 relaxation anisotropy of the articular cartilage was discovered and quantified, for the first time, at microscopic resolutions [17].

References

1. Xia Y, Callaghan PT. Study of Shear Thinning in High Polymer Solution Using Dynamic NMR Microscopy. *Macro Mol.* 1991;24(17):4777–86.
2. Xia Y, Jeffrey KR, Callaghan PT. Purpose-designed Probe for Dynamic NMR Microscopy in an Electromagnet. *Magn Reson Imaging.* 1992;10:411–26.
3. Xia Y. Static and Dynamic Imaging Using Magnetic Field Gradients [MSc thesis]. Massey University, New Zealand; 1988.
4. Tanner JE. Pulsed Field Gradients for NMR Spin-Echo Diffusion Measurements. *Rev Sci Instrum.* 1965;36(8):1086–7.
5. Anderson WA. Electrical Current Shims for Correcting Magnetic Fields. *Rev Sci Instrum.* 1961;32(3):241–50.
6. Hoult DI, Richards RE. Critical Factors in the Design of Sensitive High Resolution Nuclear Magnetic Resonance Spectrometers. *Proc R Soc Lond A.* 1975;344:311–40.
7. Mansfield P, Chapman B. Active Magnetic Screening of Coils for Static and Time-dependent Magnetic Field Generation in NMR Imaging. *J Phys E: Sci Instrum.* 1986;19:540–5.
8. Köckenberger W, Pope JM, Xia Y, Jeffrey KR, Komor E, Callaghan PT. A Non-invasive Measurement of Phloem and Xylem Water Flow in Castor Bean Seedlings by Nuclear Magnetic Resonance Microimaging. *Planta.* 1997;201:53–63.
9. Aguayo JB, Blackband SJ, Schoeniger J, Mattingly M, Hinterman M. Nuclear Magnetic Resonance Imaging in a Single Cell. *Nature.* 1986;322(10):190–1.
10. Eccles CD, Callaghan PT. High-Resolution Imaging. The NMR Microscope. *J Magn Reson.* 1986;68:393–8.
11. Johnson GA, Thompson MB, Gewalt SL, Hayes CE. Nuclear Magnetic Resonance Imaging at Microscopic Resolution. *J Magn Reson.* 1986;68:129–37.
12. McGinley JVM, Ristic M, Young IR. A Permanent MRI Magnet for Magic Angle Imaging Having Its Field Parallel to the Poles. *J Magn Reson.* 2016;271:60–7.
13. Kose K, Haishi T, Caprihan A, Fukushima E. Real-Time NMR Imaging Systems Using Personal Computers. *J Mag Reson.* 1997;124:35–41.
14. Black RD, Early TA, Roemer PB, Mueller OM, Mogro-Campero A, Turner LG, et al. A High-temperature Superconducting Receiver for Nuclear Magnetic Resonance Microscopy. *Science.* 1993;259(5096):793–5.
15. Xia Y. MRI of Articular Cartilage at Microscopic Resolution. *Bone & Joint Res.* 2013;2(1):9–17.
16. Xia Y, Farquhar T, Burton-Wurster N, Lust G. Origin of Cartilage Laminae in MRI. *J Magn Reson Imaging.* 1997;7(5):887–94.
17. Xia Y. Relaxation Anisotropy in Cartilage by NMR Microscopy (µMRI) at 14 µm Resolution. *Magn Reson Med.* 1998;39(6):941–9.

Part V

Quantitative and Creative MRI

14

Image Contrast in MRI

The most common conundrum in MRI concerns the lack of a direct proportionality between the nuclear spin density $\rho(\boldsymbol{r})$ and the image intensity $S(\boldsymbol{k})$ of the sample. According to a number of equations in the earlier chapters [e.g., $\boldsymbol{M} = \sum \boldsymbol{\mu}_i$, $M_0 \sim \exp(\hbar\gamma B_0/k_{\mathrm{B}}T)$, $M_\perp \sim M_0$, $S(t) \sim M_\perp$, and $\rho(\boldsymbol{r}) \sim S(\boldsymbol{k})$ via FT], a direct proportionality between the nuclear spin density and the signal in MRI is expected. In practice, however, such direct relationship between $\rho(\boldsymbol{r})$ and $S(\boldsymbol{k})$ often does not exist. Figure 14.1 shows three examples of this lack of proportionality. One should therefore *never* assume in ^{1}H MRI that (a) the image intensity represents the true amount of water in a specimen, and (b) a black color (or a very low intensity) in a regular gray-scale image represents nothing.

14.1 NON-TRIVIAL RELATIONSHIP BETWEEN SPIN DENSITY AND IMAGE INTENSITY

The most common cause of this non-proportionality comes from the fact that different parts of a heterogeneous specimen can have different MRI properties; for example, different relaxation times at different locations in the specimen. Using a common spin-echo-based sequence, when the echo time TE is set, for example, at 5 ms, three tissue regions in a specimen having T_2 of 50, 5, and 1 ms would have an apparent image intensity of $0.9M_0$, $0.37M_0$, and $0.0067M_0$, since the MRI intensity can be estimated by Eq. (7.4) as $S(t) = M_0\exp(-\mathrm{TE}/T_2)$. While the tissue having an intensity of $0.9M_0$ can be considered well represented, the signal from tissue having an intensity of $0.0067M_0$ would be below the noise in the image, hence appearing black in a gray-scale image. In addition, the same specimen could appear quite differently when imaged using different TEs, as shown in Figure 14.2.

An image with some non-proportionalities between the nuclear spin density and the image intensity does not necessarily contain artifacts. Often it is the contrary – an image in which different regions have been *exaggerated* in intensity – that could be more visually appealing and more useful because the exaggerated areas will not be missed and can be examined more closely. Comparing the two images in Figure 14.2, the image on the left has more non-proportionalities than the image on the right; however, the image on the left looks *better* than the image on the right, simply because the left image has stronger contrast among different tissue regions, especially between the damaged tissues (from the skin to the mesocarp) and the normal tissues. This is similar to taking portraits of human beings – a well-liked portrait would be a photo that

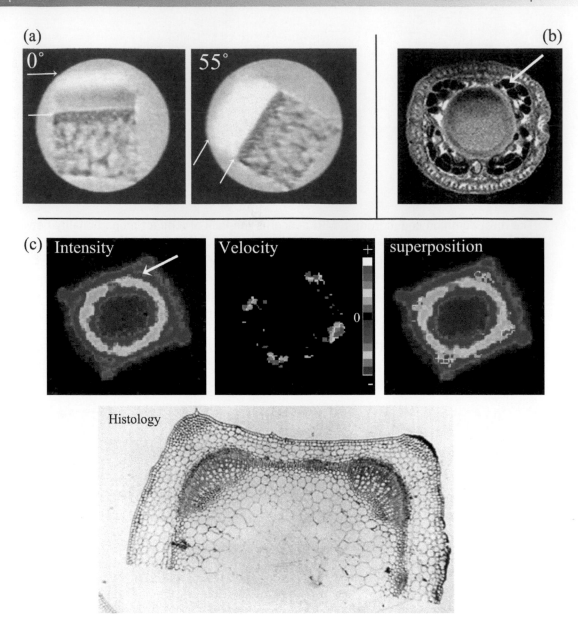

226

Figure 14.1   (a) Two 14-µm-resolution images of a cartilage and bone specimen (cartilage is between the two arrows), which was imaged with identical parameters except the orientation of the specimen in the magnetic field (vertically up) [5]. At ~ 0° orientation, the upper part of the cartilage was brighter than the lower part of the cartilage; at ~ 55°orientation, cartilage had more uniform intensity. Source: Xia [5]. (b) An axial image of a rat tail (20-µm resolution). The numerous "black holes" (arrow) contained tendons, which consist of bundled collagen fibers saturated with water. Since the tendons were in parallel with the magnetic field, they had nearly no signal in MRI by the spin-echo-based sequences. (c) The intensity image and the quantitative velocity image (with a maximum velocity of 45 µm/s) of a living plant stem (*Stachys sylvatica L.*), together with an optical image of half the stem. Source: Xia et al. [6]. Since the largest vessels of the xylem tissue are concentrated at the four corners of the stem (see the optical image and superposition image), each corner (arrow) in the intensity image should have high image intensity but actually had low image intensity [6].

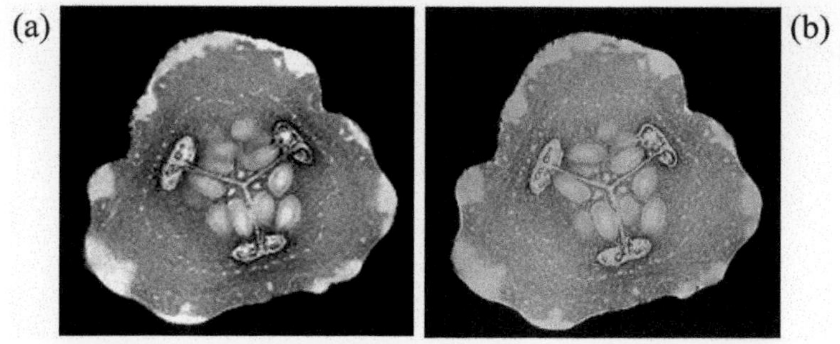

(a)   (b)

Figure 14.2 Two axial images of a cucumber acquired with a spin-echo imaging sequence at two different TEs: (a) 14.8 ms and (b) 8.2 ms. The pixel resolution of the images was 49 μm.

presents the person actually looking *better* than in reality. The same type of exaggeration is also commonly used in clinical MRI, with the terms such as T_2-weighted images or diffusion-weighted images.

227

14.2   IMAGE CONTRAST IN MRI

Image contrast can be defined as the modulation of the image intensity due to the contribution of certain molecular mechanisms other than the nuclear spin density $\rho(r)$ [1]. As shown in Figure 14.3, the time sequence of any imaging experiment consists of three sequential segments: the preparation segment, where a longitudinal magnetization is established; the excitation and evolution segment, where a transverse magnetization is established and evolved; and the imaging segment, where the signal is detected. Ideally, we would like the excitation and evolution segments to provide the image contrast and the imaging segment to map the signal faithfully. This expectation, however, may not be realized, since all images are weighted, more or less, by several intrinsic contrast factors during *all* segments of an imaging experiment, even during the detection (mapping) segment. These contrast factors are imposed on the imaging process by the molecular environment and the structure of the specimen, due to the ways the magnetization is prepared and the signal is acquired.

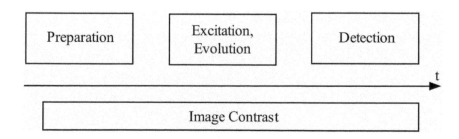

Figure 14.3 Three sequential segments of an imaging experiment, during any of which the image contrast could exist.

Instead of the straightforward Fourier relationships expressed in Eq. (11.8) and Eq. (11.9), the actual relationships between the MRI signal $S(\mathbf{k})$ and the nuclear spin density $\rho(\mathbf{r})$ are given experimentally by

$$S(\mathbf{k}) = \int \rho(\mathbf{r}) E_c(\mathbf{r}) \exp(-i2\pi \mathbf{k} \cdot \mathbf{r}) d\mathbf{r} \tag{14.1}$$

and

$$\rho(\mathbf{r}) E_c(\mathbf{r}) = \int S(\mathbf{k}) \exp(i2\pi \mathbf{k} \cdot \mathbf{r}) d\mathbf{k}, \tag{14.2}$$

where $E_c(\mathbf{r})$ is the normalized and combined contrast factor. Instead of the ideal case in MRI where a Fourier relationship exists between $S(\mathbf{k})$ and $\rho(\mathbf{r})$, the Fourier relationship in actual MRI experiments exists between $S(\mathbf{k})$ and $\rho(\mathbf{r}) E_c(\mathbf{r})$.

$E_c(\mathbf{r})$ is a combined factor that could be a product of several image contrast mechanisms, such as

$$E_c(\mathbf{r}) = E_c(v) \times E_c(D) \times E_c(T_1) \times E_c(T_2) \times E_c(\delta) \times E_c(\text{others}), \tag{14.3}$$

where

- $E_c(v)$ and $E_c(D)$ (or $E_c(\mathbf{q})$ where $\mathbf{q}$ is a dynamic reciprocal space vector, defined in Chapter 15.1) are the contrast factors representing the effect of molecular motion arising from the velocity $\mathbf{v}$ and self-diffusion $\mathbf{D}$,
- $E_c(T_1)$ is the contrast factor representing the effect of the spin-lattice relaxation,
- $E_c(T_2)$ is the contrast factor representing the effect of the spin-spin relaxation,
- $E_c(\delta)$ is the contrast factor representing the effect of chemical shift,
- $E_c(\text{others})$ is the contrast factor representing the effect of some secondary mechanisms such as the influences of magnetic susceptibility, temperature, blood oxygen level dependency, and so forth, regardless of whether one can express it as a straightforward equation or not.

The interest in MRI contrast factors is therefore two-fold (Table 14.1): (a) to appreciate the modulation to the image intensity due to these contrasts in order to understand a true image of spin density, and (b) to gain beneficial information via these contrast mechanisms that are sensitive to the molecular environment of the specimen. The rest of this chapter describes the general approaches and pitfalls in the process of extracting beneficial information from image contrast. The next chapter, Chapter 15, describes in detail the individual contrast factors and their usefulness.

Table 14.1   Relationships in signals vs images in MR.

	Signal vs. Image
MRI, the ideal case	$S(\mathbf{k}) \overset{FT(k)}{\Leftrightarrow} \rho(\mathbf{r})$
MRI, in reality	$S(\mathbf{k}) \overset{FT(k)}{\Leftrightarrow} \rho(\mathbf{r}) E_c(\mathbf{r})$
Quantitative MRI	$S(\mathbf{k}) \overset{FT(k)}{\Leftrightarrow} \rho(\mathbf{r}) E_c(\mathbf{r}), \ E_c(\mathbf{r}) \overset{?}{\Leftrightarrow} ?$

14.3 HOW TO OBTAIN USEFUL INFORMATION FROM IMAGE CONTRAST?

By choosing appropriate pulse sequences and parameters, we can emphasize the contribution of one particular contrast mechanism and find a way to extract information about it from experiments. In general, image contrast can be extracted from the images either qualitatively or quantitatively.

14.3.1 Qualitative Approach to Extract Image Contrast

The qualitative approach to extract image contrast is to make the pulse sequence sensitive to a particular contrast mechanism so that the image intensity is weighted by the influence of that mechanism. This approach is widely used in medical diagnosis where, for example, a relaxation-weighted image is often used to study abnormalities in soft tissues and organs, similar to the highlight of the damaged tissue in the cucumber image (Figure 14.2). This qualitative approach has the advantage that it is straightforward and time-efficient and requires little or no post-acquisition image process. This type of contrast-weighted approach does not necessarily have to be qualitative; it is just hard to make it quantitative. Since it is based on one single image, the lack of a reference image that contains no contrast or a second point in the contrast mechanism makes the information in the image mostly *qualitative*.

The use of this approach in high-resolution MRI (i.e., µMRI) should be handled with care. This is because, in comparison with medical applications where the magnetic field is low (1-3 Tesla), µMRI often employs a much higher polarizing field B_0 (e.g., 7 Tesla) and large imaging gradients (in order to achieve high spatial resolution in tens of microns). The image intensity at high spatial resolution could be influenced more strongly by some other factors such as susceptibility variation due to structure inhomogeneity in the sample (Chapter 15.4.1).

14.3.2 Quantitative Approaches to Extract Image Contrast

There are three approaches by which contrast information can be extracted quantitatively: the one-shot approach, the two-image approach, and the multiple-slice approach.

The quantitative one-shot approach aims to establish a direct relationship between the image contrast and image intensity in a single acquired image. In this type of one-shot scheme, the pulse sequence has to be made purposefully so that only the nuclear spins that are responsible for a particular contrast mechanism contribute to the final signal. For example, Figure 14.4 shows that the moving spins in a flowing fluid can be imaged and quantified by a "time-of-flight" imaging sequence, which utilizes the *freshness* of the spins within the imaging slice. Another example for this kind of quantitative one-shot scheme is in chemical shift imaging, where a chemical shift imaging sequence can be designed to select only one particular resonant frequency (Chapter 15.3). Numerically, the one-shot approach assumes that the values of the image voxels quantitatively represent the particular image contrast, either linearly or following some mathematical equations, with an unspoken assumption that a zero value in the image intensity means no contrast.

The quantitative two-image approach aims to establish a direct relationship between the two images obtained with different weightings of image contrast. The final image is computed from the two images on a pixel-by-pixel basis via some mathematical calculation such as subtraction,

229

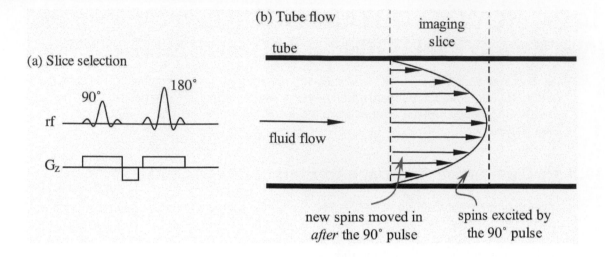

Figure 14.4 Quantitative one-shot imaging using the concept of time of flight. (a) A slice selection sequence. (b) For a steady flow in the tube, the velocity profile should be parabolic, with the highest velocity at the center of the tube. An image slice can be established axially across a tube by the first 90° soft pulse. By the time the soft 180° pulse arrives, the water in the fastest flow area has already left the imaging slice, which is replaced by the fresh water that has not been excited by the soft 90° pulse. This velocity profile should represent the amount of fresh water within the imaging slice at the time of the soft 180° pulse. Since fresh water does not contribute to the MRI signal, the image intensity across the tube should not be uniform. The intensity variation can be converted into a quantitative velocity image, provided that the flow has low Reynolds numbers (i.e., no turbulence).

division, or a trigonometric operation. This approach essentially fits a straight or curved line with the use of two data points, which accommodates the possibility of a non-zero image value at the "zero contrast," that is, to account for the influences of any intrinsic mechanism or background noise. Most of the attenuation-based contrast factors (relaxation, self-diffusion) could be extracted in this manner rather straightforwardly. In addition, even some complex fluid flows have been quantitatively imaged using a two-image approach where the velocity values were quantified via non-linear fitting by an arctan calculation [2].

The quantitative multiple-slice approach acquires multiple images, where each image is obtained at a different contrast condition [3]. This set of contrast conditions represents a progressive amount of a particular contrast along a contrast direction (Figure 14.5). Each 2D image therefore represents a "slice" in that particular contrast direction. For any complex function (e.g., velocity in a flowing fluid), both real and imaginary (quadrature) images need to be saved separately (i.e., they cannot be combined via a modulus operation like in the situation of relaxation and diffusion processes). The final quantitative image or images are computed from the multiple weighted images on a pixel-by-pixel basis via some mathematical calculation such as least-squares fitting or Fourier transformation. The least-squares fitting version of this multiple-slice approach is similar to the fitting of a straight line or curve using multiple data points, hence able to withstand the influence of experimental noise and an occasional bad data point. Chapter 15 has several examples of this multiple-slice approach.

Comparing these quantitative methods, the one-shot approach is obviously the most time-efficient technique but most likely the least accurate protocol since various imaging artifacts can also cause unwanted contributions to the final signal. By comparison, the two-image approach can be made reasonably accurate at a small increase of experimental time. The major

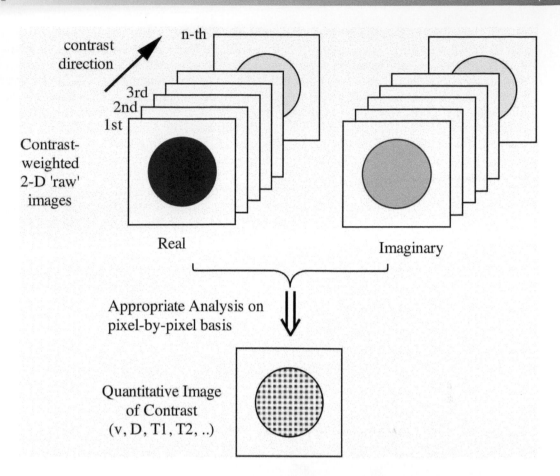

Figure 14.5 Quantitative multi-slicing approach in contrast extraction. Acquiring a set of complex image data enables the extraction of phase information (such as during velocity imaging). For the calculation of self-diffusion and relaxation, there is no need to save separately the real and imaginary images.

disadvantage of this two-image approach is that its accuracy is directly related to the SNR of the two "raw" images, since the noise in the final image could be amplified at each pixel location, especially when the two images are subtracted from each other (*check any first-year physics lab manual to see how the measurement errors combine in post-acquisition calculation*). The multiple-slicing approach can withstand the influence of noise and hence produce very accurate maps of the molecular environment at each pixel of the image comprehensively. Due to the nature of the technique, it can be time-consuming.

14.4 MAGNETIZATION-PREPARED SEQUENCES IN QUANTITATIVE MRI

Many of the imaging sequences described in Chapter 12 can be used in the two-image or multiple-slice approaches to extract some image contrast (e.g., T_2 relaxation time) by simply repeating the sequence with different variables. For example, one can carry out a quantitative T_2

imaging experiment by repeating several times the 2D FT imaging sequence shown in Figure 12.10 or Figure 12.11, each time with a different TE. This kind of approach is generally valid as long as the contributions from other image-intensity-attenuating factors, such as self-diffusion in this case, do not play an important part in the signal attenuation. As can be seen from Figure 14.6, an increase of TE in 2D FT imaging would increase the interval between the mapping gradient pulses, as well as the interval between the two slice-selection gradient pulses. Both pairs of gradient pulses could act like the diffusion-encoding gradient pair in a pulsed-gradient spin-echo (PGSE) sequence (cf. Chapter 15.1), resulting in an inaccurate calculation of T_2. This situation is worse in NMR microscopy since a large mapping gradient is required. Special care should be taken to ensure that the increment of TE does not increase the spacing of gradient pulses.

One approach to minimize unwanted influences during quantitative imaging utilizes the concept of magnetization-prepared sequences [4], shown in Figure 14.7 with a built-in spin-echo T_2 contrast [5]. This type of magnetization-prepared sequence has two well-separated time segments: a leading contrast segment that prepares the magnetization and a subsequent imaging segment that provides the spatial resolution; both segments can be said to have an echo time. In this example, during a series of T_2-weighted imaging experiments, only the timing of the leading segment is altered (i.e., only TEc is altered), while the timing of the

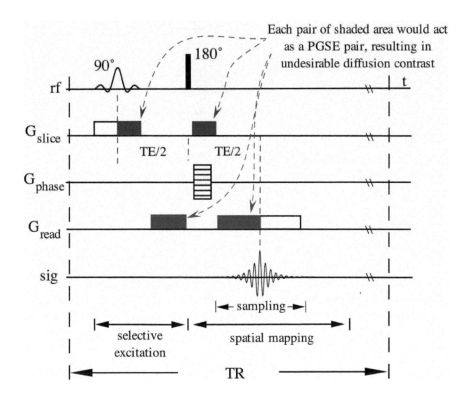

Figure 14.6 PGSE-like effect due to the slice gradients (blue) and the read gradients (red) in imaging. In a quantitative T_2 experiment using this type of sequence, for example, this PGSE-like effect will be different for each TE, which is an artifact that is difficult to eliminate subsequently in any quantitative calculation.

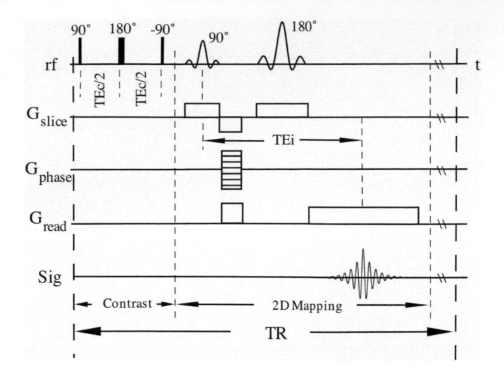

Figure 14.7 Magnetization-prepared T_2 imaging sequence. This type of imaging sequence has two totally separated time segments, one segment for the image contrast where the timings can be changed according to the amount of image contrast needed, and the other segment for 2D mapping where all timings are kept constant. This type of magnetization-prepared quantitative imaging sequences can eliminate the inaccuracy in the quantitative calculation of image contrast due to effects such as the pseudo-PGSE pair discussed in Figure 14.6.

imaging segment is kept constant in all imaging experiments (i.e., TEi is kept constant). Since there are no gradient pulses in the leading contrast segment, the intrinsic diffusion weighting and the T_2 weighting during the imaging echo time are both constant. Their effects therefore do not influence the subsequent relaxation calculation [1]. Consequently, T_2 can be determined unambiguously. One can also design this type of magnetization-prepared sequence for other types of contrast, such as different relaxation times and self-diffusion imaging, by exchanging only the leading contrast segment. See Chapter 15 for more examples.

In addition to the advantage of being able to determine more accurately the physical quantities, this type of magnetization-prepared imaging sequence can also obtain the physical quantities with less dependence on imaging resolution. This is because the magnitude of the pseudo-PGSE gradients (shaded in Figure 14.6) depends upon the imaging resolution – the higher the resolution, the higher the imaging gradients. Therefore, it could become challenging to compare the results from two studies in the literature that were acquired with different imaging resolution. In contrast, magnetization-prepared sequences can obtain the relaxation quantification with minimal influence from the imaging gradients, hence are not only more accurate but also are less resolution-dependent and sequence-dependent (cf. Chapter 15 for examples).

References

1. Xia Y. Contrast in NMR Imaging and Microscopy. *Concepts in Magn Reson*. 1996;8(3):205–25.
2. Xia Y, Callaghan PT. "One-shot" Velocity Microscopy: NMR Imaging of Motion Using a Single Phase-encoding Step. *Magn Reson Med*. 1992;23(1):138–53.
3. Xia Y. Dynamic NMR Microscopy [Ph. D. dissertation, vol. 1]. Massey University, New Zealand; 1992.
4. Haase A, Brandl M, Kuchenbrod E, Link A. Magnetization-prepared NMR Microscopy. *J Magn Reson A*. 1993;105:230–3.
5. Xia Y. Relaxation Anisotropy in Cartilage by NMR Microscopy (µMRI) at 14 µm Resolution. *Magn Reson Med*. 1998;39(6):941–9.
6. Xia Y, Sarafis V, Campbell EO, Callaghan PT. Non Invasive Imaging of Water Flow in Plants by NMR Microscopy. *Protoplasma*. 1993;173:170–6.

15

Quantitative MRI

In Chapter 14, we discussed the non-trivial relationship between the spin density in a specimen and the image intensity as measured in MRI. This type of unanticipated image manipulation can often be referred to as *artifacts* in MRI since they mess up the goal of imaging: to have a faithful representation of the specimen. All of these artifacts have their origin in physics principles. At the same time, the same physics principles that give us image artifacts can also give us useful information in the form of image contrast. Whether a particular physics principle is called the source of an image artifact or the origin of image contrast depends on whether it messes up the image or reveals useful information. In this chapter, the beneficial aspects of these sources and causes are discussed, with the proper equations and sequences. In these discussions, we limit ourselves to proton MRI, which essentially deals with water-containing specimens (biological or not).

Figure 15.1 summarizes the sources of image contrast in MRI. For each spin species (e.g., ^{1}H, ^{31}P), there are a number of primary sources of image contrast, where their influence on the image can be traced back to the Bloch equations, either the original equations or the extended equations. The secondary sources of contrast are defined as the additional physical parameters that can be obtained via the primary sources. For example, self-diffusion is temperature-dependent, which provides an opportunity to study the temperature distribution in an object non-destructively and non-invasively. In addition, one can also combine two different sources of contrast into one experiment, such as to image the velocity of a particular component that has been selected via the relaxation or chemical shift mechanism before the onset of the velocity-imaging sequence; for example, using the chemical shift to isolate the specific oil/fat-rich fluid in some plants (e.g., pine trees or rubber trees) and then imaging the velocity of this particular oil/fat-rich fluid.

15.1 QUANTITATIVE IMAGING OF VELOCITY *V* AND MOLECULAR DIFFUSION *D*

MRI is extremely useful in detecting two types of molecular motion. First, the self-diffusion coefficient D is characteristic of the random Brownian motion in liquid-containing objects, where self-diffusion of molecules leads to an irreversible loss of the echo signal [1]. Second, velocity v describes a translational motion of the molecules in a fluid flow; a molecule with a velocity v will

Essential Concepts in MRI: Physics, Instrumentation, Spectroscopy, and Imaging, First Edition. Yang Xia.
© 2022 John Wiley & Sons Ltd. Published 2022 by John Wiley & Sons Ltd.

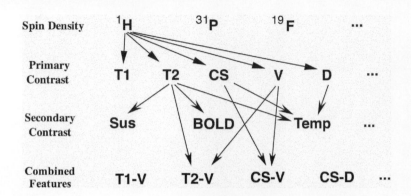

Figure 15.1 Summary of image contrasts. The lines relate some of the relationships discussed in this chapter. In addition, one can also combine two different contrasts into one experiment, such as to image the velocity of a particular component that has been selected via the relaxation or chemical shift mechanism before the onset of the velocity imaging sequence. The same relationships can also be used in other nuclear species. (BOLD, blood oxygen level dependent; CS, chemical shift; D, self-diffusion; Sus, magnetic susceptibility; T_1, longitudinal relaxation; T_2, transverse relaxation; Temp, temperature; V, velocity).

move to a new position over a specific time interval in the image pulse sequence, which also leads to signal loss due to outflow. Both types of motion can provide invaluable information about the liquid or liquid-containing specimen – velocity is associated with the mass flow of the fluid, and the self-diffusion coefficient reflects the microscopic structure of the molecules and their surroundings.

15.1.1 PGSE Sequence

In order to probe the motion of nuclear spins using MRI, the usual imaging sequence needs to be made motion-sensitive. A precise motion-sensitive algorithm uses the narrow pulsed-gradient spin-echo (PGSE) sequence [2, 3]. Figure 15.2a shows the classical form of the PGSE sequence in its spin-echo version, where the dephasing and refocusing of the nuclear spins occur between two strong and identical gradient pulses. (Note that these PGSE sequences are introduced in this imaging chapter. The same sequences can be used in the spectroscopy mode, much like the ways that T_1 and T_2 of bulk samples are measured by the sequences in Chapter 7.) The normalized echo amplitude for the PGSE sequence is given by

$$
\begin{aligned}
E_c(\boldsymbol{g}, \delta, \Delta) &= \exp(i\gamma\delta\boldsymbol{g}\cdot\boldsymbol{v}\Delta - \gamma^2\delta^2 g^2 D\Delta) \\
&= \exp(i\gamma\delta\boldsymbol{g}\cdot\boldsymbol{v}\Delta)\exp(-\gamma^2\delta^2 g^2 D\Delta),
\end{aligned}
\tag{15.1}
$$

where Δ is the separation between the two gradient pulses, δ is the duration of the gradient pulses, g is the magnitude of the gradient pulses, $\boldsymbol{v}$ is the velocity of the fluid (a vector quantity), and D is the self-diffusion coefficient of the liquid. An exact analysis for Brownian motion shows that Δ may be replaced by $(\Delta - \delta/3)$ in Eq. (15.1).

Note that the first term in the above equation that senses the velocity is complex, which means in order to distinguish the velocity vector (i.e., the flow direction), one needs to keep

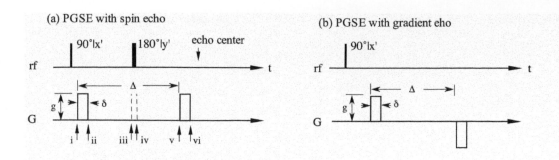

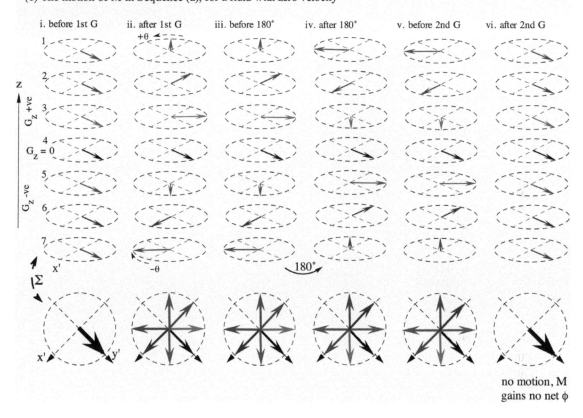

Figure 15.2 The PGSE pulse sequences in the spin-echo form (a) and the gradient-echo form (b). The motion of the magnetization for the spin-echo form of the PGSE sequence is illustrated without fluid flow in (c) and with fluid flow in (d). In (c) and (d), seven rows of vectors (1–7) mark the magnetization components in seven sequential physical locations along the z axis, where *row 4* is at the center of the PGSE gradient G_z, which has a zero-gradient value; while the six columns of vectors (*events i–vi*) mark the orientation of these M components at six events along the axis of time, indicated by the six arrows in (a). See text for explanation. Note that the reduction of the net magnetization due to T_2 relaxation and diffusion is not illustrated in the schematic for simplicity. Note also that the application of the 180° rf pulse, on the y' axis, does not change the orientation of the magnetization components on the $+y'$ and $-y'$ axis; hence, the black M component in (c) and the black and green M components in (d) stay where they are under the 180° rf pulse.

(d) The motion of M in Sequence (a), for a fluid with a finite velocity, along the z axis

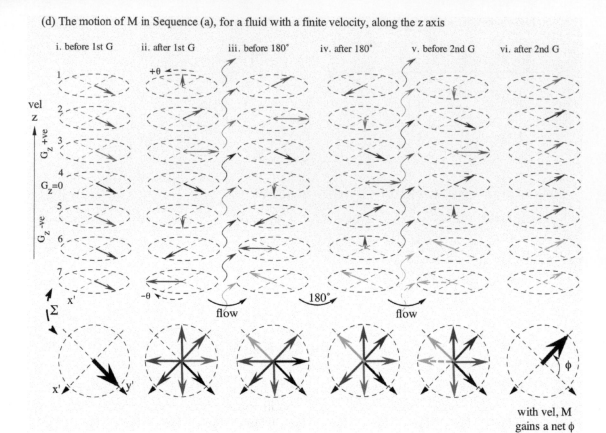

Figure 15.2   (Continued)

the complex form of the FID signal (i.e., keep the real and imaginary signals separate in data acquisition, as shown explicitly in Figure 14.5). In contrast, the second term in the above equation that senses the random Brownian motion is real, which means that one can take the usual modulus approach to process the raw data, since self-diffusion is determined by the attenuation of the signal.

The detection of the motion of magnetization in this sequence can be appreciated by comparing the winding and unwinding schematics in Figure 15.2c (zero velocity) and Figure 15.2d (a finite velocity). In Figure 15.2c, there is no translational flow. The application of the first gradient pulse (*event ii*) imposes a location-dependent phase shift to the *M* components in the specimen (cf. Figures 11.8 and 12.2b) – where the red M_{row1} experiences the largest G_z gradient and hence possesses the largest positive phase shift, the black M_{row4} receives no phase shift (since it is at the center of G_z), and the blue M_{row7} receives the largest negative phase shift. Since there is no flow, all *M* components stay where they are physically between *event ii* and *event iii* and between *event iv* and *event v*. The 180° rf pulse inverts the phases. The application of the second gradient pulse re-phases all M components to their original orientation.

In Figure 15.2d, by comparison, there is a translational flow characterized by a velocity along the +z direction (i.e., the direction of the gradient pulses); consequently, the molecules will move themselves to a new physical location where the gradient is *different* (in the schematic, moving upwards constantly and uniformly). Between *event ii* and *event iii*, the red M moves away to an even higher location, and the green M moves into *row 7* from below. Between *event iv* and *event v*, the flow continues upwards, with the pink M moving away to a higher location and a green-dashed M moving into *row 7*. The application of the second gradient pulse would not be able to re-phase all M components to their original orientations as in *event i*; instead, the sum of the magnetization will gain a net phase shift ϕ that is proportional to the velocity of the fluid flow. Note that between *event iii* and *event iv*, the black and green M components are not tipped by the $180°|_{y'}$ rf pulse since they are parallel to the $+y'$ and $-y'$ axis.

This PGSE sequence can also be designed in its gradient-echo version, shown in Figure 15.2b. Since the gradient echo only refocuses the spins that have been dephased by the gradient itself, the dephasing of the magnetization due to the inhomogeneous effects are not refocused by the gradient echo. Since there is no hard 180° rf pulse between the gradient pulses, this gradient-echo version of the PGSE could be the preferred choice in situations where a hard rf pulse is not used (e.g., multi-slice imaging in clinical scanners).

Note that Figure 15.2 only illustrates the effect of the translational motion in fluid flow on the magnetization; the schematic does not include the dephasing and re-phasing events of the magnetization associated with the transverse relaxation for simplicity. To illustrate the motion of the transverse relaxation in these schematics, each colored vector (i.e., each component of the magnetization) in the schematics should have been drawn in the same way as in Figure 7.6a, that is, each colored magnetization component in Figure 15.2 is further fanning out before the 180° rf pulse and re-phased after the 180° rf pulse. To include the attenuation effect of T_2 relaxation or diffusion on the final magnetization, the final vectors in column *vi* should be shorter than the original vectors in column *i*.

15.1.2 *q*-Space Imaging

Following a similar approach used in the spatial coding in MRI where a reciprocal vector $\boldsymbol{k}$ is defined as $(2\pi)^{-1}\gamma\boldsymbol{G}t$, a dynamic reciprocal space vector, $\boldsymbol{q}$, can be defined as [4]

$$\boldsymbol{q}=(2\pi)^{-1}\gamma\boldsymbol{g}\delta, \qquad\qquad (15.2)$$

where $\boldsymbol{g}$ is the PGSE gradient (a vector quantity) and δ is the duration of the PGSE gradient pulses. A dimensional analysis yields the dimension of $\boldsymbol{q}$ to be m^{-1}, which is a spatial frequency of a wave or wavenumber (i.e., the number of wave cycles per unit distance).

A wave can be measured in both time (in seconds) or distance (in meters), as shown in Figure 15.3. When a wave is measured in time, one complete waveform is said to have one period T (in seconds), which is the inverse of the frequency in hertz, that is, $f = 1/T$. When a wave is measured in distance, one complete waveform is said to have one wavelength λ (in meters), which is the inverse of the wavenumber κ, that is, $\kappa=2\pi/\lambda$ (in rad/m when κ is the angular wavenumber), or $\kappa=1/\lambda$ (in 1/m when κ is the linear wavenumber).

239

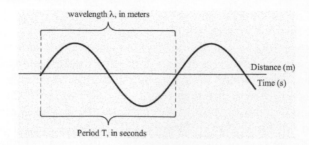

Figure 15.3 A sinusoidal wave can be characterized either by distance traveled or time evolved, with wavelength in meters and period in seconds.

When the width of the gradient pulse is narrow, and when both translational flow and random Brownian motion occur, the result can be derived as a simple product of a phase shift term and an amplitude attenuation term in the Fourier domain [4, 5], as the normalized contrast factor:

$$E_c(\boldsymbol{q}) = \exp(i2\pi qv\Delta)\exp(-4\pi^2 q^2 D\Delta), \tag{15.3}$$

where v is the value of the velocity in the direction of $\boldsymbol{q}$, D is the self-diffusion coefficient, and Δ is the separation of the two PGSE pulses.

When the dynamic dimension conjugate to $\boldsymbol{q}$ is defined as Z and the relevant components of the diffusion tensor and the velocity vector in the Z direction are D and $\boldsymbol{v}$, $E_c(\boldsymbol{q})$ has the form of an oscillatory function in $\boldsymbol{q}$ space modulated by a Gaussian decay, which is the characteristics of Brownian motion and shown in Figure 15.4 [6]. Quantitative imaging experiments can be carried out by repeating an imaging sequence with a successive increment of the $\boldsymbol{q}$ gradient in a number of discrete steps, from zero to some maximum value g_m. Each step corresponds to a "slice of contrast" in $\boldsymbol{q}$ space (cf. Figure 14.5), where one pair of complex images (one in-phase and one quadrature-phase) is reconstructed using the usual $\boldsymbol{k}$-space reconstruction algorithm. Each slice in the contrast direction (here in $\boldsymbol{q}$ space) is weighted progressively by $E_c(\boldsymbol{q})$ as the stepping up of $\boldsymbol{g}$. Subsequent analysis along the particular $\boldsymbol{q}$ direction using the multiple $\boldsymbol{q}$-contrasted slices produces comprehensive information about molecular motion at each pixel of the image, shown as $\mathcal{F}[E(\boldsymbol{q}, \boldsymbol{r}, \Delta)]$ in Figure 15.4.

The PGSE sequence shown in Figure 15.2 can be combined with almost any spatial mapping sequence, either 2D or 3D mapping. Figure 15.5a shows an example of the insertion of the PGSE sequence in the middle of a 2D FT imaging sequence (Figure 12.11a), which has been used in many successful experiments [4]. One can also adapt the concept of the magnetization preparation (Chapter 14.4) and add the PGSE sequence to the front of a 2D imaging sequence (Figure 15.5b). These motion-sensitive sequences are capable of imaging fluid flow and Brownian motion simultaneously. Figure 15.6 shows one set of experimental data in MRI of velocity and diffusion, when a cross-sectional slice was selected for a circular tube [4]. The experimental velocity profiles were cylindrically symmetric and confirmed the expected Poiseuille flow behavior.

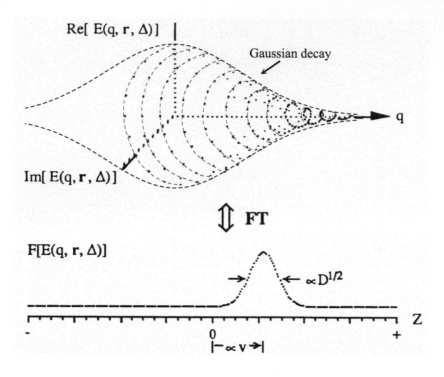

Figure 15.4   The dynamic dimension conjugate to *q* is defined as *Z*, and the relevant components of the diffusion tensor and the velocity vector in the *Z* direction are *D* and **v** [6]. The normalized contrast factor $E(q)$ has the form of an oscillatory function of *q* modulated by a Gaussian decay. A Fourier transform of this contrast factor will be a Gaussian position shifted along the *Z* axis, which is proportional to the velocity in the translational flow. The full width at half maximum of the Gaussian peak is proportional to the self-diffusion of the fluid. Source: Xia [6].

Note that the motion-sensitive gradient **g** can be applied in any arbitrary direction in the 3D space, which determines the direction at which self-diffusion and flow is measured. Figures 15.7 and 15.8 show a different set of experimental data [7] in which the transverse velocity was quantified around a junction of two glass tubes that had a sudden change in the tube diameter (Figure 15.7). One can compare with the *q* slices between Figures 15.6a and 15.8a, where Figure 15.6a is the *q* slices when the longitudinal flow was measured in a constant-diameter tube, while Figure 15.8a is the *q* slices when the transverse flow was measured around the junction of two tubes. The amplitude of **g** together with the duration of the gradient pulses δ determines the *q*-vector amplitude, *q*. When the spatial mapping is 3D and **g** is only along one particular direction, the overall imaging process is four dimensional. More commonly, a 2D imaging sequence is combined with a single **g** direction, which makes the experiment (Figures 15.6 and 15.8) three dimensional.

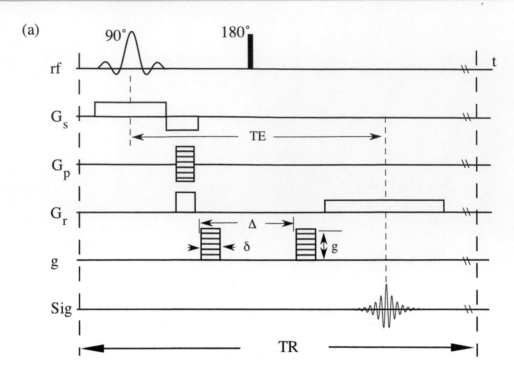

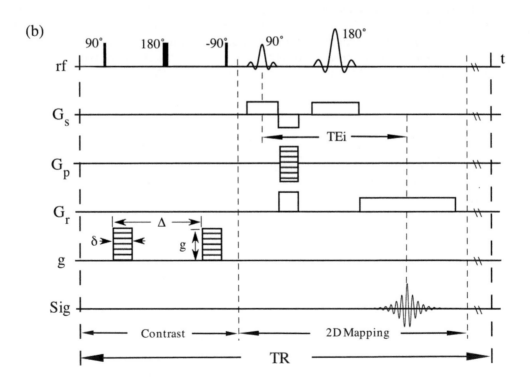

Figure 15.5 (a) A 2D imaging sequence that employs the spin-echo version of the PGSE sequence (Figure 15.2a), which is capable of imaging velocity and self-diffusion simultaneously. (b) A 2D flow-sensitive imaging sequence that uses the magnetization-prepared concept.

(a) q-space slices

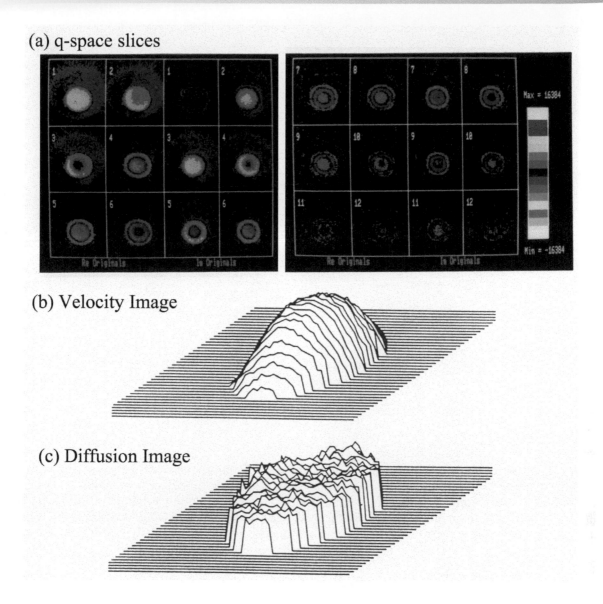

(b) Velocity Image

243

(c) Diffusion Image

Figure 15.6 A set of 2D *q* space MRI data that imaged the water flow in a single plastic tube, with the slice selection axial/longitudinal [65]. (a) The first 12 *q* slices, which are separately saved as real images and imaginary images. Since the first *q* slice has the PGSE gradient set at zero, the first real-image was an ordinary 2D image of the tube, and the first imaginary image contained only the noise. Starting from the second *q* slice, PGSE gradient was increased in a number of steps, so the flow was encoded as oscillations, centered about the axis of the tube since the tube flow is symmetrical Poiseuille flow. A subsequent pixel-by-pixel calculation, extracted from the raw data, of two quantitative images, (b) a 2D velocity map and (c) a 2D self-diffusion map. Source: Callaghan et al. [4].

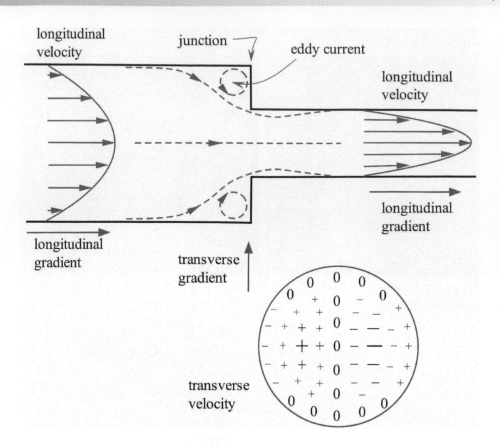

Figure 15.7 A configuration of two glass tubes forming a junction (the small tube had a 1.8-mm internal diameter; the large tube had a 2.9-mm internal diameter). The water flow, which had the Reynolds numbers from 12 to 20, was imaged at a 20 μm/pixel resolution. When the **q**-space gradient is along the length of the tubes, the parabolic Poiseuille's flows can be imaged (shown as the profiles inside the tubes). When the **q**-space gradient is in a single transverse direction, the centrally symmetric transverse velocity can be mapped out. Source: Based on Xia et al. [7].

15.1.3 Imaging Fluid Flow

Being able to image the velocity distribution and profile of a fluid flow can yield a tremendous amount of information about the flow system, including various tube flows in big and small devices [4, 8, 9], delicate vascular flows in plants [10, 11], and pulsatile blood flows in humans and animals. One can simply use the phase part of the PGSE equation [the first part in Eq. (15.1) and Eq. (15.3)] for any quantitative measurement to determine the velocity profile and volume flow rate. Since MRI velocity measurement relies on the internal properties of the fluid, and hence is totally non-invasive and non-destructive, the quantitative determination of velocity by MRI is much more accurate than other available methods such as doping the fluid with small particles or dye.

In general, there are two categories of flows, those in which the spin density $\rho(r)$ is independent of time and those in which $\rho(r)$ is dependent on time.

- Fluid flows with time-independent spin density can be found in various tube configurations with homogeneous fluids in steady state [7, 9]. Since $\rho(r)$ is time-independent, most of the

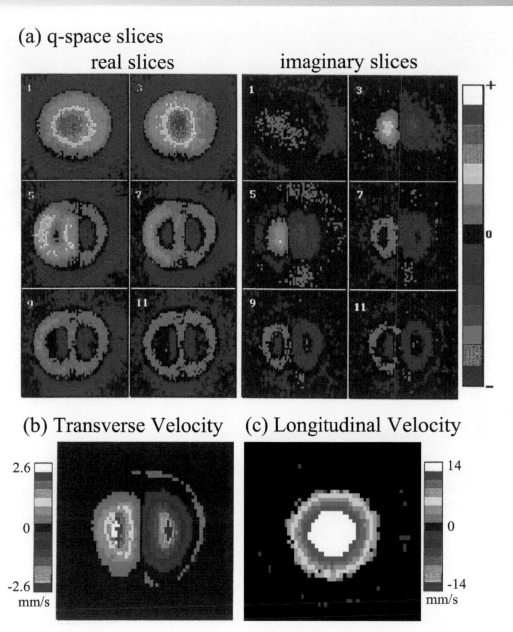

Figure 15.8 (a) A set of 2D *q*-space MRI data that imaged the eddy currents in water flow at a junction of two glass tubes as in Figure 15.7. When the *q*-space gradient was in a single transverse direction, different *q* slices exhibited a left-right symmetry. (b) The transverse velocity map at the junction, which is symmetric about the center of the *q*-space gradient. (c) The longitudinal velocity map at the junction, which has a Poiseuille's profile in the small tube at the center while the background (i.e., the large tube) has low negative velocities (very dark blue colors), indicating the negative eddy current [7]. The 2D *q*-space MRI data in the longitudinal direction are similar to the data shown in Figure 15.6. Source: Xia et al. [7].

imaging methods discussed in the previous sections can be applied to image both velocity and self-diffusion simultaneously, including the one shot, phase encoding [12], and *q*-space imaging [5]. Various flow systems have been investigated [9, 13, 14], using both Newtonian and non-Newtonian fluids [5, 8]. Another type of flow is the vascular flow in xylem and phloem of a live plant [10, 11]. Although it is never certain that vascular flows in plants are truly steady state, it is slow enough for the duration of the imaging experiment so it can be considered as steady state. Many successful experiments have been carried out to measure the mean velocity of vascular flow, which can be as low as several microns per second [15].

- Fluid flows with time-dependent spin density can be blood flows in animals and humans, where the motion in arteries and veins is irregular (or pulsatile/periodic). Since $\rho(\boldsymbol{r})$ is time-dependent, the only solution in MRI is to either image the specimen fast or image it while gated if the time-dependence is periodic.

- A clinical MRI technique that uses the time-of-flight principle (Figure 14.4) in blood flow imaging is called arterial spin labeling (ASL) [16, 17], which capitalizes on the fact an increase in the neural activity in the brain will increase the local cerebral blood flow in the network of capillary vessels. ASL uses an rf pulse (usually a 180° inversion pulse) to label/tag the magnetization of the upstream blood that inflows into the arterials of the brain, which is measured/imaged downstream after a short delay inside the brain. Within the limit of T_1 relaxation, the rf-tagged magnetization can generate an MRI signal, different than the not-tagged magnetization (ASL is essentially an inversion recovery experiment as discussed in Chapter 7.2). ASL is an indirect measurement of neural activity, which can also be measured via the oxygenation consumption in functional MRI (Sections 15.4.2 and 16.4). Since the labeling effect of ASL is weak, repeated acquisition together with image and statistical analysis is needed.

15.1.4   Imaging Molecular Diffusion

Self-diffusion is characteristic of the thermal motion of the molecules, the Brownian motion. All self-diffusion measurements are based on signal attenuation due to incoherent random motion [18], which can be implemented in MRI by utilizing the amplitude decay part of the PGSE equation [the second part in Eq. (15.1) and Eq. (15.3)]. Similar to the velocity measurement by MRI, the diffusion MRI relies on the internal properties of the fluid and therefore is totally non-invasive and non-destructive [8]. Since the basic principle behind a self-diffusion measurement relies on signal attenuation due to the unrecoverable loss of phase coherence caused by the Brownian motion, there is no reliable method to quantify diffusion of molecules in any specimen with time-dependent spin density. Examples of fluid flows with time-dependent spin density include blood flows in animals and humans, where the motion is irregular (or pulsatile/periodic).

In the medical literature of diffusion, MRI in the absence of translational flow, a "*b*-factor" is commonly used to include all terms in Eq. (15.3) except *D*, as the following:

$$E_c(D) = \exp(-bD) \tag{15.4}$$

where

$$b = \gamma^2 \delta^2 g^2 (\Delta - \delta/3). \tag{15.5}$$

The use of this "*b*-factor" in diffusion MRI is convenient in qualitative comparison of the MRI sensitivity to diffusion among different studies in the literature. However, the use of a single parameter *b* can bury some subtle differences among different studies. This is because the same *b*-factor can be obtained by different combinations of the gradient magnitude *g*, diffusion time Δ, and pulse duration δ where each term actually senses diffusion *differently*.

There are a few cautionary notes in the experimental measurement of diffusion using MRI. First, the term self-diffusion should, strictly speaking, be reserved for the random Brownian motion in non-viscous aqueous solutions where there is no chemical potential gradient in the system (both the solution and its environment). For example, you can put a small drop of liquid ink into a bucket of water and watch the ink spreading from its original location to the rest of the water; this process is a diffusion process but not a self-diffusion process because there is a concentration gradient of the ink that drives the process. In ^{1}H NMR and MRI when the sample is not an aqueous solution, the measured value of diffusion should be named more appropriately as the apparent diffusion coefficient (ADC, D_{app}), which is usually slower than the diffusion of water at the same temperature. A number of "obstructive effects" can slow down molecular diffusion. For example, the measured diffusion values were found reduced by some large molar mass, and hence less-mobile macromolecules [e.g., poly(ethylene oxide)] in solutions [8], and by some immobile and structured biomolecules in tissues (e.g., collagen fibers in articular cartilage and other connective tissues) [19]. See Chapter 16.1.3 for the measurement of restricted diffusion and the diffusion tensor.

Second, apart from the Brownian motion, there are other factors that may cause signal attenuation in the pulse sequence designed to detect self-diffusion. These include perfusion or micro-circulation of molecules in the sample [20] and the existence of velocity shear such as in laminar flows [5]. These influences could become dominant in any quantitative self-diffusion measurement and should be carefully addressed during the experiments – for example, the use of a double PGSE sequence can cancel the phase shifts due to velocity so that the self-diffusion can be measured accurately in the presence of velocity shear [8]. At the same time, image attenuation due to relaxation is likely not a major issue, since the signal attenuation due to the diffusion gradient is commonly much larger than the effect of relaxation (unless the T_2 of the specimen is short and TE of the imaging sequence is long).

Finally, the measurement of diffusion much slower than that of ordinary water at room temperature ($\sim 2 \times 10^{-9}$ m^2/s) should be handled carefully because the tiny movement/vibrations of the sample or probe due to the induced eddy current in the surrounding metals will ruin the chance of accurate diffusion measurement, a well-known fact in conventional pulsed-gradient NMR experiments. In addition, the mismatch of the PGSE gradient pulses (due to electronic noise) will also cause inaccurate measurement (cf. Chapter 13.2).

15.2 QUANTITATIVE IMAGING OF RELAXATION TIMES T_1, T_2, $T_{1\rho}$

T_1 and T_2 relaxation times in the Bloch equations (Chapter 2) describe the evolution of a nuclear spin system after being disturbed from its thermal equilibrium, one longitudinally and one transversely. T_1 and T_2 relaxation are two of the most common contrast factors that influence the image intensity in MRI. They occur *naturally* in *all* MRI experiments, regardless of whether one pays attention to them or not. They are also the most common causes of image artifacts in MRI. In addition, $T_{1\rho}$ relaxation time (Chapter 7.6) has been increasingly used during the last two decades in biomedical research.

Before we start on quantitative MRI of relaxation, we make a note on a common interpretation or wording in the literature of biomedical MRI relaxation studies, "*we show $T_1/T_2/T_{1\rho}$ measures whatever macromolecule.*" No, they do not. Different relaxation times in ^{1}H NMR and MRI reflect the *dynamic motion of water molecules* in the tissue in different frequency ranges (cf. Chapter 3.7). Assuming the dipolar interaction plays an important role in spin relaxation in biological tissues, the T_1 relaxation process requires the frequency spectrum of the randomly fluctuating molecular motion to have frequency components that are able to stimulate transitions between energy levels at the Larmor frequency. In contrast, the T_2 relaxation process can occur under the influence of static or slowly fluctuating magnetic fields in tens of hertz [21]. The frequency range of the molecular motion sensitive to the $T_{1\rho}$ relaxation mechanism would be intermediate, around tens of kilohertz.

247

These motional characteristics of water molecules in the tissue are modulated by the interactions between water and neighboring macromolecules, by the anatomic and histological structures of the tissue in which the water molecules are measured, and sometimes by the orientation of the tissue in the magnetic field. Whenever a certain degradation or lesion in the tissue leads to any reduction or increase of water or other macromolecules, or any conformational change of macromolecules, the motional characteristics of water will change because of the changes in the molecular interactions and/or structure, which will lead to the changes in the relaxation times. But the relaxation times themselves from ^{1}H MRI do not *directly* measure any macromolecule in the tissue.

15.2.1   Magic Angle Effect and Relaxation Anisotropy

A subtle feature of the relaxation time measurement in biological systems is the possible dependence of their values on the physical orientation of the specimen in the $\boldsymbol{B}_0$ field (i.e., the magnet), in which case the relaxation times are said to have anisotropy. Anisotropy offers a powerful way to examine the ultrastructure of biological tissue, often related to the orientation and organization of macromolecules in the tissue. This relaxation anisotropy can cause some connective tissues (tendon, ligament, articular cartilage, etc.) in MRI to have high/low intensity in the images, depending upon the local orientation to $\boldsymbol{B}_0$, a phenomenon called *the magic angle effect* in the clinical MRI literature [22]. The main cause of the relaxation anisotropy is the dipolar interaction between the water molecules and the macromolecules (which are less mobile and hence hard to be seen in MRI) in the tissue. In addition, the exchange between different molecular environments and cross relaxation between protein macromolecules and water are important aspects of the relaxation mechanism in biological materials.

Figure 15.9 shows a set of quantitative relaxation MRI experiments in articular cartilage [19], where T_1 relaxation was found to be isotropic (Figure 15.9c) while T_2 relaxation was found to have a strong orientational dependence (Figure 15.9d) that varied as $(3\cos^2\theta - 1)$, the geometrical factor that dominates the non-zero dipolar Hamiltonian (cf. Chapter 4.1). This difference in the anisotropic characteristics between T_1 and T_2 is a clear indication of slow macromolecular motion in cartilage, likely related to the highly constrained motion of swollen proteoglycans in the collagen matrix. The features of these relaxation profiles have been quantitatively verified to be consistent with the anatomic structure of articular cartilage (drawn schematically in Figure 15.9a and b) by a number of correlation studies between μMRI and optical imaging [23, 24]. These relaxation anisotropies are responsible for the orientational dependence of articular cartilage in MRI shown in Figure 14.1a and the disappearance of the tendon bundles in rat tail in the MRI image shown in Figure 14.1b. When using MRI to image relaxation times in some biological tissues such as articular cartilage, tendon, ligaments, meniscus, and muscles, one should always pay attention to the possibility of the magic angle effect in the images and the influence of the anisotropy on quantitative calculation.

15.2.2 T_1 Relaxation Time

T_1 relaxation time varies in different biological tissues or between normal and abnormal tissues, so that a knowledge of T_1 distribution in a specimen is useful in the differentiation of different tissue types. Since the longitudinal magnetization is determined jointly by the T_1 relaxation in the specimen and the repetition time of the experiment, T_1 contrast occurs in an image when the repetition time TR of the experiment is not long enough for the magnetization to fully recover its thermal equilibrium before the start of the next rf pulse, that is, anytime TR$<5\times T_1$. T_1 contrast can be incorporated into an MRI experiment simply by repeating a spin-echo sequence at different TR (Figure 15.10a). In this situation, the normalized image contrast is given approximately by

$$E_c(T_1) = 1 - \exp(-\text{TR}/T_1), \tag{15.6}$$

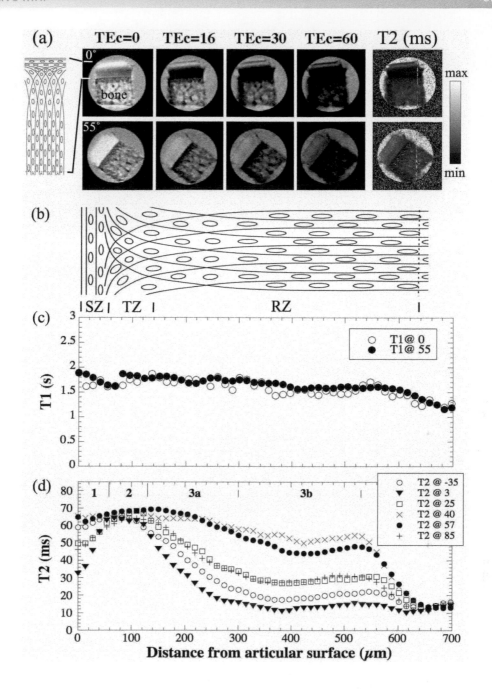

Figure 15.9 (a) The T_2-weighted proton images and the calculated T_2 images for a cartilage-bone plug oriented at ~0° and ~55° to B_0 (oriented vertically up), respectively. The image intensity is sensitive to the orientational angle of the sample. (b) A schematic drawing showing the orientation of collagen fibers (lines) and chondrocytes (ovals) in cartilage (not drawn to scale), which is also shown as a small version on the left-hand side of panel (a). (c) 1D cross-sectional profiles of T_1 taken from the 2D T_1 images of cartilage, where no orientational dependence of T_1 was observed. (d) The cross-sectional profiles of T_2 for the 2D cartilage images, which exhibited three distinct regions (1–3). These three MRI regions were later confirmed to be equivalent to the three histological zones (SZ, superficial zone; TZ, transitional zone; RZ, radial zone) by polarized light microscopy. Source: Reproduced with permission from Xia [19].

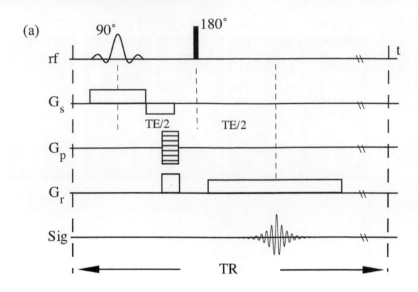

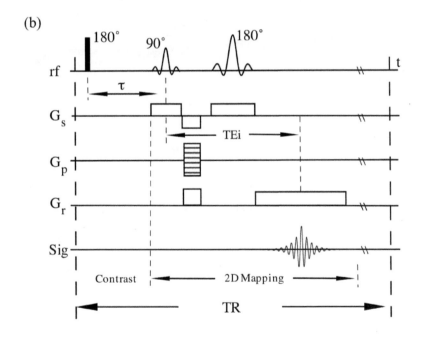

Figure 15.10 (a) A 2D imaging sequence that can quantify T_1 by repeating it at a fixed TE but different TRs. (b) A 2D T_1 imaging sequence that uses the magnetization-prepared concept, which has an inversion recovery sequence before the onset of 2D imaging. A quantitative T_1 experiment can be carried out by repeating the sequence at different τ, while TEi and TR are both fixed.

which is the normalized version of Eq. (7.3). An alternative method to apply T_1 contrast is to utilize the concept of magnetization preparation (Chapter 14.4) by incorporating the inversion-recovery sequence (Figure 7.3) prior to the slice excitation of a 2D imaging sequence, as in Figure 15.10b. Utilizing the zero-crossing point at $0.693\,T_1$, it can suppress the component with a characteristic T_1 value. In this situation, the normalized image contrast function becomes

$$E_c(\boldsymbol{r}) = 1 - 2\exp(-\tau/T_1) + \exp[-(TR - \tau)/T_1] \qquad (15.7)$$

where the first two terms are due to the inversion-recovery sequence [i.e., Eq. (7.1)], while the last term reflects the recovery of the magnetization due to T_1 relaxation. If TR is $\gg T_1$, the effect of the last term is limited.

15.2.3 T_2 Relaxation Time

T_2 relaxation is an extremely sensitive parameter in biomedical research. T_2 contrast occurs in all MRI experiments due to signal decay during the echo time TE in an MRI scan. Since TE is finite and commonly 5–20 ms (it is extremely difficult to make TE much less than 5 ms without causing severe image artifacts), T_2 relaxation contrast is the most common source of contrast or artifact in MRI. The magnitude of T_2 contrast depends on the size of TE and the T_2 relaxation of the sample. Since a heterogeneous sample is assumed to have a distribution of T_2 relaxation times over different structures of the specimen, the contrast due to T_2 is also heterogeneous, that is, not uniform over the entire image.

One can repeat an imaging experiment at a fixed TR but with different TE (Figure 15.11a). In this situation, the normalized T_2 image contrast function can be normalized from Eq. (7.4) as

$$E_c(T_2) = \exp(-TE/T_2), \qquad (15.8)$$

provided that TR is larger than T_1 to ensure that the T_2 contrast is dominant.

The sequence such as in Figure 15.11a could be sensitive to the effect of pseudo-PGSE gradients that we have discussed in Chapter 14.4, which cause the quantitative T_2 values to be dependent upon a number of experimental parameters such as the image resolution and matrix size, hence difficult to be compared with the T_2 results from a different study. A group of more accurate sequences for quantitative T_2 imaging uses the magnetization-prepared concept (Chapter 14.4), where the T_2 contrast for the quantitative calculation is determined by the attenuation that occurs only in the leading contrast segment that does not have any gradient pulse. One version of the magnetization-prepared T_2 sequence using the spin-echo sequence has been shown previously in Figure 14.7. One can also incorporate the Carr–Purcell–Meiboom–Gill (CPMG) sequence (Figure 7.7) for 2D imaging using the concept of magnetization preparation, as shown in Figure 15.11b.

15.2.4 $T_{1\rho}$ Relaxation Time

As shown in Figure 7.9, $T_{1\rho}$ relaxation occurs when a $\boldsymbol{B}_1$ field is applied to the transverse plane immediately after the magnetization is tipped into the transverse plane in such a direction that locks the magnetization to the direction of the $\boldsymbol{B}_1$ field. Since the B_1 field is the only field in the rotating frame, the magnetization is said to be spin-locked in the rotating frame. However, the magnitude of the magnetization was established under the external polarizing field $\boldsymbol{B}_0$, which is usually far bigger than $\boldsymbol{B}_1$. Hence, the magnetization will reduce in magnitude in the transverse plane without the presence of dephasing. This reduction of $\boldsymbol{M}$ can be measured with the use of

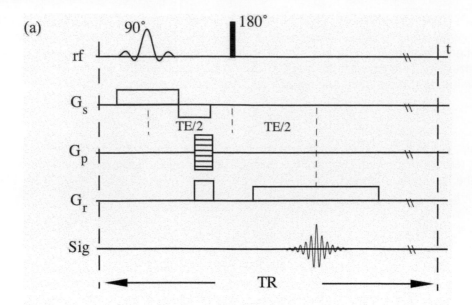

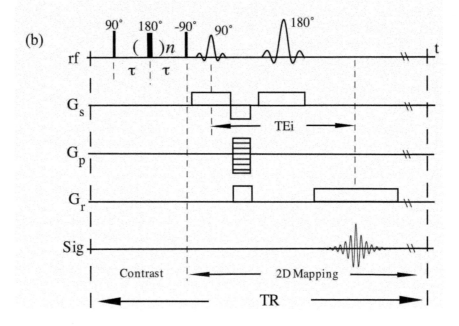

Figure 15.11 (a) A 2D imaging sequence that can quantify T_2 by repeating it at a fixed TR but different TEs. (b) A 2D T_2 imaging sequence that uses the magnetization-prepared concept, which has a CPMG sequence before the onset of 2D imaging. A quantitative T_2 experiment can be carried out by repeating the sequence at different n (which varies the number of 180° pulses), while τ, TEi, and TR are all fixed.

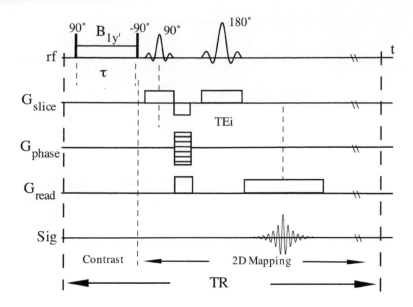

Figure 15.12   A 2D $T_{1\rho}$ imaging sequence that uses the magnetization-prepared concept, which has a spin-lock sequence before the onset of 2D imaging. A quantitative $T_{1\rho}$ experiment can be carried out by repeating the sequence at different spin-lock frequencies, while TEi and TR are both fixed.

a pulse sequence such as in Figure 15.12, which is also based on the concept of magnetization preparation. Commonly, the sequence is repeated several times, each with a different duration τ (or less commonly, different magnitude) of the spin-lock field, so that the reduction of magnetization under $T_{1\rho}$ can be measured experimentally.

Equation (7.6) implies that the normalized contrast factor due to $T_{1\rho}$ is

$$E_c(T_{1\rho}) = \exp(-\tau / T_{1\rho}). \tag{15.9}$$

The usefulness of $T_{1\rho}$ in scientific research lies in its sensitivity to molecular motion, which is in a frequency range between T_1 and T_2 (cf. Chapter 3.7). $T_{1\rho}$ senses slow (but not static) motional interactions between the confined water molecules and the macromolecules.

A unique feature (or complication) of $T_{1\rho}$ relaxation is its dependence on the strength of the spin-lock field, a phenomenon termed as $T_{1\rho}$ dispersion and shown in Figure 15.13 [25]. When the power of the spin-lock field is set to zero, $T_{1\rho}$ relaxation is identical to T_2 relaxation, which in connective tissues has all of the features of T_2 anisotropy such as the magic angle effect (see the similarity of the $T_{1\rho}$ (0 Hz) profile at 0° in Figure 15.13a and the T_2 profile at 3° in Figure 15.9d). With an increase of the spin-lock strength, $T_{1\rho}$ becomes less anisotropic and has higher values. With a sufficiently high-spin-lock field (e.g., over 2000 Hz), the influence of the dipolar interaction to spin relaxation is sufficiently minimized to yield the measurement of an isotropic $T_{1\rho}$ (see the similarity between two $T_{1\rho}$ profiles at 2000 Hz in Figures 15.13a and 15.13b, and compare them with the T_2 profile at 57° in Figure 15.9d). Note that in most clinical MRI scanners, this "sufficient spin-lock power" condition is never met, since most clinical MRI scanners can only have the highest spin-lock field less than 800 Hz (for safety reasons). The $T_{1\rho}$ value and profiles from the clinical MRI scanners are therefore not isotropic and still subject to the influence of the dipolar interaction (implying a smaller magic angle effect [26]).

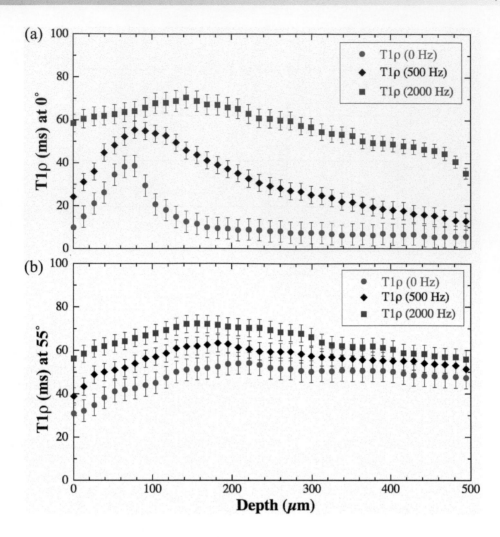

Figure 15.13 $T_{1\rho}$ dispersion in articular cartilage – where the $T_{1\rho}$ values depend upon the power of the spin-lock frequencies. Source: Wang and Xia [25].

15.3 QUANTITATIVE IMAGING OF CHEMICAL SHIFT δ

Chemical shift is the most important parameter in NMR to identify the structure of molecules. For ^{1}H NMR, the frequency reference is commonly TMS, $Si(CH_3)_4$, and the proton in pure water (i.e., no solvent) has a chemical shift of about 4.7 ppm [27]. In proton MRI when water is the dominant nucleus in the sample, this proton chemical shift at 4.7 ppm is simply ignored because it represents zero displacement of the image object if the resonant frequency is set exactly at the water frequency. If a sample contains more than one chemical species, the situation will be different.

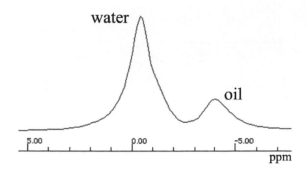

Figure 15.14 A 1D NMR spectrum of a fresh acorn, which has both water and oil components (~3 ppm apart in chemical shifts).

When the chemical environment is considered in MRI experiments, the relationship between the resonant frequency and the spatial position becomes

$$\omega = \gamma(B_0 + \boldsymbol{G} \cdot \boldsymbol{r} + \delta_i B_0 \times 10^{-6})\boldsymbol{k}, \qquad (15.10)$$

where δ_i represent different chemical shifts. Without the middle term, Eq. (15.10) is the precise equation in NMR spectroscopy [i.e., Eq. (4.7)] that is used to quantify the chemical shift information. Without the last term, Eq. (15.10) becomes the standard relationship when an MRI experiment is described [i.e., Eq. (11.3)]. When all three terms are considered and when there are two different chemical shifts ($i = 2$), a usual imaging sequence will reconstruct two position-shifted objects in a single image simultaneously (provided the spectrometer has high enough resolution and the difference in the two chemical shifts is sufficiently large), resulting in a super-position of these two images. A well-known example in chemical shift imaging is the peak from fat, whose chemical shift is about 3 ppm upfield from that of water, as shown in Figure 15.14.

Since a spatial distribution of the chemical shift information in a specimen contains extremely valuable information about the specimen [28], a number of MRI techniques have been developed in chemical shift imaging. The most comprehensive approach is the chemical shift imaging (CSI) technique [29, 30], where the frequency spread due to the chemical shifts becomes an extra dimension in the imaging experiment. Hence the digital array in CSI is in principle four dimensional: three spatial dimensions (x, y, z) and one chemical dimension (δ_i), with the normalized contrast factor as

$$E_c(\boldsymbol{r}, f) = \int g(\boldsymbol{r}, f) \exp(i2\pi f t) df, \qquad (15.11)$$

where $g(\boldsymbol{r}, f)$ is the spectral profile function at $\boldsymbol{r}$. The exact function of $g(\boldsymbol{r}, f)$ can only be written down in a few cases. For example, the spectral profile from a sample containing simple liquids bears a Lorentzian line shape. In most cases, a Fourier transform is used to find out the line shape of $g(\boldsymbol{r}, f)$, as it is done in NMR spectroscopy.

This type of 4D imaging experiment is of course time-consuming, in particular since one likely needs high resolution (i.e., high number of data points) in the chemical shift dimension. In practice, one often reduces the digital array size and/or numbers of dimensions so that the experiment can be completed within a reasonable amount of experimental time. There are two extremes in the data size/dimension reduction in CSI experiments.

One extreme in CSI is to reduce the spatial dimension to a single voxel (the region of interest, ROI) but to obtain the full chemical shift information from the selected volume (Figure 15.15),

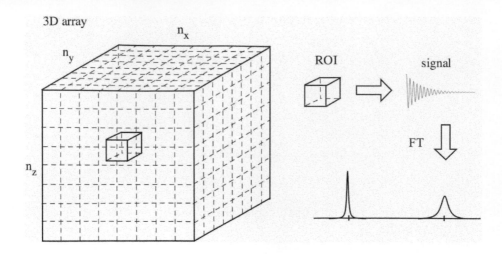

Figure 15.15   A schematic diagram for chemical shift spectroscopic imaging. By selecting a small voxel from a 3D object, one can get the full chemical information from that selected volume. It is a very useful experimental approach in clinical diagnostics.

which is also termed as localized or volume-selected spectroscopy. This approach, which is no longer a true imaging experiment, is useful in medical diagnosis and *in vivo* biological studies, such as the study of proton and phosphorus metabolism in brain and muscles [31, 32]. See Chapter 12.9 for a note about this type of volume-selective sequences.

Examples of this volume-selected spectroscopic imaging include ISIS [33], DRESS [34], and SPACE [35]. A notable success in the application of localized spectroscopy is the study of phosphorus metabolism. One can monitor quantitatively the phosphates of energy metabolism: ATP, phosphocreatine, sugar phosphate, inorganic phosphate, and so forth. There is rich information in a ^{31}P spectrum, such as the inorganic phosphate shift that is determined by the local pH, in which case the splitting of the lines is influenced by the intra- and extracellular pH. Hence changes in health and disease, in exercise and at rest, can be followed biochemically *in vivo*. Another area of interest and increasing study in recent years is the function and biochemistry of brain. Using NMR imaging based on proton and phosphorus localized spectroscopy, brain development and function can be imaged, analyzed, and interpreted structurally and biochemically.

The other extreme in CSI is to reduce the observation window on the chemical shift spectrum to a single shift (the shift of interest, SOI) while the spatial information of the image is retained. CHEmical Shift Selective (CHESS) imaging [36–38] is an example of such an approach that produces an entire image of one pre-selected chemical shift component in the sample, using either the pre-saturation technique to null one unwanted signal (Figure 15.16a), or the pre-selection technique to select one desired signal (Figure 15.16b). In both sequences, the first pulse is a soft sinc pulse *without* the simultaneous application of any gradient, which excites a narrow range of frequencies (cf. Figure 12.3a). Hence this first soft pulse is not a slice selection pulse but a chemical shift selection pulse for the entire sample. In Figure 15.16a, the frequency of the first soft 90° pulse is set at the undesired spins in the specimen. The second 90° pulse, this time together with a gradient pulse, does the usual slice selection, which turns all spins within that slice by 90°. Since the undesired spins have been tipped 90° by the first soft pulse already, the undesired spins will be turned to the z axis upon the second 90° pulse.

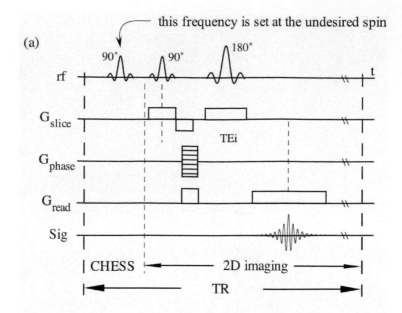

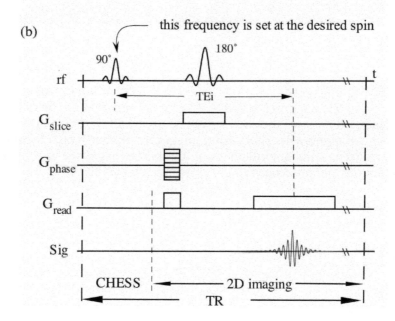

Figure 15.16 Two CHEmical Shift Selective (CHESS) imaging sequences that use the magnetization-prepared concept, which has a soft rf pulse (i.e., a narrow-frequency band) before the onset of 2D imaging. (a) The soft selected pulse is set at the frequency of the undesired spin, which can null its signal. (b) The soft selected pulse is set at the frequency of the desired spin, which can prepare it for signal acquisition.

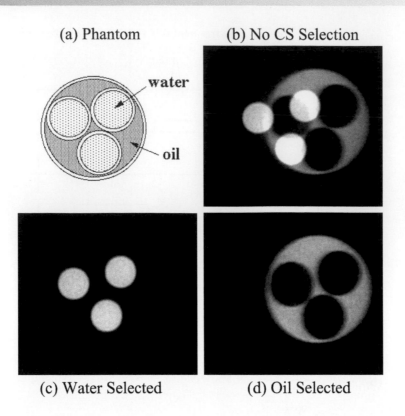

(a) Phantom   (b) No CS Selection

(c) Water Selected   (d) Oil Selected

Figure 15.17 A CHESS imaging example. (a) A phantom of oil and water in tubes, consisting of three small glass capillaries in a 4-mm NMR tube. The small capillaries are filled with water while the 4-mm tube is filled with ordinary oil. The chemical shifts of oil and water are about 3 ppm apart. (b) Conventional spin-echo imaging where both components are imaged simultaneously. (c) Water selected. (d) Oil selected. The resonant frequency of the spectrometer was unchanged during the three experiments. Source: Reproduced with permission from Xia [68].

The acquired signal therefore comes from all components in the slice except the undesired spins. In Figure 15.16b, the frequency of the first soft 90° pulse is set at the desired spins in the specimen, which turns the desired spins to the transverse plane (while all other spins are unaffected by this soft 90° pulse and remain in parallel with the z axis). The 180° slice selection pulse forms a slice of this desired spin with the use of a gradient, which can be mapped out by the usual k-space imaging.

CHESS is much more straightforward in implementing and less time-consuming than a full-featured CSI, if one knows, *a priori*, what chemical shift to monitor [38]. Figure 15.17 shows an example of a tube phantom of oil and water, where the resonance frequency of the spectrometer was unchanged during the three experiments. Hence the oil image is displaced along the read direction (horizontal) relative to the water image (cf. Figure 15.14). One can also center the spectrometer frequency to whatever chemical shift component is selected. Figure 15.18 shows two examples of CHESS applications, where each image was acquired on the selected resonance frequency so that the object in the image has zero physical displacement, which enables the identification of the oil-rich (3 ppm away from water) and sugar-rich (1 ppm away from water) anatomic structures in the overall specimen.

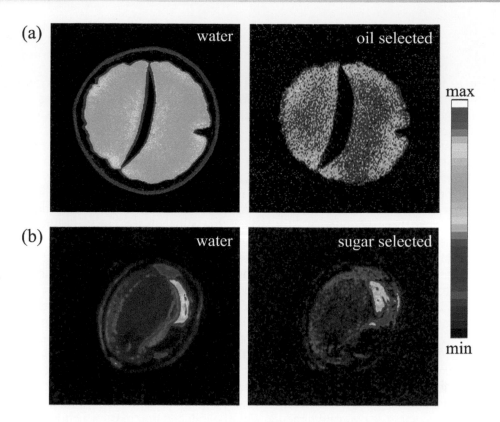

Figure 15.18   Two CHESS imaging examples, using (a) a fresh acorn and (b) a sweet pea as the specimens. The 1D NMR spectrum of the acorn has been shown in Figure 15.14.

15.4   SECONDARY IMAGE CONTRASTS IN MRI

The contrast factors discussed in Section 15.3 can be considered as "primary contrasts," since each of them can be included directly into the Bloch equation. For example, the self-diffusion term can be included as the Bloch–Torrey equation [39]. In addition to primary contrasts, image intensity can also be manipulated by numerous other functions or mechanisms, which act via the effects of the primary contrasts, hence are termed as "secondary contrasts" in this book (Figure 15.1). Spatially resolved images of these functions or mechanisms can follow in a similar manner as we discussed for the primary contrasts. A general trend is that any quantitative determination of a secondary contrast requires much longer experimental time, since each slice in the secondary contrast needs to be a quantitative image of a primary contrast. A few examples are briefly discussed in this section, but there are likely numerous other examples in which some physical quantities of a specimen can be measured non-invasively and non-destructively.

15.4.1   Susceptibility

While the external origin of the magnetization $\boldsymbol{M}$ is the applied magnetic field $\boldsymbol{B}_0$, the intrinsic origin of $\boldsymbol{M}$ is due to the magnetic susceptibility, χ_m, as

$$M = \frac{\chi_m}{\mu_0(1+\chi_m)} B, \tag{15.12}$$

where μ_0 is the absolute permeability ($4\pi \times 10^{-7} Hm^{-1}$). χ_m is a dimensionless number, positive for paramagnetic materials and negative for diamagnetic materials (such as biological and organic materials), due to the atomic structures of the material that can be explained using quantum mechanics.

Since samples in MRI are heterogeneous, χ_m could be different in different microstructures inside a specimen. At the boundaries of the local internal microstructures between themselves or with air, a difference in two neighboring χ_m, even just a few ppm, will induce a local magnetic field gradient across the boundary. These induced internal magnetic field gradients will degrade the homogeneity of the applied field gradients [40, 41], not only at the boundaries but also in the space near these boundaries. This means that although the susceptibility differences are *local* at the boundaries, their effects extend into the nearby media. As a consequence, additional T_2^* relaxation would be induced in these regions, which results in image distortions and artifacts. Since the magnetic susceptibility is field-dependent, the higher the field, the bigger the effect. Clearly the image distortion is the worst at microscopic resolution.

Figure 15.19 is an example where severe image distortions can be seen at the interface between air and water, since air is paramagnetic (the mass susceptibility is about $+3.6 \times 10^{-7}$ m³/kg, due to the presence of oxygen) and water is diamagnetic (the mass susceptibility is about -9.0×10^{-6} m³/kg). This difference in magnetic susceptibility creates a local field gradient across the boundary at the water surface (the meniscus), which degrades the homogeneity of the applied field gradient in imaging, degrading the water deep inside the tube. The artifact can appear differently, depending upon the direction of the imaging gradient (i.e., whether the imaging gradient is in the same or opposite direction with the induced gradient), which is influenced by the curvature of the concave meniscus of the liquid surface. This artifactual distortion is also responsible for the nearly black color (i.e., no signal) in medical MRI of lung, where the large amount of air in the lung makes the lung appear with a low intensity (black color) in any spin-echo-based MRI. The use of a noble gas to fill the voids could facilitate the study of these types of air-rich structures [42], which will be discussed more extensively in Chapter 16.5.

15.4.2 BOLD

BOLD stands for Blood Oxygenation Level Dependent. The mechanism of BOLD contrast is the following. By giving up oxygen, the hemoglobin iron in blood undergoes a change of its spin state from diamagnetic low spin (S = 0) in the oxygenated state to paramagnetic high spin (S = 2) in the deoxygenated state. An increase in paramagnetic iron content in blood increases the difference in bulk magnetic susceptibilities among the blood and the surrounding tissue. It also changes the T_1 of the sample significantly. Hence, the coherence in the NMR signal decays quickly.

Early studies [43] show that BOLD contrast in proton NMR depends strongly on the T_2 relaxation of the blood water, and the content of paramagnetic deoxyhemoglobin in red cells was responsible for the T_2 change. In MRI, however, blood water in microvessels is often difficult to image in soft tissue such as brain. To enhance the effect of the BOLD contrast, the effect of T_2^* was explored by using gradient-echo imaging at high magnetic fields (up to 7 Tesla) [44–46]. These studies show that in addition to the contributions from the blood water, the effect of BOLD contrast can be enhanced by including contributions from the perfused tissue water *around* the blood vessels in gradient-echo experiments. In other words, while a spin-echo signal is attenuated only by the effect of T_2, a gradient echo is additionally attenuated by dephasing due to

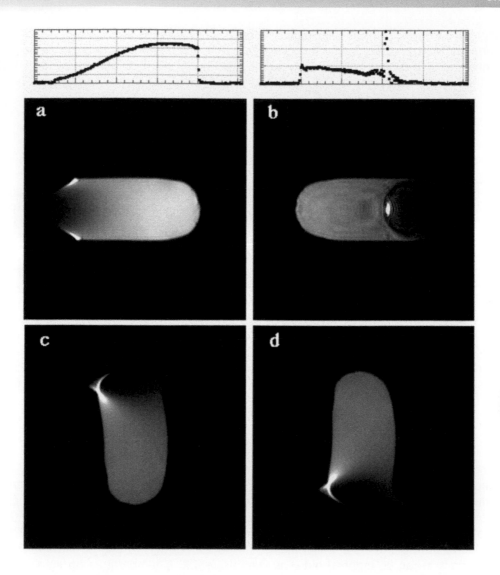

Figure 15.19 Effect of susceptibility artifact. The sample is water in 4-mm NMR tube. The length of the water is short compared with the height of the rf coil (which ensures the uniformity of the rf field over the sample space). The four images were acquired with the same sample location, same pulse sequence, and same parameters but different gradient directions, using a vertical-bore superconducting magnet (where the z axis is in parallel with the tube length). The slice is selected vertically across the middle of the NMR tube. (a) The phase gradient is x, pointing vertically upwards in the image; and the read gradient is z, pointing to the right. (b) The same as (a) except the read gradient points to the left. (c) The phase gradient is z, pointing vertically upwards; and the read gradient is x, pointing to the right. (d) The same as (c) except the phase gradient points downwards. The two profiles above (a) and (b) are drawn through the centers of (a) and (b), respectively. The artifacts are due to the curved meniscus of water surface in the tube: the local gradients due to the susceptibility difference at the water/air interface distort the linearity of the imaging gradients. Source: Reproduced with permission from Xia [68].

microscopic static field inhomogeneities. Because of the extensive network of blood microcapillaries in the brain tissue, the imaging signal from the brain reflects blood oxygenation via the enhanced BOLD contrast. BOLD is the physical mechanism for functional MRI (fMRI), which will be further discussed in Chapter 16.4.

15.4.3   Temperature

The early development of imaging temperature distributions using MRI was motivated by the clinical application in hyperthermia cancer therapy, during which the cancerous tissue is heated to just over 42°C while keeping the surrounding healthy tissue to a temperature less than 40°C [47]. This temperature mapping technique turned MRI into a gigantic thermometer, expensive but totally non-invasive and with 3D spatial resolution. Further studies show that such an MRI monitoring technique can guide thermal surgery where thermal profiles of the heated tissue can be imaged before it cools using a fast imaging sequence [48]. In addition to clinical applications, temperature mapping studies can also be used in non-biological materials such as solids and polymers. Besides mapping temperature, other thermal phenomena can also be mapped via their temperature-dependence. For example, the mapping of Rayleigh convection has been demonstrated using a flow-sensitive imaging sequence [49].

The most versatile temperature-sensitive MRI scheme utilizes self-diffusion of molecules, which measures random thermal motion. The theoretical relationship of diffusion vs temperature can be derived from the Stokes–Einstein equation that relates viscosity to translational self-diffusion. Assuming the activation energy E_a is a constant, the Stokes–Einstein equation can be differentiated [50, 51] to give

$$\frac{D_T - D_0}{D_0} = (\frac{E_a}{k_B T_0})(\frac{T - T_0}{T_0}), \tag{15.13}$$

where D_T is the measured self-diffusion coefficient at temperature T (in Kelvin), D_0 is a reference diffusion coefficient at the reference temperature T_0, E_a is the activation energy that measures the amount of energy required to break hydrogen bonds, and k_B is the Boltzmann constant. Note that $(k_B T_0)$ is about 4.41×10^{-21} Joule at room temperature, and $[E_a/(k_B T_0)]$ is a factor that determines the sensitivity of the temperature measurement using self-diffusion. For a series of multiple-imaging experiments, the factor $[E_a/(k_B T_0)]$ can be normalized so that the temperature can be mapped out. For example, a 0.2 °C temperature difference can be determined over 0.3 cm^3 regions using polyacrylamide gel and the two-image approach [52]. A practical estimation for the temperature-dependence of self-diffusion in water can utilize the data of Mills [53], as in Figure 15.20. These data show that every 1 °C temperature change would result in approximately a 2.4% change in the diffusion coefficient within the room temperature range, almost linearly.

In addition to using self-diffusion to map out the temperature, relaxation times are also known to be temperature-dependent [54–56]. Since nuclear spin relaxation is caused by the distribution of local interactions experienced by the nuclear spins, both T_1 and T_2 relaxation times could exhibit temperature-dependence. For pure, degassed water, the T_1 value changes at a rate of about 100 ms/°C between 0 and 100 °C [50], and the T_2 value changes at a rate about 74 ms/°C in the range of 30–50 °C [57].

Due to the temperature-dependence of the nuclear screening constant, chemical shift can also exhibit temperature-dependence [58]. Early experiments demonstrated the temperature-dependence of chemical shift for a variety of well-characterized compounds, such as water, methanol, and ethylene glycol [57, 59]. For example, the OH peak of ethylene glycol shifts toward the methyl peak at a rate about 2 Hz/°C between 27 °C and 77 °C at 300-MHz proton frequency [57]. This temperature-dependent frequency shift can be detected either directly by

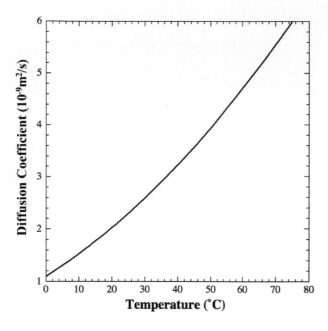

Figure 15.20 A plot of the self-diffusion of water as a function of temperature. Source: Based on Mills [53].

a chemical shift selected imaging sequence [59] or through the additional phase shift in a gradient-echo sequence [60].

15.4.4 The Use of Contrast Agents

Image contrast in MRI can also be established by injecting contrast agents into the veins of humans and animals or soaking tissues in a bath of contrast agent. A family of common contrast agents in MRI is based on gadolinium ions. Since Gd^{3+} is a paramagnetic ion that can shorten the T_1 relaxation significantly, and a low quantity of gadolinium diethylene triamine penta-acetic acid $(Gd(DTPA)^{2-})$ is usually harmless to healthy humans, it is possible to map the local concentration of Gd ions, based on the Donnan equilibrium theory between the infused Gd ions and the native distribution of local charges in tissue [61, 62],

$$[Gd]_t = \frac{1}{R}(\frac{1}{T_{1Gd}} - \frac{1}{T_1}), \tag{15.14a}$$

$$FCD = [Na^+]_b (\sqrt{\frac{[Gd]_t}{[Gd]_b}} - \sqrt{\frac{[Gd]_b}{[Gd]_t}}), \tag{15.14b}$$

$$[GAG]_t = FCD \times 502.5 / (-2), \tag{15.14c}$$

where T_1 and T_{1Gd} are the T_1 values in the tissue as measured before and after the Gd administration, respectively; $[Gd]_t$ and $[Gd]_b$ are the Gd concentrations in the tissue and the bath

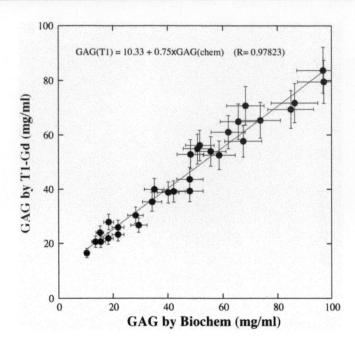

$$GAG(T1) = 10.33 + 0.75 \times GAG(chem) \quad (R = 0.97823)$$

Figure 15.21   Quantitative correlation of the GAG concentrations by an MRI method and by a biochemical assay. The straight line is the linear fit of the data. Source: Zheng and Xia [63].

(soaking solution), respectively; R is the relaxivity of the Gd ions in saline; $[Na^+]_b$ is the sodium concentration in the bath; FCD is the fixed charge density; and $[GAG]_t$ is the glycosaminoglycans (GAGs) concentration in tissue. The two constants in Eq. (15.14c), "-2" and "502.5," come from the estimates that there are 2 moles of negative charges per mole of disaccharide with a molecular weight of 502.5 g/mole. By monitoring the changes in T_1 relaxation time, one can determine the local concentration of macromolecules, such as the negatively charged proteoglycan molecules in solutions (Figure 15.21) and in tissues [63]. There are many uses of the contrast agents in MRI, in both clinical diagnostics and laboratory research.

15.5   POTENTIAL ISSUES AND PRACTICAL STRATEGIES IN QUANTITATIVE MRI

Since all methods to detect image contrast require additional post-acquisition calculation, the algorithms and approaches in these calculations are critically important for extracting useful information. There are essentially two general approaches in contrast extraction (Figure 14.5): having a complex form of contrast (as in velocity imaging) that requires an additional Fourier transform and having attenuation-based contrast (as in diffusion and relaxation imaging) that requires a least-squares fit.

15.5.1   How Quantitative Is This Quantitative Result?

The quantitative calculation of self-diffusion and relaxation times depends upon the signal attenuation, which is commonly obtained from the slope of a least-squares fit through the data. Although one would like to see that only the desired contrast causes the attenuation in a series of images, the reality is that many other factors could also cause the signal attenuation. A single algorithm cannot determine which portion of the attenuation is from which source of contrast. In general, unexpected T_2 weighting in MRI is common for all images when a finite TE is used in signal acquisition. In addition, unexpected T_1 weighting can occur when a short repetition time TR is used in imaging, which results in insufficiently restored magnetization and hence low image intensity. These differences in magnetization recovery would cause more harm than the visualization error when different parts of the specimen have different T_1 and T_2 relaxation times, hence different parts of the tissue could be weighted differently.

Figure 15.22 compares the two T_2 relaxation profiles in a piece of articular cartilage. One experiment was done by a magnetization-prepared spin-echo sequence (Figure 15.11b), while the other was done by a typical gradient-echo-based quantitative sequence (Figure 12.11b), at the same 26-μm/pixel resolution [64]. Although the profiles looked similar, quantitative T_2 values by the gradient-echo sequence were about 1/3 of that by the magnetization-prepared spin-echo sequence. Surely, the gradient-echo sequence of the T_2 profiles was from the images that were further attenuated by the pseudo-PGSE effect (cf. Figure 14.6). The same study also showed the laminar appearance of articular cartilage can have opposite intensity patterns in the deep part of the tissue, depending on whether the image is T_1 weighted or T_2 weighted. These types of inaccuracy and variation, which are field-dependent, sequence-dependent, and resolution-dependent, contribute to the difficulty in comparing results from different studies and different laboratories in the literature, since the sequences and parameters were unlikely to be the same.

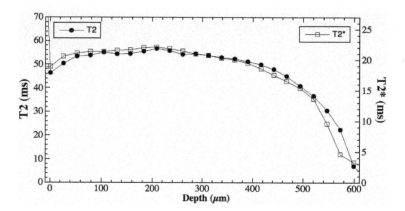

Figure 15.22 T_2 and T_2* profiles in articular cartilage by the magnetization-prepared MRI sequence and the gradient-echo MRI sequence. Source: Modified from Xia and Zheng [64].

15.5.2   Issues and Strategies in Discrete Fourier Transform

The velocity signal in q-space formalism (Section 15.1) is inherently complex, which requires saving the real and imaginary FID images separately for discrete Fourier transform (Figure 15.6). The transform utilized the discrete form of the normalized contrast factor [5] for velocity and diffusion in the PGSE sequence, where Eq. (15.1) can be written as

$$E_c(n, r, \Delta) = \exp(i\gamma\delta v \Delta (g_m/n_D)n) \exp(-\gamma^2\delta^2 (g_m/n_D)^2 n^2 D\Delta), \tag{15.15}$$

where g_m is the maximum gradient used in the velocity and diffusion contrast imaging, and n_D is the number of steps in the contrast direction (n in Figure 14.5). Hence, $(g_m/n_D)n$ marks the nth-step on the contrast direction.

The quantitative calculation of velocity in q-space MRI is inherently more robust than the calculation of diffusion and relaxation in MRI because of its use of phase shift rather than intensity attenuation. There are nevertheless a number of issues that are unique to the quantitative calculation in q-space MRI, which are briefly discussed here [6, 65].

- **Gradient-dependent phase shifts:** Any imperfection in the PGSE gradient [g in Eq. (15.1)] could cause additional phase shift in the FID signal. An analysis has shown that this phase shift does not present a major problem to any attenuation-based quantitative experiment, provided that it is consistently reproducible. For velocity imaging where the phase shift is quantified as arising from fluid flow, and when the magnitude of the PGSE gradient is different in each q-space slice, one must devise a means of distinguishing flow-related phase shifts from spectrometer artifacts. Some additional measurement of phase shift on stationary fluid can be implemented; for example, to compensate the phase shift due to the fluid flow.
- **The influence of the slice selection gradient:** In Chapter 14.4, the magnetization-prepared sequences are developed to provide more quantitative and resolution-independent measurement in MRI. One of the potential artifacts is the influence of the slice selection gradient if the 180° rf pulse is incorporated (cf. Figure 12.5), which will also generate a fixed phase shift due to the fluid flow, common to each q-space slice. It is clear that this effect depends on the distance moved by the spins during the slice excitation period, which would be location-dependent; for example, a Newtonian fluid flowing through a tube has a parabolic velocity profile (Poiseuille flow). Special compensation for this issue is needed in the quantitative calculation of the velocity by MRI.
- **Digital broadening and baseline artifacts in the Fourier domain:** Prior to the Fourier transformation in q space, it is common to zero-fill in q space from n_D+1 to N, where N is an integer much larger than n_D. This is equivalent to data interpolation in the conjugate velocity space and helps to reduce the effect of quantization errors in the computation of the dynamic displacement profile, a process identical to the zero-filling in NMR spectroscopy (cf. Chapter 6.8). The zero-filling also helps to satisfy the need to have 2^n data points when performing a fast Fourier transform. Since zero-filling corresponds to the multiplication of the q-space signal by a step function (Figure 6.11), a sinc convolution is necessarily introduced in the velocity space, and artifactual broadening will result. The effect is not important if the dynamic displacement profile is sufficiently broader than the sinc function, a condition equivalent to requiring that the signal amplitude has significantly decayed at the maximum gradient value g_m corresponding to the last data point n_D before the onset of zero-filling [5]. Compare the three sets of data in Figure 6.11 for the effect of zero-filling and truncation.
- **Systematic errors in diffusion measured in the presence of a velocity shear:** In the pixel-by-pixel calculation and analysis of dynamic displacement profile, it is assumed that the broadening arises from Brownian motion (cf. Figure 15.4). In many experiments, however, a local velocity shear exists at a single pixel, which could introduce additional "broadening" to the

dynamic displacement profile [8]. This type of velocity shear inside individual pixels would lead to a distribution in phases within the pixel that will emulate the effect of diffusion in the direction of the velocity and so lead to a systematic error in the estimation of D. It can be shown that artifactual enhancements to the apparent self-diffusion coefficient are not important when the velocity shear between two adjacent pixels is much smaller than $(D/\Delta)^{1/2}$. For more accurate determination of self-diffusion in the presence of a velocity shear, a double PGSE imaging sequence has been used in polymer tube flow where the phase shifts due to velocity are cancelled [8].

15.5.3　Issues and Strategies in Least-squares Fitting

Most of the multiple-image approaches (Figure 14.5) to extract diffusion and relaxation use the least-squares fitting method, which minimizes the square of the error between the original data and the values predicted by an equation. This method is well understood and requires little computation power. The major weakness of the method is its sensitivity to noise and outliers in the data. Consider a quantitative T_2 experiment as an example to illustrate potential issues and strategies in the least-squares method (the discussion would be equally applicable to other attenuation-based quantitative calculations including the calculation of diffusion). We first re-write Eq. (15.8) as

$$\text{Sig}(t_i) = \text{Sig}(t_0)\exp(-t_i/T_2) \tag{15.16}$$

where t_i are several time delays associated with several specific T_2 weightings that are selected for the imaging experiment, $\text{Sig}(t_i)$ is the measured image intensity at a particular t_i, and $\text{Sig}(t_0)$ is the image intensity at the shortest T_2 weighting. In theory, one wants the shortest t_i to be at zero, in order to catch the signal as soon as possible; in actual experiments, the shortest t_i is likely on the order of μs in spectroscopy and ms in imaging, given the needs of the instrument operation.

Visualize the data using a linear plot: The best way to judge if the data are good or not is to visualize it. Since Eq. (15.16) is an exponential decay function, we could easily plot it as it is. However, the human eye is not sensitive to a small deviation away from exponential decay, which may be caused by multi-component relaxation (Section 16.2) or by non-physical behavior of the data due to instrumental artifacts. A much better approach is to take the natural logarithm for both sides of Eq. (15.16), which turns the exponential equation into a linear equation,

$$\ln\big(\text{Sig}(t_i)/\text{Sig}(t_0)\big) = -t_i/T_2. \tag{15.17}$$

Alternative to using Eq. (15.17) to convert your data into a linear form, (a) you could plot the exponential decay data in Eq. (15.16) on semi-log graph paper, if you can still find it; or (b) you can use software if your graphing software can plot the amplitude of the data in a natural logarithm scale. Figure 15.23 shows the exponential and linear formats of a data generated by the sum of two exponential decays, $5\times\exp(-t/10)+5\times\exp(-t/60)$, that is, one T_2 at 10 ms and one T_2 at 60 ms, both having the same maximum amplitude. Although it is visually challenging to recognize the double exponentials on the exponential plot by Eq. (15.16) in Figure 15.23a, the linear plot by Eq. (15.17) in Figure 15.23b shows clearly the data can be fitted by two straight lines with two different slopes, which demonstrates the double exponential decays (i.e., two T_2s) in this set of data.

Recognize the noise floor in the data: Practical experimental data always contain some level of noise, which presents itself in the magnitude images as positive and random fluctuations, a "floor" in the data. (The noise floor would become all positive fluctuations once you take the modulus in the data.) For an exponential decay due to diffusion and relaxation, one can reach the noise floor if the contrast weighting is excessively long with respect to the relaxation or diffusion values. Figure 15.24 shows the same data as in Figure 15.23, now with added random noise at 20% of the maximum amplitude. The noise floor could influence the least-squares fitting by skewing the results of the regression. Comparing the data between Figures 15.23 and 15.24, it is clear that the data with values below 20% of the peak amplitude or obtained beyond 100 ms in time are dominated by noise. If all the data were used for the fit, the slope of the

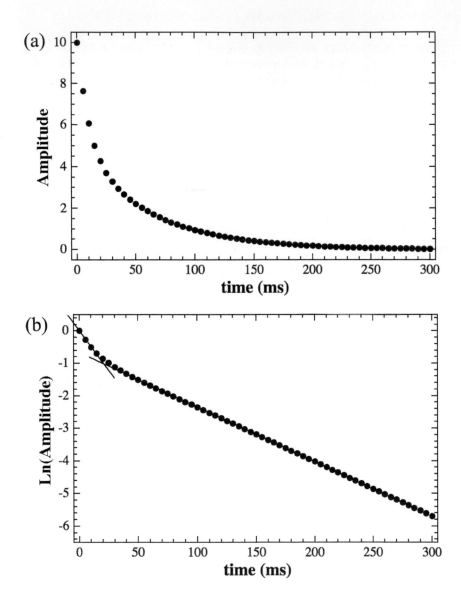

Figure 15.23 A simulated noiseless double-exponential decay, shown in a linear amplitude plot (a) and a natural logarithm plot (b).

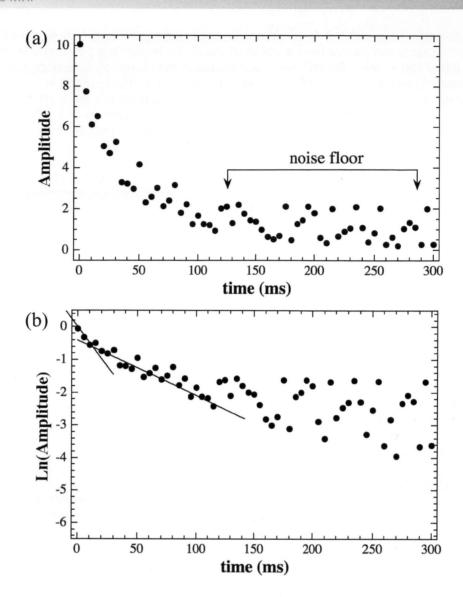

Figure 15.24 The same simulated double-exponential decay as in Figure 15.23, now with 20% random noise added, shown in a linear amplitude plot (a) and a natural logarithm plot (b).

fitted line would contain a large error. In this situation, one must judge where the noise becomes dominant and truncate the data (i.e., discard the rest of the data). The two fitted lines in the linear plot in Figure 15.24b have the same slopes as in Figure 15.23b. Without recognizing the noise floor in the data, the slope of the second fitted line in Figure 15.24b, hence a T_2 value, would be very different from the true value.

Optimize (i.e., compromise on) the T_2 weightings: In any real experiment, one only has a finite amount of experimental time. For the multi-slice contrast extraction based on Figure 14.5, one can only select a handful of T_2 weightings for any single experiment. A tricky situation may arise when the same sample contains some short T_2s in one region (e.g., 10 ms) and some long T_2s in another region (e.g., 60 ms). One such sample is the relaxation anisotropy plots in articular cartilage, as shown in Figure 15.9d. When the specimen is oriented at 3° in the magnetic field, T_2 is about 60 ms for tissue at about 100 μm below the surface but about 10 ms for tissue from 300 to 600 μm [19]. How do we select just several T_2 weightings (e.g., 3–5) to adequately sample both short-T_2 and long-T_2 regions in one quantitative imaging experiment? It is very likely that the short-T_2 region has already reached the noise floor when the long-T_2 region is just becoming sufficiently weighted or attenuated. The only sensible approach, instead of a large number of T_2 weightings, is to select a few short weightings to take care of the short component and a few more long weightings to take care of the long components, and check the influence of the noise floor (see above) in the quantification.

Use the amplitude weighting, and add an anchor point at the far end of the signal decay: There are two practical strategies when experimental noise and limited contrast slices pose experimental challenges; they are (a) weight the amplitudes of the data points, and (b) add fictional data points at the far end of the weightings.

The first strategy recognizes the fact that the initial points (either low PGSE gradient values in diffusion measurement or short TE values in relaxation measurement) are intrinsically more reliable, due to their lower % influence from the experimental noise. In any actual imaging calculation, the number of data points are large (e.g., a 128×128 image has 16,384 data points); it is therefore not possible to visually inspect the signal decay one by one at *all* pixel locations as in Figure 15.24. An automatic way in the image contrast calculation is to incorporate a data weighting step in the least-squares fitting process, which gives more weight to higher amplitude data points. The implementation of this strategy in the least-squares fitting routine can be found in *Numerical Recipes* [66] and in practical research [67].

The second strategy recognizes that for a sample with a finite T_2 (say, 50 ms), the signal must become zero when attenuated by the relaxation at some long time, say 500 ms. So to fit the data from a set of actual T_2 weightings (e.g., 2, 8, 16, and 40 ms), one could add one or two fictional data points around 500 ms with the amplitude value set to the background noise. These fictional values act as the anchor points to pin the "tail" in the least-squares fitting to the "floor." A comparison from two practical T_2 imaging experiments using the same cartilage specimen is shown in Figure 15.25, which essentially recognizes the experimental data to be

$$\text{Sig}(t_i) = \text{Sig}(t_0)\exp(-t_i/T_2) + \text{noise}(t_i), \tag{15.18}$$

which has the generic form that can be used in the least-squares fitting, $y = a \times \exp(-t_i/T_2) + b$.

One T_2 imaging experiment in Figure 15.25 had five actual contrast weightings (TE = 2, 4, 10, 30, 70 ms), which were analyzed with and without two artificial noise anchors extracted from the background of the actual image and placed at TE = 500 and 1000 ms. A second T_2 imaging experiment had seven contrast weightings (TE = 2, 4, 10, 30, 70, 500, 1000 ms). Since the T_2 of cartilage is within 10–100 ms, there was no signal left in the cartilage portion of the images at TE = 500 and 1000 ms. However, the actual noise from the two long-TE images was used in the analysis. As one can see, the artificially long-TE analysis and the actual long-TE analysis both yield largely consistent results, both quite different from the fitting of the 5-TE experiment without artificial anchor. These strategies have all been used extensively over the years in our experimental work.

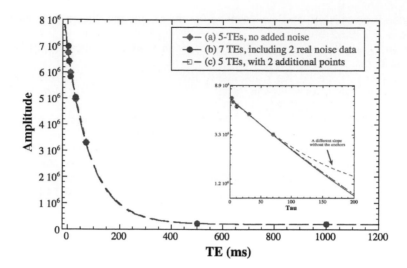

Figure 15.25 Fitting decay data with a single exponential. (a) A five-TE experimental plot (TE = 2, 4, 10, 30, 70 ms), which has a fitted T_2 of 77.81 ms. (b) A seven-TE experimental plot (TE = 2, 4, 10, 30, 70, 500, 1000 ms), which has a fitted T_2 of 90.98 ms. (c) The same five-TE data as in (a) with two noise values added in the data analysis at the far end of the exponential decay (TE = 500, 1000 ms), which has a fitted T_2 of 93.32 ms. The insert shows more clearly the difference in fitting.

References

1. Hahn EL. Spin Echoes. *Phys Rev.* 1950;80:580–94.
2. Stejskal EO, Tanner JE. Spin Diffusion Measurements: Spin Echoes in the Presence of a Time-Dependent Field Gradient. *J Chem Phys.* 1965;42(1):288–92.
3. Stejskal EO. Use of Spin Echoes in a Pulsed Magnetic-Field Gradient to Study Anisotropic Restricted Diffusion and Flow. *J Chem Phys.* 1965;43(10):3597–603.
4. Callaghan PT, Eccles CD, Xia Y. NMR Microscopy of Dynamic Displacements: k-space and q-space Imaging. *J Phys E: Sci Instrum.* 1988;21:820–2.
5. Callaghan PT, Xia Y. Velocity and Diffusion Imaging in Dynamic NMR Microscopy. *J Magn Reson.* 1991;91:326–52.
6. Xia Y. Dynamic NMR Microscopy [Ph. D. dissertation]. Massey University, New Zealand; 1992.
7. Xia Y, Jeffrey KR, Callaghan PT. Imaging Velocity Profiles: Flow Through an Abrupt Contraction and Expansion. *AIChE J.* 1992;38(9):1408–20.
8. Xia Y, Callaghan PT. Study of Shear Thinning in High Polymer Solution Using Dynamic NMR Microscopy. *Macro Mol.* 1991;24(17):4777–86.
9. Xia Y, Callaghan PT. Imaging the Velocity Profiles in Tubeless Siphon Flow by NMR Microscopy. *J Magn Reson.* 2003;164(2):365–8.
10. Jenner CF, Xia Y, Eccles CD, Callaghan PT. Circulation of Water Within Wheat Grain Revealed by Nuclear Magnetic Resonance Micro-imaging. *Nature.* 1988;336:399.
11. Xia Y, Sarafis V, Campbell EO, Callaghan PT. Non Invasive Imaging of Water Flow in Plants by NMR Microscopy. *Protoplasma.* 1993;173:170–6.
12. Xia Y, Callaghan PT. "One-Shot" Velocity Microscopy: NMR Imaging of Motion Using a Single Phase-encoding Step. *Magn Reson Med.* 1992;23(1):138–53.

13. Callaghan P. Rheo-NMR: Nuclear Magnetic Resonance and the Rheology of Complex Fluids. *Rep Prog Phys*. 1999;62:599–668.

14. Callaghan PT, Xia Y. Nuclear Magnetic Resonance Imaging and Velocimetry of Fano Flow. *J Phys: Condens Matter*. 2004;16:4177–92.

15. Köckenberger W, Pope JM, Xia Y, Jeffrey KR, Komor E, Callaghan PT. A Non-invasive Measurement of Phloem and Xylem Water Flow in Castor Bean Seedlings by Nuclear Magnetic Resonance Microimaging. *Planta*. 1997;201:53–63.

16. Williams DS, Detre JA, Leigh JS, Koretsky AP. Magnetic Resonance Imaging of Perfusion Using Spin Inversion of Arterial Water. *Proc Natl Acad Sci*. 1992;89(1):212–6.

17. Wong EC. An Introduction to ASL Labeling Techniques. *J Magn Reson Imaging*. 2014;40:1–10.

18. Carr HY, Purcell EM. Effects of Diffusion on Free Precession in Nuclear Magnetic Resonance Experiments. *Phys Rev*. 1954;94(3):630–8.

19. Xia Y. Relaxation Anisotropy in Cartilage by NMR Microscopy (μMRI) at 14 μm Resolution. *Magn Reson Med*. 1998;39(6):941–9.

20. Le Bihan D, Breton E, Lallemand D, Aubin M-L, Vignaud J, Laval-Jeantet M. Separation of Diffusion and Prefusion in Intravoxel Incoherent Motion MR Imaging. *Radiology*. 1988;168:497–505.

21. Fullerton GD, Cameron IL. Relaxation of Biological Tissues. In: Wehrli FW, Shaw D, Kneeland JB, editors. *Biomedical Magnetic Resonance Imaging – Principles, Methodology, and Applications*. New York: VCH; 1988. pp. 115–55.

22. Xia Y. Magic Angle Effect in MRI of Articular Cartilage – A Review. *Invest Radiol*. 2000;35(10):602–21.

23. Xia Y, Moody J, Burton-Wurster N, Lust G. Quantitative In Situ Correlation Between Microscopic MRI and Polarized Light Microscopy Studies of Articular Cartilage. *Osteoarthritis Cartilage*. 2001;9(5):393–406.

24. Xia Y, Moody J, Alhadlaq H. Orientational Dependence of T2 Relaxation in Articular Cartilage: A Microscopic MRI (μMRI) Study. *Magn Reson Med*. 2002;48(3):460–9.

25. Wang N, Xia Y. Depth and Orientational Dependencies of MRI T2 and T1rho Sensitivities Towards Trypsin Degradation and Gd-DTPA(2-) Presence in Articular Cartilage at Microscopic Resolution. *Magn Reson Imaging*. 2012;30(3):361–70.

26. Xia Y. MRI of Articular Cartilage at Microscopic Resolution. *Bone and Joint Res*. 2013;2(1):9–17.

27. Shaw D. *Fourier Transform NMR Spectroscopy*. Amsterdam: Elsevier; 1976.

28. Rumpel H, Pope JM. Chemical Shift Imaging in Nuclear Magnetic Resonance: A Comparison of Methods. *Concepts in Magn Reson*. 1993;5:43–55.

29. Maudsley AA, Hilal SK, Perman WH, Simon HE. Spatially Resolved High Resolution Spectroscopy by "Four-Dimensional" NMR. *J Magn Reson*. 1983;51:147–52.

30. Martin JF, Wade CG. Chemical-Shift Encoding in NMR Images. *J Magn Reson*. 1985;61:153–7.

31. Hu J, Jiang Q, Xia Y, Zuo C. High Spatial Resolution In Vivo 2D 1H Magnetic Resonance Spectroscopic Imaging of Human Muscles with a Band-selective Technique. *Magn Reson Imaging*. 2001;19(8):1091–6.

32. Hu J, Xia Y, Feng W, Xuan Y, Shen Y, Haacke EM, et al. Orientational Dependence of Trimethyl Ammonium Signal in Human Muscles by (1)H Magnetic Resonance Spectroscopic Imaging. *Magn Reson Imaging*. 2005;23(1):97–104.

33. Ordidge RJ, Connelly A, Lohman JAB. Image-Selective *In Vivo* Spectroscopy (ISIS). A New Technique for Spatially Selective NMR Spectroscopy. *J Magn Reson*. 1986;66:283–94.

34. Bottomley PA, Foster TH, Darrow RD. Depth-resolved Surface-coil Spectroscopy (DRESS) for In Vivo 1H, 31P, and 13C NMR. *J Magn Reson*. 1984;59:338–43.

35. Doddrell DM, Brooks WM, Bulsing JM, Field J, Irving MG, Baddeley H. Spatial and Chemical-shift-encoded Excitation. SPACE, a New Technique for Volume-selected NMR Spectroscopy. *J Magn Reson*. 1986;68:367–72.

36. Bottomley PA, Foster TH, Leue WM. *In Vivo* Nuclear Magnetic Resonance Chemical Shift Imaging by Selective Irradiation. *Proc Natl Acad Sci USA*. 1984;81:6856–60.

37. Haase A, Frahm J, Hänicke W, Matthaei D. ^{1}H NMR Chemical Shift Selective (CHESS) Imaging. *Phys Med Biol*. 1985;30(4):341–4.

38. Xia Y, Jelinski LW. Imaging Low-concentration Metabolites in the Presence of a Large Background Signal. *J Magn Reson Ser B*. 1995;107:1–9.

39. Torrey HC. Bloch Equations with Diffusion Terms. *Phys Rev*. 1956;104(3):563–5.

272

40. Kärger J, Pfeifer H, Rudtsch S. The Influence of Internal Magnetic Field Gradients on NMR Self-Diffusion Measurements of Molecules Adsorbed on Microporous Crystallites. *J Magn Reson*. 1989;85:381–7.

41. Callaghan PT. Susceptibility-Limited Resolution in Nuclear Magnetic Resonance Microscopy. *J Magn Reson*. 1990;87:304–18.

42. Albert MS, Cates GD, Drlehuys B, Happer W, Saam B, Springer CSJ, et al. Biological Magnetic Resonance Imaging Using Laser-polarized 129 Xe. *Nature*. 1994;370:199–201.

43. Thulborn KR, Waterton JC, Matthews PM, Radda GK. Oxygenation Dependence of the Transverse Relaxation Time of Water Protons in Whole Blood at High Field. *Biochimica et Biophysica Acta*. 1982;714:265–70.

44. Ogawa S, Lee TM. Magnetic Resonance Imaging of Blood Vessels at High Fields: In Vivo and In Vitro Measurements and Image Simulation. *Magn Reson Med*. 1990;16:9–18.

45. Turner R, Le-Bihan D, Moonen CTW, Despres D, Frank J. Echo-Planar Time Course MRI of Cat Brain Oxygenation Changes. *Magn Reson Med*. 1991;22(1):159–66.

46. Ogawa SJ, Tank DW, Menon R, Ellemann JM, Kim SG, Merkle H, et al. Intrinsic Signal Changes Accompanying Sensory Stimulation: Functional Brain Mapping with Magnetic Resonance Imaging. *Proc Natl Acad Sci USA*. 1992;89:5951–5.

47. van der Zee J. Heating the Patient: A Promising Approach? *Ann Oncol*. 2002;13:1173–84.

48. Cline HE, Hynynen K, Hardy CJ, Watkins RD, Schenck JF, Jolesz FA. MR Temperature Mapping of Focused Ultrasound Surgery. *Magn Reson Med*. 1994;31:628–36.

49. Gibbs SJ, Carpenter TA, Hall LD. Magnetic Resonance Imaging of Thermal Convection. *J Magn Reson A*. 1993;105:209–14.

50. Simpson JH, Carr HY. Diffusion and Nuclear Spin Relaxation in Water. *Phys Rev*. 1958;111(5):1201–2.

51. Le Bihan D, Delannoy J, Levin RL. Temperature Mapping with MR Imaging of Molecular Diffusion: Application to Hyperthermia. *Radiology*. 1989;171:853–7.

52. Zhang Y, Samulski TV, Joines WT, Mattiello J, Levin RL, LeBihan D. On the Accuracy of Noninvasive Thermometry Using Molecular Diffusion Magnetic Resonance Imaging. *Int J Hyperthermia*. 1992;8(2):263–74.

53. Mills R. Self-Diffusion in Normal and Heavy Water in the Range 1–45°. *J Phys Chem*. 1973;77(5):685–8.

54. Parker DL, Smith V, Sheldon P, Crooks LE, Fussell L. Temperature Distribution Measurements in Two-dimensional NMR Imaging. *Med Phys*. 1983;10(3):321–5.

55. Dickinson RJ, Hall AS, Hind AJ, Young IR. Measurement of Changes in Tissue Temperature Using MR Imaging. *J Comput Assist Tomogr*. 1986;10(3):468–72.

56. Doran SJ, Carpenter TA, Hall LD. Noninvasive Measurement of Temperature Distribution with High Spatial Resolution Using Quantitative Imaging of NMR Relaxation Times. *Rev Sci Instrum*. 1994;65(7):2231–7.

57. Knüttel B, Juretschke HP. Temperature Measurements by Nuclear Magnetic Resonance and Its Possible Use as a Means of In Vivo Noninvasive Temperature Measurement and for Hyperthermia Treatment Assessment. *Recent Results in Cancer Research*. 1986;101:109–18.

58. Hindman JC. Proton Resonance Shift of Water in the Gas and Liquid States. *J Chem Phys*. 1966;44(12):4582–92.

59. Hall LD, Talagala SL. Mapping of pH and Temperature Distribution Using Chemical-Shift-Resolved Tomography. *J Magn Reson*. 1985;65:501–5.

60. De Poorter J, De Wagter C, De Deene Y, Thomsen C, Stahlberg F, Achten E. The Proton-resonance-frequency-shift Method Compared with Molecular Diffusion for Quantitative Measurement of Two-dimensional Time-dependent Temperature Distribution in a Phantom. *J Magn Reson B*. 1994;103:234–41.

61. Maroudas A, Muir H, Wingham J. The Correlation of Fixed Negative Charge with Glycosaminoglycan Content of Human Articular Cartilage. *Biochim Biophys Acta*. 1969;177:492–500.

62. Bashir A, Gray ML, Burstein D. Gd-DTPA2- as a Measure of Cartilage Degradation. *Magn Reson Med*. 1996;36(5):665–73.

63. Zheng S, Xia Y. The Impact of the Relaxivity Definition on the Quantitative Measurement of Glycosaminoglycans in Cartilage by the MRI dGEMRIC Method. *Magn Reson Med*. 2010;63(1):25–32.

273

64. Xia Y, Zheng S. Reversed Laminar Appearance of Articular Cartilage by T1-weighting in 3D Fat-suppressed Spoiled Gradient Recalled Echo (SPGR) Imaging. *J Magn Reson Imaging*. 2010;32(3):733–7.

65. Xia Y. Static and Dynamic Imaging Using Magnetic Field Gradients [M.Sc thesis]. Massey University, New Zealand; 1988.

66. Press WH, Flannery BP, Teukolsky SA, Vetterling WT. *Numerical Recipes*. Cambridge: Cambridge University Press; 1989.

67. Xia Y. Dynamic NMR Microscopy Vol 2 (Software) [Ph. D. Dissertation]. Massey University, New Zealand; 1992.

68. Xia Y. Contrast in NMR Imaging and Microscopy. *Concepts in Magn Reson*. 1996;8(3):205–25.

16

Advanced Topics in Quantitative MRI

Most quantitative MRI parameters discussed in Chapter 15 have been treated as a scalar quantity, that is, a number. If close attention is paid to the characteristics of these quantitative parameters, especially when careful measurements are made to study biological specimens (e.g., brain, musculoskeletal system), one can often notice more complicated characteristics of these parameters beyond the simple scalar form. Some of the non-scalar characteristics include anisotropy, dispersion, tensor, and multi-components. The intrinsic origin for these non-scalar characteristics of MRI parameters is commonly the tissue properties, associated with the morphological and micro-/meso-/macroscopic molecular structures of the tissue. The extrinsic origin for these non-scalar characteristics is associated with the ways we carry out measurements in MRI.

Figure 16.1 provides an overview to the "measurement evolution" of MRI parameters, as we pay closer and closer attention to the details of our measurement. The original and still the most common MRI images are the magnitude/intensity images (Figure 16.1a), which are made of numerical numbers organized as a 2D or 3D array, perhaps with a ± sign in front of the number. We can use these intensity images to explore primary image contrast and generate quantitative images that represent the voxel-averaged molecular environment (Figure 16.1b), which are the images of relaxation, velocity, diffusion, chemical shift, and so forth as described in Chapter 15. These quantitative images are still made of scalar quantities (technically, velocity is a vector), which are isotropic and have no measurement dependence, that is, each number can be visualized as a sphere. When we use advanced knowledge of physics and physiology and anatomy to examine these sources of quantitative image contrast, we notice the value of these parameters often has dependence on some orientation in space or a certain experimental method, that is, these quantitative parameters are in fact anisotropic in nature and are best visualized as an ellipsoid (Figure 16.1c). The non-scalar characteristics of these parameters can be linked to the micro-/meso-/macroscopic anatomic structures in the specimen (e.g., the networks of neurons and microscopic blood vessels in brain, the networks of collagen fibers in connective tissues, the network of vascular tubes in plants). One can describe the non-scalar features of these quantitative parameters in MRI as being an anisotropy property, a dispersion property, or a tensor property. Finally, since a single image voxel, no matter how small, must contain numerous macromolecules and some quadrillions of water molecules, we can divide the population of the water molecules within *a single voxel* into several discrete pools of sub-populations, each molecular pool having some distinctly different characteristics. This is the situation of multi-component MRI parameters (Figure 16.1d), which applies in principle to all of the MRI parameters we have

Essential Concepts in MRI: Physics, Instrumentation, Spectroscopy, and Imaging, First Edition. Yang Xia.
© 2022 John Wiley & Sons Ltd. Published 2022 by John Wiley & Sons Ltd.

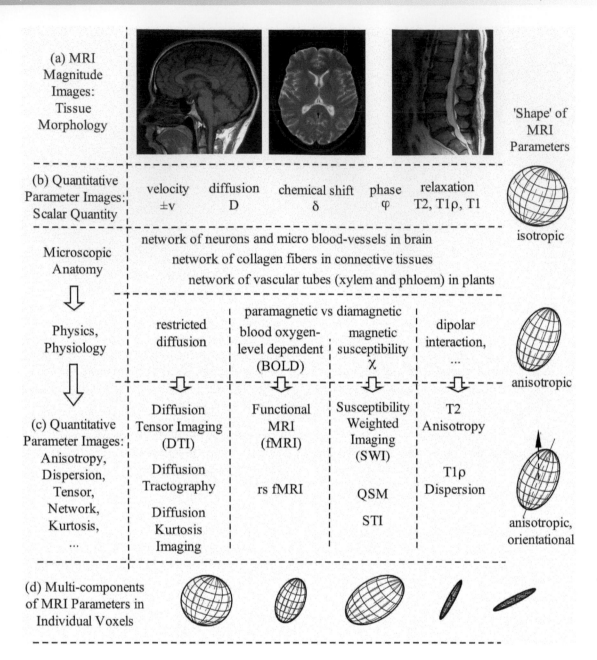

Figure 16.1 An evolutional view of MRI images. (a) MRI magnitude images, which show the tissue morphology and structure. With the advance of the technology, quantitative parameter images (b) can be generated where the parameters are treated as scalar quantity (shaped as a sphere). With further application of physics and physiology, the non-scalar properties of the quantitative parameters can be studied (c) – such properties can be visualized as an ellipsoid, with different shapes and orientations. An even closer examination can review the multi-component nature of the MRI parameters, which reflects the existence of multiple molecular pools within the individual voxels (d). (rs fMRI, resting-state functional MRI; QSM, quantitative susceptibility mapping; STI, susceptibility tensor imaging).

276

discussed. This chapter briefly discusses a few advanced topics in quantitative MRI, when additional information can be extracted from the non-scalar characteristics and less-used features of the MRI measurements.

16.1   ANISOTROPY AND TENSOR PROPERTIES IN QUANTITATIVE MRI

Anisotropy and tensor properties in quantitative MRI refer to the fact that the values of many measured MRI parameters vary according to the conditions of the measurement as well as the hierarchical anatomic structures in the specimen.

16.1.1   Anisotropy and T_2 Anisotropy

Anisotropy is the property of being directionally dependent. A well-known example is the T_2 anisotropy (cf. Figure 15.9) in articular cartilage, which is a thin layer of connective tissue that covers the opposing end-surfaces of two long bones in a synovial joint (e.g., a knee, a hip). When the same block of tissue is measured in MRI with the tissue block set at different orientations with respect to the polarizing magnetic field B_0, different relaxation values can be measured at the same tissue location. The intrinsic origin of this anisotropy is the 3D orientation of the collagen fibers (Figure 16.2b) in cartilage. Starting from the articular surface, noncalcified cartilage is commonly sub-divided along its thickness into three sub-tissue zones based on the local orientation of its collagen fibers: the superficial zone (SZ) where the fibers are mostly parallel to the surface, the transitional zone (TZ) where the fibers are mostly random, and the radial zone (RZ) where the fibers are oriented perpendicularly to the surface [1, 2]. Other major molecular components in cartilage include water and proteoglycan, which have no preferred orientation and structure in aqueous solutions [3–5].

Although water, which is being measured in NMR and MRI, does not possess motional or orientational anisotropy in a simple environment, it can become anisotropic in tissues due to its close interactions or bindings with the collagen fibers. This water–fiber interaction results in a non-zero averaging of the spin Hamiltonian that describes the nuclear dipolar interaction [6]. In Chapter 4.1, we showed that the dipolar Hamiltonian contains a geometrical factor $(3\cos^2\theta - 1)$, where θ is the angle between the position vector joining the two spins and the external magnetic field B_0 (Figure 16.2a). When $\theta = 54.74°$ (the magic angle in NMR and MRI; see Chapters 4.1 and 15.2), the dipolar Hamiltonian becomes zero and hence has no influence on T_2 relaxation, which gives cartilage its most uniform and highest intensity. When $\theta \neq 54.74°$, T_2 in cartilage is being attenuated by the dipolar Hamiltonian, which gives the image its tissue-depth dependent and tissue-orientation dependent appearance (Figure 15.9a). This orientational dependent influence to T_2 relaxation of tissue water gives rise to the variable and laminated appearance of articular cartilage in MRI, which is called the magic angle effect in cartilage MRI literature [6–8]. Although μMRI does not have the resolution to identify individual collagen fibers, many μMRI studies have demonstrated that the local fibril architecture in cartilage modulates the anisotropic characteristics of the T_2 relaxation, which can be modeled by an ellipse schematic (Figure 16.2c) [9]. Figure 16.3 shows a set of MRI results where the cartilage blocks were rotated in the magnet over a 360° angular space, where the geometrical factor in the dipolar Hamiltonian can be minimized at four particular angles [10]. Similar relaxation anisotropy should exist in any animal and human tissue that contains a network of organized fibril structures, including tendon and ligament, meniscus, spinal disc, muscles, membrane, and so forth.

277

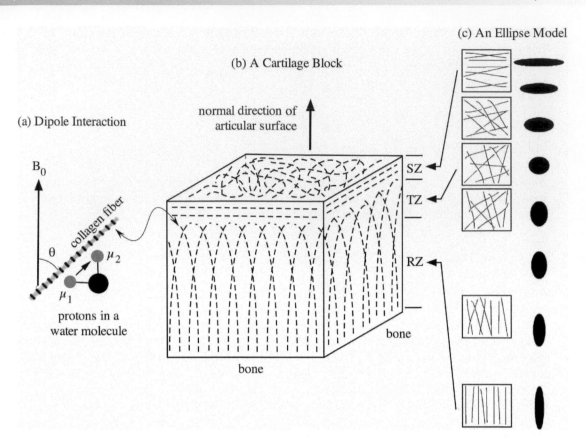

Figure 16.2 (a) Dipolar interaction between two protons (μ_1 and μ_2) in a water molecule that is bound to a collagen fiber (not to scale). Each proton generates a local dipolar field (as shown in Figure 4.1a) and experiences a small contribution of magnetic fields from its (many) neighbor protons. (b) The orientation of the collagen fibers in a block of articular cartilage, donated by the dashed lines in the tissue. SZ, TZ, and RZ refer to three histological zones of articular cartilage: superficial zone, transitional zone, and radial zone, respectively. (c) Schematic illustration of the averaging effect of collagen fibers in a single imaging voxel at different tissue depths in cartilage, whose anisotropy behaviors can be visualized graphically by an ellipse model in this 2D schematic [9]. Source: Xia [9].

16.1.2 Dispersion and $T_{1\rho}$ Dispersion

Dispersion in NMR and MRI refers to the fact that some measured values depend upon the experimental conditions. A well-known example is the dispersion of $T_{1\rho}$ relaxation, which has been discussed in Chapter 15.2.4. In NMR (Figure 7.9) and MRI (Figure 15.12), $T_{1\rho}$ is measured with the application of a $\boldsymbol{B}_1$ field to the direction of the transverse magnetization after it has been tipped to the transverse plane, which spin-locks the magnetization to the transverse plane. Although no spin-spin relaxation occurs when the $\boldsymbol{B}_1$ field is turned on, the magnetization shrinks in magnitude due to the fact that $|\boldsymbol{B}_1| << |\boldsymbol{B}_0|$. $T_{1\rho}$ dispersion occurs in the NMR and

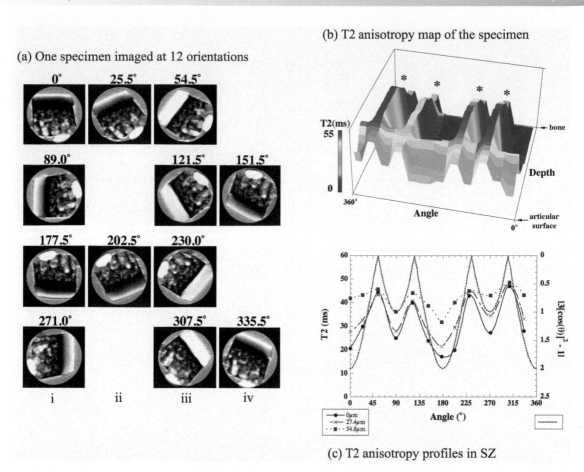

(a) One specimen imaged at 12 orientations

(b) T2 anisotropy map of the specimen

(c) T2 anisotropy profiles in SZ

Figure 16.3 (a) The proton images of one cartilage-bone specimen imaged at 12 angles, with the B_0 direction pointing vertically upwards. The four images in the first column (*i*) were at approximately 90° apart, corresponding to the four peaks in the curve of $(3\cos^2\theta - 1)$. The four images in the third column (*iii*) were imaged at approximately the four magic angles for the radial zone fibrils. The images in the second and forth columns (*ii*, *iv*) were imaged at the angles close to the four magic angles for the superficial zone fibrils. (b) The quantitative map of T_2 anisotropy for the specimen shown as a 3D surface image. The four stars indicate the nominal orientations of the magic angles (for the radial zone fibrils), where the T_2 profiles are mostly uniform and have the highest values (similar to the profiles of 40° and 57° in Figure 15.9d). (c) The T_2 anisotropy cross sections in the superficial zone, together with the curve of $|(3\cos^2\theta - 1)|$. A close correlation between the local minima in $|(3\cos^2\theta - 1)|$ and the local maxima in T_2 can be seen. Source: Xia et al. [10].

MRI measurement [11] since the values of $T_{1\rho}$ depend on the strength of the spin-lock field, as shown in Figure 15.13.

Figure 16.4 shows an orientational study of $T_{1\rho}$ dispersion using bovine nasal cartilage, which has a simpler fibril structure than articular cartilage (in the specimens, the fibers were in parallel with the short length of the block). One can see that the greatest relaxation difference as the function of specimen orientation occurs when the samples contain no gadolinium contrast agent and the spin-lock field is zero (which are T_2 values in Figures 16.4a and 16.4c). This difference

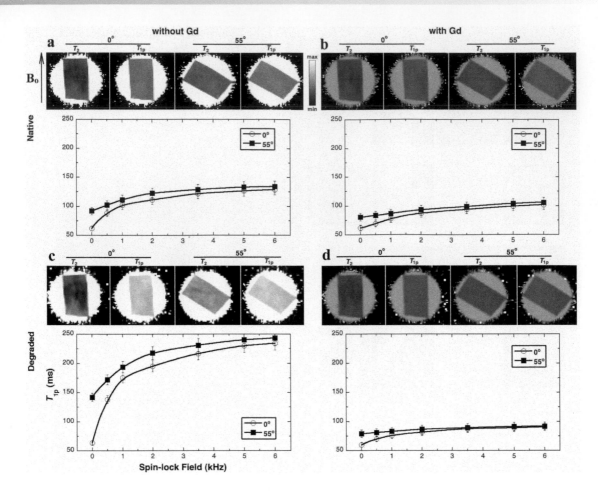

Figure 16.4 $T_{1\rho}$ dispersion plots of the bovine nasal cartilage specimens with and without Gd-DTPA^{2-} at 0° (open circles) and the magic angle (solid squares) for (a) native specimens, (b) native specimens with Gd-DTPA^{2-}, (c) trypsin-degraded specimens, and (d) trypsin-degraded specimens with Gd-DTPA^{2-}. The collagen fibers were in parallel with the short length of the specimens. All images were plotted with the same intensity limits (0–400 ms) in the usual gray scale. Source: Wang and Xia [12].

decreases when tissue is subject to an increasingly larger spin-lock field or is treated with gadolinium ions. One can also see that $T_{1\rho}$ relaxation is a function of the spin-lock field: the larger the spin-lock field, the longer the $T_{1\rho}$ relaxation. The use of the gadolinium contrast agent reduces the magic angle effect as well as the sensitivity of $T_{1\rho}$ to tissue degradation (compare Figures 16.4c and 16.4d) [12].

In addition to $T_{1\rho}$ dispersion, it is also possible to investigate the dispersion behaviors of other MRI parameters [13, 14], such as T_1 dispersion [15], which is the dependence of T_1 values with the external magnetic field. (T_1 values get longer as the field strength increases; the dependence of T_1 values on the field strength varies in different tissues.) Special instrumentation is needed to incorporate T_1 dispersion into MRI [16].

16.1.3 Tensor and Diffusion Tensor Imaging

A tensor is a mathematical quantity that describes the multi-directionality of an object or a measurement. For a generic object in the usual 3D space, a 3 × 3 matrix (see Appendix 1) can adequately describe the characteristics of a tensor. In NMR and MRI, the most common tensor measurement is of diffusion in biological tissues.

In pure liquids and unstructured/randomly structured materials, water molecules are in random Brownian motion, which is free and un-obstructed (only limited by the wall of the container). The behaviors of such diffusion process should be isotropic regardless of diffusion time (Figure 16.5a). Quantitatively, the mean-square displacement in a 1D diffusion process can be estimated as

$$<\left(x - x_0\right)^2 >^{0.5} = \sqrt{2Dt},\qquad(16.1)$$

where the numerical factor 2 is for 1D diffusion, and t is the diffusion time. In NMR and MRI experiments when the PGSE sequence is used to probe the diffusion process, t is Δ in the PGSE sequence (Figure 15.2), that is, the time between the two diffusion-sensitive gradient pulses.

In biological tissues, the morphological structures at some micro-/meso-/macroscopic levels such as neurons, blood vessels, and vascular tubes can become obstacles to free diffusion, which act to restrict the diffusion process toward a certain space or direction [17–19]. In an NMR or MRI measurement in which the PGSE sequence is used (Figure 15.2), the molecules are tagged by the first PGSE gradient and measured after the second PGSE gradient. When there is no obstacle in the environment, the water molecules are in thermal random motion, which is called free diffusion (Figure 16.5a). When there are obstacles in the path of the diffusion molecules, the tagged molecules could hit a "wall" and bounce back during the diffusion time Δ, which leads to the diffusion values being attenuated (Figure 16.5b). This diffusion process is called restricted diffusion. In addition to obstacles, transient hydrogen bonding can occur between macromolecules and diffusion water [20], which also restricts the diffusion. Since these obstacles and bonding have some preferred local orientations and structures, the molecular diffusion becomes highly geometry- and size-dependent, hence a tensor property.

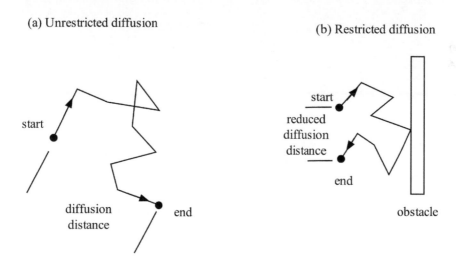

Figure 16.5   Schematics of free diffusion (a) and restricted diffusion (b), where the diffusion distance as specified in Eq. (16.1) is reduced due to the obstacles.

Figure 16.6 shows a set of quantitative diffusion imaging experiments, where the structure of the vascular vessels and the skin of asparagus can be visualized by the quantitative diffusion maps that illustrate restricted diffusion. In this experiment, the diffusion distance was about 5 μm, which was sufficient to sense the sizes (diameters) of the vascular vessels as well as the structure of the inner skin of the specimen. For specimens such as asparagus, the structural orientation is relatively simple. Two or three ordinary diffusion-sensitive experiments where the diffusion gradient is set at different orthogonal directions can reveal the location and structure of these obstacles. This type of restricted diffusion experiment is a useful tool to probe non-invasively the microscopic textures in biological tissues.

When the orientational structures of the tissue become complicated, such as the networks of axon bundles in brain, diffusion can be treated as a tensor property in MRI and measured by diffusion tensor imaging (DTI) [21]. In MRI of the brain, each imaging voxel contains many axons, whose voxel-averaged diffusivity should in general have a tensor property. The simplest situation would have the majority of the axons oriented along a single direction, which can be

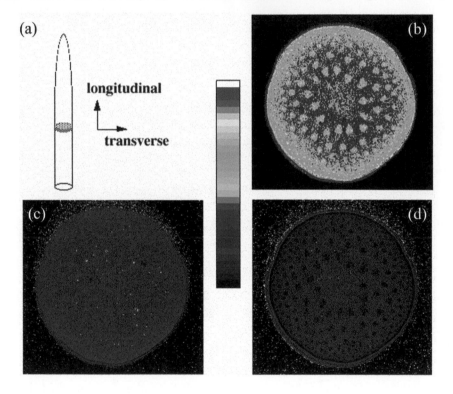

Figure 16.6   An example of a restricted diffusion experiment by microscopic MRI on a 7T instrument. (a) The orientations of two diffusion gradients, one longitudinal and one transverse. (b) The proton image of the specimen (imaging voxel size = 39 μm × 39 μm × 1 mm), where each highlighted region inside the stalk of asparagus is a bundle of vascular vessels, which transports fluids vertically along the stem of the plant. (c) and (d) Quantitative diffusion maps with the diffusion gradient at the longitudinal direction (c) and transverse direction (d). Both quantitative images were plotted using the same diffusion scale. Note that in (d) when the PGSE gradient was set transversely, the diffusion values at the bundles of the vascular vessels are lower than the background tissue, which illustrated the restricted diffusion in the vessel areas. In contrast, when the PGSE gradient was set longitudinally in (c), diffusion along the length of the stem did not experience any restriction, which yielded a uniform map.

modeled as a single ellipsoid shown in Figure 16.1. Studies have found that water molecules diffuse more easily along the axon bundle (the long axis of the ellipsoid) than perpendicular to the axon bundle [22]. In real tissue, there are likely numerous "shapes" for the voxel-averaged diffusivity tensor. For example, when two bundles of axons cross each other in the voxel, one would need to fuse two ellipsoids into one complicated shape, depending upon the relative angle between the two bundles. Different anisotropic properties have been identified in different regions of the brain, for example, between the white matter and the gray matter, as well as between healthy and lesioned tissues [23].

Mathematically, one can write down the tensor property in DTI by a 3×3 matrix (Appendix 1), as

$$\boldsymbol{D} = \begin{bmatrix} D_{xx} & D_{xy} & D_{xz} \\ D_{yx} & D_{yy} & D_{yz} \\ D_{zx} & D_{zy} & D_{zz} \end{bmatrix}. \tag{16.2}$$

When the medium of diffusion is isotropic, $D_{xx} = D_{yy} = D_{zz} = D$ and $D_{xy} = D_{xz} = D_{yz} = 0$. To accurately determine the shape and orientation of the diffusion tensor property in brain, many diffusion experiments are needed. During each diffusion-sensitive experiment one varies the combination of the gradient direction (G_x, G_y, G_z), gradient strength (g), and gradient diffusion time Δ (cf. Figure 15.2). Since there are six independent elements in the diffusion tensor (note that $D_{xy} = D_{yx}$, $D_{xz} = D_{zx}$, and $D_{zy} = D_{yz}$), one needs at least six gradient directions to carry out a tensor measurement. (One likely needs much more than six independent experiments, given the complex geometry of the axon network in brain.) From these diffusion measurements, it is possible to determine three diffusion coefficients (λ_1, λ_2, λ_3) along three principal directions of this ellipsoid by the diagonalization of the diffusion tensor. A generic 3D ellipsoid can have three anisotropic length properties (λ_1, λ_2, λ_3) and three orientation properties (θ_1, θ_2, θ_3). From these calculations, one can determine the mean diffusivity,

$$D = \frac{D_{xx} + D_{yy} + D_{zz}}{3}, \tag{16.3}$$

and the fractional anisotropy,

$$FA = \sqrt{\frac{3}{2}} \frac{\sqrt{(D_{xx} - D)^2 + (D_{yy} - D)^2 + (D_{zz} - D)^2}}{\sqrt{D_{xx}^2 + D_{yy}^2 + D_{zz}^2}}, \tag{16.4}$$

which is a number from 0 (isotropic) to 1 (extreme anisotropic). The 3D maps of these final parameters, which are often color-coded (Figure 16.7), can be extremely valuable in the understanding of brain functions and diagnostics of brain disease [24]. Based on similar methodology, diffusion tractography has been developed, which links the principal diffusion at each voxel to form a single line, to indicate the trajectory orientations of the axon bundles [25].

16.1.4 Kurtosis and Diffusion Kurtosis Imaging

We mentioned in Chapter 15.1 that the self-diffusion coefficient is characteristic of the random Brownian motion in water-containing objects, which assumes a Gaussian distribution function (Figures 15.4 and 16.8a) and can be described by Eq. (15.1). More commonly in clinical MRI when the formalism of b-factor is used, Eq. (15.4) is used to describe the mono-exponential attenuation of the water signal by molecular diffusion. It is easy to see that by taking the natural logarithm on Eq. (15.4), the signal attenuation due to diffusion is expected to follow a straight line in the semi-logarithmic plot, with the slope of the straight line as the diffusion coefficient

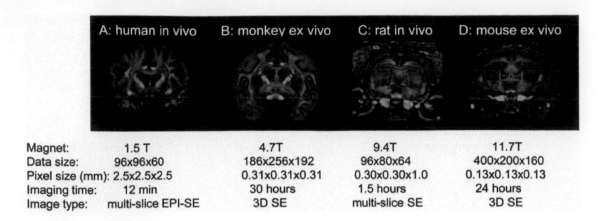

Magnet:	1.5 T	4.7T	9.4T	11.7T
Data size:	96x96x60	186x256x192	96x80x64	400x200x160
Pixel size (mm):	2.5x2.5x2.5	0.31x0.31x0.31	0.30x0.30x1.0	0.13x0.13x0.13
Imaging time:	12 min	30 hours	1.5 hours	24 hours
Image type:	multi-slice EPI-SE	3D SE	multi-slice SE	3D SE

Figure 16.7 Examples of brain DTI of different animals and imaging parameters. The colors indicate the orientation of the ellipsoid (red x, green y, blue z). Source: Mori and Zhang [24].

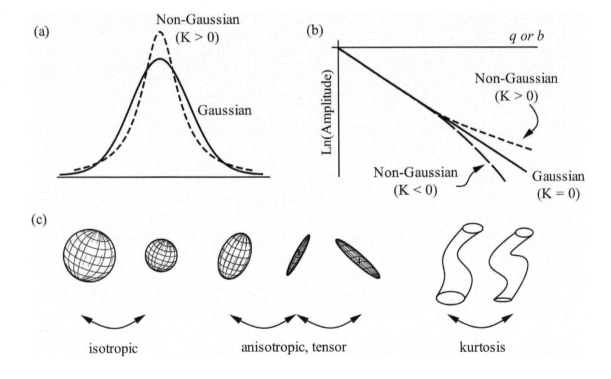

Figure 16.8 The Brownian motion of water molecules has a Gaussian distribution function in probability (a) and can be measured in MRI diffusion experiments by fitting the data to a straight line in a semi-logarithmic plot (b). Any deviation from the straight line implies a non-Gaussian distribution, which can be described by its kurtosis value (K = 0 for Gaussian; K > 0 is commonly seen in MRI of biological tissues; K < 0 is uncommon in MRI diffusion experiments). (c) The evolution of the MRI diffusion experiments, from isotropic to anisotropic/tensor properties. The recognition of the kurtosis properties of diffusion comes from the heterogeneities of the microscopic structural environment in brain.

(Figure 16.8b). In MRI diffusion experiments of biological tissues at high *b*-factors or high *q*-values [Eq. (15.3)], however, the deviations from the straight line in these plots have been noticed, which would correspond to non-Gaussian distribution of water diffusion. Note that the deviations only occur at high *b*-factors or high *q*-values, which implies the restriction occurs when the water molecules have large space to explore [i.e., a large mean-square displacement in the diffusion process as described in Eq. (16.1)].

The reason for the non-Gaussian attenuation of the diffusion signal in MRI is the fact that in biological tissues such as brain, the microscopic environment for water diffusion is no longer homogeneous, which is used in diffusion modeling. In reality, the axon bundles in brain have various kinks and bulges along the lengths (Figure 16.8c); these kinks and bulges can also be the crossover locations of two or more axon bundles that MRI or another imaging method is unable to resolve spatially. A study of retinal ganglion cell axons has found that the axon caliber varies by 17.5-fold (from ~0.2 μm to ~3.5 μm in diameter) and that volume (proportional to r^2) varies by ~300-fold [26]. Such variations suggest that the space for the diffusion of water molecules in biological tissues is not only highly anisotropic but also highly heterogeneous. This complex diffusion environment deviates from the idealized assumption that derives the Gaussian distribution function.

A more realistic model in MRI diffusion imaging of biological tissues employs kurtosis (K), a term in the probability theory and statistics, to describe the shape of a probability distribution function. In recent years, diffusion kurtosis imaging (DKI) has emerged as a new technique in diffusion tensor imaging of human brain [27]. When the diffusions of molecules can be considered isotropic, one measures merely the size of the circular balls (Figure 16.8c). When the diffusions of molecules become anisotropic and are influenced by the orientational structures of the axon bundles in brain, one employs the properties of anisotropy and tensor to describe the 3D orientation of the axon bundles as rods and ellipsoids. When the kinks and bulges along the lengths of the axon bundles are considered, the equation for the water diffusion in the formalism of *b*-factor can be extended [27] from Eq. (15.4) to

$$Ln(E_c) = -bD + \frac{1}{6}b^2 D^2 K, \tag{16.5}$$

where K is a dimensionless parameter that describes the kurtosis of the distribution function. A Gaussian distribution has K = 0 and returns Eq. (16.5) to Eq. (15.4), which is the signature of water diffusion in simple liquids. A distribution function with a peak width narrower than the Gaussian has positive K values (i.e., K > 0, as in Figure 16.8a), which has been found to describe the water diffusion in complex biological tissues. (K < 0 is uncommon in MRI of biological tissues.) Graphically in the semi-logarithmic plot (Figure 16.8b), a signal decay with positive kurtosis has the tail end of the data (i.e., at high *b* or *q* values) tilted above the straight line, which reflects slower diffusion when water encounters more microscopic structures (i.e., more restrictive diffusion). A number of studies have found the evaluation of the non-Gaussian distribution function can provide valuable information about the microscopic structures in brain [27]. In general, diffusion kurtosis can be treated as a tensor.

16.2 MULTI-COMPONENT NATURE IN QUANTITATIVE MRI

Not all water molecules in biological tissue have the same molecular environment because of the hierarchical organization and structure of the tissue. For example, a tissue could have water associated with the rigid tissue structures (the crystalline water), water associated with macromolecules (the hydration water), and a pool of relatively free water [28]. One can also consider

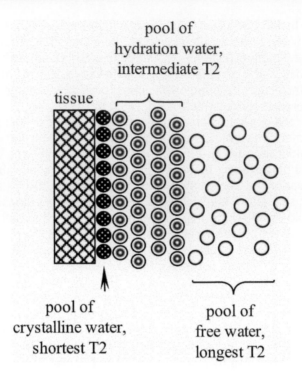

pool of
hydration water,
intermediate T2

tissue

pool of
crystalline water,
shortest T2

pool of
free water,
longest T2

Figure 16.9 The nature of multiple pools of water molecules in biological tissue, where each pool of water will have a very different molecular environment, hence different MRI properties. The pool of free water would have the longest T_2 and isotropic T_2, while the pool of crystalline water would have the shortest T_2 and anisotropic T_2.

that the multilayers of hydration water act as "a delay line" or buffer that slows down the relaxation of the protons diffusing through it to the crystalline compartment (Figure 16.9). (Ignore for the moment the exchanges among the different populations and the distribution of each population, which act to blend the populations together.)

Since the dynamics of water molecules closely reflect the macromolecules they associate or interact with, biological tissues should be considered to *always* have more than one population of water molecules, where each population has a unique environment (Figure 16.9). Consequently, the general practice in MRI of assigning only one value to T_1, T_2, or $T_{1\rho}$ at any pixel location in a complex tissue is merely an approximation. Careful MRI studies of some connective tissues such as articular cartilage and tendon at high spatial resolution have indeed found two to four relaxation times [29–32].

Figure 16.10 shows a set of quantitative $T_{1\rho}$ data in MRI from bovine nasal cartilage [33]. When the collagen fibers in nasal cartilage were oriented at the magic angle (55°) to $\boldsymbol{B}_0$, both T_2 and $T_{1\rho}$ were single component (Figure 16.10f), regardless of the spin-lock field strength in the pulse sequence. When the collagen fibers in nasal cartilage were oriented at 0° to $\boldsymbol{B}_0$, both T_2 and $T_{1\rho}$ at a spin-lock field of 500 Hz had two components (Figure 16.10e). When the spin-lock field was increased to 1000 Hz or higher, $T_{1\rho}$ relaxation in nasal cartilage became a single component, even when the specimen orientation was 0°. These results demonstrate that the specimen orientation must be considered for any multi-component relaxation analysis. Since the rapidly and slowly relaxing components can be attributed to different pools of the water population in tissue, the

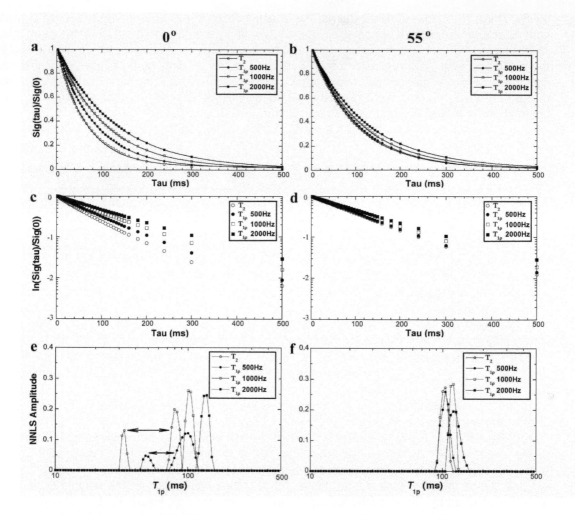

Figure 16.10 Example of multi-component $T_{1\rho}$ relaxation in bovine nasal cartilage by MRI. (a) and (b) are the normalized signal decay in the MRI experiments, where the solid line in the figure is an exponential fit with one decay-constant. (c) and (d) are the natural log plots of data shown in (a) and (b). (e) and (f) are the $T_{1\rho}$ distribution profiles by the non-negative least-squares (NNLS) method from the same data. Source: Wang and Xia [33].

ability to resolve different relaxation components could be used to quantitatively examine individual molecular components in connective tissues, including between healthy and lesioned tissues.

A few notes will help to explain the measurement of multi-component MRI. First, to resolve the distribution of relaxation in any complex tissue, one needs high resolution in the direction of relaxation, that is, the number *n* in the contrast direction in Figure 14.5 needs to be large, which increases the experimental time. Second, common MRI protocols all have a minimum echo time of at least several milliseconds, during which the molecular information is not accessible. Most imaging experiments, hence, cannot resolve the shortest relaxation components (the water molecules that tightly interact or bind with the macromolecules), which simply disappear when we are waiting during TE, unless the sequence is the ultra-short echo (UTE) type [34] (cf. Chapter

12.8). Third, the size and orientation of the imaging voxel needs to be small in a multi-component relaxation experiment, since any low-resolution image could have its signal from an imaging voxel that is large enough to contain several different anatomic structures, which would lead to the exhibition of several relaxation components in the tissue. Only by high-resolution MRI can one measure multi-component relaxation coexisting *intrinsically* in the tissue. Finally, the discussion of multi-component relaxation could equally be applied to other sources of image contrast, such as self-diffusion and chemical shift.

16.3 QUANTITATIVE PHASE INFORMATION IN THE FID DATA – SWI AND QSM

We mentioned in the beginning of this book that the FID data in NMR and MRI is complex, consisting of a real component and an imaginary component (Chapter 2.13). This complex signal enables us to determine the phase of the magnetization in the transverse plane. In routine NMR spectroscopy, this phase information is used in the post-acquisition process to generate an in-phase spectrum (cf. Figure 6.12), which is the NMR spectrum being displayed on computer monitors and saved in hardcopies. In MRI, only the velocity measurement by the PGSE sequence retains the phase information, which is used in the post-acquisition process to determine the direction of the velocity by Fourier transform (Chapter 15.1). All quantitative measurements of relaxation and diffusion discussed in Chapter 15 utilize the attenuation of the FID signal amplitude, during which the phase information is simply discarded. Are there other uses of the phase information in the FID data?

Yes. The phase of the magnetization in the transverse plane is an extremely sensitive measure of the magnetization, which can be influenced by the inhomogeneities of the magnetic field as well as manipulated by various nuclear interactions (e.g., dipolar interaction, chemical shift, scalar interaction). As discussed in Chapter 15.4.1, the homogeneity of the magnetic field can also be perturbed by the magnetic susceptibility differences (e.g., Figure 15.19). This perturbation to the imaging field by the local susceptibility was first noticed in the mid-1980s in clinical MRI [35, 36], when the phase changes in the FID data of human brain were attributed to the variations in the tissue's magnetic susceptibility, related to the concentration of local paramagnetic spins associated with deoxyhemoglobin, methemoglobin, free ferric iron, hemosiderin, and so on. Since a majority of the tissues are diamagnetic, the differences between the paramagnetic components and diamagnetic tissues, which are associated with the oxygenation situation of the blood, become a very powerful mechanism for generating extra image contrast (see also Chapter 15.4.2).

By noting $\varphi = \omega t = \gamma B t$ and $t = \text{TE}$, one can write the extra phase due to the magnetic susceptibility as

$$\Delta\varphi = \Delta\omega t = \gamma \Delta\chi B_0 \text{TE}, \tag{16.6}$$

where χ is the magnetic susceptibility as defined in Chapter 15.4.1. Equation (16.6) shows that this phase term relates to the local susceptibility in tissue, as well as the length of TE, which can be long in the gradient-echo and multi-echo MRI experiments. Two highly sensitive methodologies have been developed in MRI to systematically utilize the phase information in the FID data; one method is called susceptibility weighted imaging (SWI) [37] and the other is termed as quantitative susceptibility mapping (QSM) [38].

SWI aims to enhance the susceptibility-induced image contrast. It uses the gradient-echo sequence and can generate a pair of images, the SWI filtered phase image and the merged SWI magnitude image [39]. These phase-enhanced images emphasize the differences among the magnetic susceptibilities in tissue, which can be localized to the vast network of blood vessels in human brain. Hence SWI can improve the visualization of microscopic blood vessels in brain. Some studies have shown that SWI is 3–6 times more sensitive than conventional T_2*-weighted gradient-echo sequences in detecting the hemorrhagic lesions in diffuse axonal injury [40].

One limitation to the direct use of the phase information measured in gradient-echo MRI is that the phase depends on imaging parameters, is orientation-dependent, and is non-local. As we have seen in Figure 15.19, the difference in the magnetic susceptibility creates a local field gradient that can extend into the surrounding space. One could treat the susceptibility-induced field as if it comes from a magnetic dipole, which has a known distribution of its field into the nearby space (Figure 4.1a). Hence the phase value measured at any voxel location in SWI contains both local tissue properties and surrounding tissue properties. By solving the inverse problem from magnetic field to the susceptibility source [38, 41], QSM aims to map quantitatively the local tissue property. Figure 16.11 compares a set of images without and with the phase information (the magnitude image, the SWI fused image, and the QSM image). QSM is a mathematically intensive method, which has been used extensively in recent years to diagnose a number of neurodegenerative diseases, including Alzheimer's disease, Parkinson's disease, Huntington's disease, and so forth [38]. In some tissues, one can even treat the magnetic susceptibility as an anisotropic property (following similar steps as in the case of the diffusion tensor), which measures the phases at different orientations with respect to B_0 (i.e., the sample needs to be rotated in B_0). Such an MRI method is called susceptibility tensor imaging (STI) [42].

A final note on these phase-based imaging contrast methods is their need for an extensive network of microscopic blood vessels in the tissue. For example, we have found no image contrast in articular cartilage by SWI, since mature and healthy cartilage is an avascular tissue.

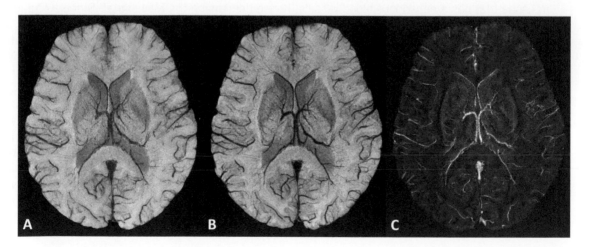

Figure 16.11 (a) An original magnitude image of brain by a gradient-echo sequence, (b) the processed SWI magnitude image, after phase mask multiplication with the original SWI magnitude image, and (c) the QSM image. The imaging parameters for this data are: B_0 = 3T, TE/TR = 22.5 ms/36 ms, flip angle = 20°, bandwidth = 100 Hz/pixel, resolution = 0.22 × 0.44 × 1 mm^3. All images are minimum/maximum projected over 8 slices or 8 mm. Source: Images courtesy of Sagar Buch, Ph.D. and E. Mark Haacke, Ph.D. in the MR Research Facility at Wayne State University, Detroit, Michigan.

16.4 FUNCTIONAL MRI (fMRI)

In Chapter 4.2, we mentioned that diamagnetic materials have a negative bulk susceptibility χ and an induced magnetic field antiparallel to $\boldsymbol{B}_0$; in contrast, paramagnetic materials have a positive bulk susceptibility and an induced magnetic field parallel to $\boldsymbol{B}_0$. Since most water-rich biological tissues are diamagnetic, a diamagnetic material has been assumed in most of this book (e.g., in all discussions about chemical shift). Our first discussion on paramagnetic material was in Chapter 15.4.2, when the hemoglobin iron in blood will change its spin state from diamagnetic low spin (S = 0) in the oxygenated state to paramagnetic high spin (S = 2) in the deoxygenated state. This spin conversion to paramagnetic iron will increase the difference in magnetic susceptibilities among the blood and the surrounding tissue, hence changing a number of MRI properties.

In the last section (Section 16.3), the concentration change of local paramagnetic spins in the network of blood vessels was related to the variations in the tissue's magnetic susceptibility, which can be mapped out by the phase information as saved by the complex FID data. By using the gradient-echo sequences that enhance the FID decay beyond the T_2 relaxation, a number of phase-based MRI techniques have been developed, such as SWI, QSM, and STI, which have been used extensively to study mainly the brain and its diseases. The same techniques can also be used to study other non-brain tissues, as long as there is a network of microscopic vessels.

In this brief section, the concentration change of local paramagnetic spins in the network of blood vessels was related to a dynamic supply-and-demand process in the brain, called BOLD contrast (Blood Oxygenation Level Dependent), which we have discussed in Chapter 15.4.2 [43]. The blood oxygenation level varies when there is a change in the dynamic equilibrium between the normal supply and demand of oxygen to the brain tissue. Oxygenated hemoglobin is diamagnetic, which appears in MRI just like the rest of the brain tissue. In contrast, deoxygenated hemoglobin is paramagnetic, which acts as an endogenous contrast agent to increase the T_2* contrast in tissue that has a network of blood vessels, hence decreasing the MRI signal. Since a change in the demand of oxygen can come from an increase of the neuronal activity in the brain, which can be triggered by some external visual or audio stimulation or some repetitive motion, the association between the brain activity and MRI signal change becomes the scientific basis for functional MRI (fMRI) [44–47].

A functional brain imaging experiment usually requires a number of high-speed gradient-echo-based imaging sequences with different stimulating conditions, which need to be optimized not only for spatial resolution but also for temporal resolution. Extensive image and statistical analyses are needed subsequently, usually on a voxel-by-voxel basis to compare the background image and the stimulated image, in order to identify the stimulus-correlated changes in the voxel intensities. A positive confirmation is indicated by superimposing the confirmed voxel as a highlight on a high-resolution "anatomical" image of the subject. In addition to the response of the brain to external stimulation, fMRI has also been used to study brain's "background" activities in a resting or task-less state (rs fMRI) [48, 49], which can be further extended to study the regional connectivity of baseline brain activities, hence any changes in psychological conditions.

16.5 OPTICAL PUMPING AND HYPERPOLARIZATION IN MRI

In Chapter 3, we showed that the size of the magnetization in thermal equilibrium is proportional to the ratio $\hbar\gamma B_0/k_B T$, which is given by Eq. (3.14). For protons in a magnetic field B_0 of 7 Tesla and at room temperature (T = 300 K), $k_B T$ is over 4 orders of magnitude bigger than

290

$\hbar\gamma B_0$, which makes the ratio (i.e., the magnetization in thermal equilibrium) tiny. All NMR and MRI experiments that we have discussed so far in this book utilize the thermal magnetization. Is it possible to go beyond the thermal polarization? The answer is yes.

Hyperpolarization benefits from the principle of optical pumping. By collisional spin exchange with laser optically pumped rubidium (Rb) vapor [50–52], one can increase the population difference in noble gases (e.g., ^{3}He and ^{129}Xe, which are the only two noble gas isotopes with spin-1/2 nuclei) from a net polarization of only a few in 10^6 to 10–30%! This astonishing increase in the spin polarization more than overcomes the spin density difference between the gas and liquid states, consequently permitting high-resolution NMR and MRI experiments of gaseous spaces.

The initial motivation of using the hyperpolarized gas in MRI is to image the lung [53, 54]. Conventional MRI of lungs is difficult because of the several hundred million tiny air pockets called the alveoli in the lung, which is the microscopic system that exchanges oxygen and carbon dioxide molecules to and from the blood. As we have seen from Figure 15.19, the interface between air and water/tissue has a large difference in magnetic susceptibilities, which significantly degrades the image quality and increases the T_2^* relaxation. So conventional MRI of lung using regular imaging sequences generates no image of the lung (Figure 16.12a). In this situation, one could fill the alveoli with the noble gas (of course mixed with oxygen), then acquire the signal from the hyperpolarized gas. These two noble gases are harmless to inhale and can stay hyperpolarized for long times (from several minutes to hours). Figure 16.12 shows the comparison of conversional ^{1}H MRI and hyperpolarized ^{129}Xe gas MRI of the lung – the diagnostic value of the hyperpolarized ^{129}Xe gas MRI is clear [55]. A wide range of pulmonology disorders can be investigated by hyperpolarized gas MRI, including from the recent COVID-19 patients who were found to have a higher rate of ventilation defects and longer gas-blood exchange time compared with healthy individuals [56].

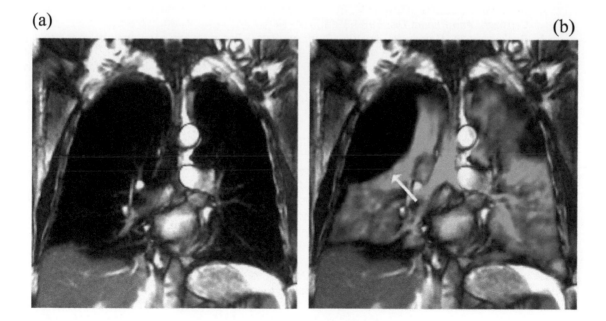

(a)

(b)

Figure 16.12 The coronal mid-lung ^{1}H MR image without (a) and with (b) hyperpolarized ^{129}Xe Gas contrast (green overlay), which nicely demonstrate the lack of ventilation in a large right upper lobe bullae (arrow) in this patient. Source: Roos et al. [55].

Between the two noble gases ^{3}He and ^{129}Xe, ^{3}He is more sensitive because it has a larger gyromagnetic ratio. However, ^{3}He has very poor solubility in water and lipids (it is also scarcer and more expensive). ^{129}Xe is lipophilic and has a solubility in blood an order of magnitude greater than that of helium. ^{129}Xe also shows very large chemical shift differences (ca. 200 ppm) between the gas and tissue-dissolved environments [57], consequently permitting differential studies of gas and tissue-dissolved phases. Another interesting feature is that the signal-to-noise ratio in ^{3}He and ^{129}Xe NMR does not depend upon the resonant frequency ω (assuming the thermal noise in the sample dominates in experiments). This is because the hyperpolarized gases are no longer polarized by the magnet. Consequently, low-field NMR systems can be just as sensitive as high-field systems in hyperpolarized gas NMR. Hyperpolarized noble gas NMR and MRI have found a range of entirely new applications with temporal and spatial resolution comparable to proton NMR and MRI. In addition, a different technology that transfers spin polarization from electrons to nuclei, which was discovered by physicists in the 1950s and called dynamic nuclear polarization (DNP) [58], has been used to hyperpolarize other nuclear species, such as ^{13}C in injectable solutions. Hyperpolarized ^{13}C MRI can monitor specific metabolic activities such as the metabolism of pyruvate to lactate, which is enhanced in cancerous tissues [59].

References

1. Weiss C, Rosenberg L, Helfet AJ. An Ultrastructural Study of Normal Young Adult Human Articular Cartilage. *J Bone Joint Surgery*. 1968;50 A(4):663–74.

2. Jeffery AK, Blunn GW, Archer CW, Bentley G. Three-dimensional Collagen Architecture in Bovine Articular Cartilage. *J Bone Joint Surgery*. 1991;73 B(5):795–801.

3. Venn M, Maroudas A. Chemical Composition and Swelling of Normal and Osteoarthritic Femoral Head Cartilage. *Ann Rheum Dis*. 1977;36(2):121–9.

4. Maroudas A, Bayliss MT, Venn M. Further Studies on the Composition of Human Femoral Head Cartilage. *Ann Rheum Dis*. 1980;39(5):514–34.

5. Buckwalter JA, Mankin HJ. Articular Cartilage. Part I: Tissue Design and Chondrocyte-matrix Interactions. *J Bone Joint Surgery (Am)*. 1997;79:600–11.

6. Xia Y. Relaxation Anisotropy in Cartilage by NMR Microscopy (µMRI) at 14 µm Resolution. *Magn Reson Med*. 1998;39(6):941–9.

7. Rubenstein JD, Kim JK, Morava-Protzner I, Stanchev PL, Henkelman RM. Effects of Collagen Orientation on MR Imaging Characteristics of Bovine Articular Cartilage. *Radiology*. 1993;188(1):219–26.

8. Xia Y. Magic Angle Effect in MRI of Articular Cartilage – A Review. *Invest Radiol*. 2000;35(10):602–21.

9. Xia Y. Averaged and Depth-Dependent Anisotropy of Articular Cartilage by Microscopic Imaging. *Semin Arthritis Rheum*. 2008;37(5):317–27.

10. Xia Y, Moody J, Alhadlaq H. Orientational Dependence of T2 Relaxation in Articular Cartilage: A Microscopic MRI (µMRI) Study. *Magn Reson Med*. 2002;48(3):460–9.

11. Borthakur A, Wheaton AJ, Gougoutas AJ, Akella SV, Regatte RR, Charagundla SR, et al. In Vivo Measurement of T1rho Dispersion in the Human Brain at 1.5 Tesla. *J Magn Reson Imaging*. 2004;19(4):403–9.

12. Wang N, Xia Y. Orientational Dependent Sensitivities of T2 and T1rho Towards Trypsin Degradation and Gd-DTPA (2-) Presence in Bovine Nasal Cartilage. *MAGMA*. 2012;25(4):297–304.

13. Kimmich R, Winter F, Nusser W, Spohn K. Interactions and Fluctuations Deduced from Proton Field-cycling Relaxation Spectroscopy of Polypeptides, DNA, Muscles, and Algae. *J Magn Reson*. 1986;68(2):263–82.

14. Koenig SH, Schillinger WE. Nuclear Magnetic Relaxation Dispersion in Protein Solutions. I. Apotransferrin. *J Biol Chem*. 1969;244(12):3283–9.

15. Lurie DJ, Aime S, Baroni S, Booth NA, Broche LM, Choi C-H, et al. Fast Field-cycling Magnetic Resonance Imaging. *C R Physique*. 2010;11:136–48.

16. Ungersma SE, Matter NI, Hardy JW, Venook RD, Macovski A, Conolly SM, et al. Magnetic Resonance Imaging with T1 Dispersion Contrast. *Magn Reson Med*. 2006;55(6):1362–71.

17. Tanner JE, Stejskal EO. Restricted Self-diffusion of Protons in Colloidal Systems by the Pulsed-gradient, Spin-echo Method. *J Chem Phys*. 1968;49(4):1768–77.

18. Merboldt K-D, Hänicke W, Frahm J. NMR Imaging of Restricted Diffusion. *Ber Bunsenges Phys Chem*. 1987;91:1124–6.

19. D'Orazio F, Bhattacharja S, Halperin WP, Gerhardt R. Enhanced Self-diffusion of Water in Restricted Geometry. *Phys Rev Lett*. 1989;63(1):43–6.

20. Fullerton GD. The Magic Angle Effect in NMR and MRI of Cartilage. In: Xia Y, Momot KI, editors. *Biophysics and Biochemistry of Cartilage by NMR and MRI*. Cambridge: Royal Society of Chemistry; 2017. pp. 109–44.

21. Basser PJ, Mattiello J, LeBihan D. MR Diffusion Tensor Spectroscopy and Imaging. *Biophys J*. 1994;66:259–67.

22. Chenevert TL, Brunberg JA, Pips JS. Anisotropic Diffusion in Human White Matter: Demonstration with MR Technique In Vivo. *Radiology*. 1990;177:401–5.

23. Cavaliere C, Aiello M, Di Perri C, Fernandez-Espejo D, Owen AM, Soddu A. Diffusion Tensor Imaging and White Matter Abnormalities in Patients with Disorders of Consciousness. *Front Hum Neurosci*. 2014;8:1028.

24. Mori S, Zhang JY. Principles of Diffusion Tensor Imaging and Its Applications to Basic Neuroscience Research. *Neuron*. 2006;51:527–39.

25. Basser PJ, Pajevic S, Pierpaoli C, Duda J, Aldroubi A. In Vivo Fiber-tractography in Human Brain Using Diffusion Tensor MRI (DT-MRI) Data. *Magn Reson Med*. 2000;44(4):625–32.

26. Perge JA, Koch K, Miller R, Sterling P, Balasubramanian V. How the Optic Nerve Allocates Space, Energy Capacity, and Information. *J Neurosci*. 2009;29(24):7917–28.

27. Jensen JH, Helpern JA, Ramani A, Lu HZ, Kaczynski K. Diffusional Kurtosis Imaging: The Quantification of Non-Gaussian Water Diffusion by Means of Magnetic Resonance Imaging. *Magn Reson Med*. 2005;53(6):1432–40.

28. Fullerton G, Potter J, Dornbluth N. NMR Relaxation of Protons in Tissues and Other Macromolecular Water Solutions. *Magn Reson Imaging*. 1982;1:209–28.

29. Zheng S, Xia Y. Multi-components of T2 Relaxation in Ex Vivo Cartilage and Tendon. *J Magn Reson*. 2009;198(2):188–96.

30. Reiter DA, Lin PC, Fishbein KW, Spencer RG. Multicomponent T2 Relaxation Analysis in Cartilage. *Magn Reson Med*. 2009;61(4):803–9.

31. Zheng S, Xia Y. On the Measurement of Multi-component T2 Relaxation in Cartilage by MR Spectroscopy and Imaging. *Magn Reson Imaging*. 2010;28:537–45.

32. Wang N, Xia Y. Experimental Issues in the Measurement of Multi-component Relaxation Times in Articular Cartilage by Microscopic MRI. *J Magn Reson*. 2013;235:15–25.

33. Wang N, Xia Y. Dependencies of Multi-component T2 and T1rho Relaxation on the Anisotropy of Collagen Fibrils in Bovine Nasal Cartilage. *J Magn Reson*. 2011;212:124–32.

34. Mahar R, Batool S, Badar F, Xia Y. Quantitative Measurement of T2, T1rho and T1 Relaxation Times in Articular Cartilage and Cartilage-bone Interface by SE and UTE Imaging at Microscopic Resolution. *J Magn Reson*. 2018;297:76–85.

35. Lüdeke KM, Röschmann P, Tischler R. Susceptibility Artefacts in NMR Imaging. *Magn Reson Imaging*. 1985;3(4):329–43.

36. Young IR, Khenia S, Thomas DG, Davis CH, Gadian DG, Cox IJ, et al. Clinical Magnetic Susceptibility Mapping of the Brain. *J Comput Assist Tomogr*. 1987;11(1):2–6.

37. Haacke EM, Xu YB, Cheng Y-CN, Reichenbach JR. Susceptibility Weighted Imaging (SWI). *Magn Reson Med*. 2004;52:612–8.

38. Liu CL, Wei HJ, Gong NJ, Cronin M, Dibb R, Decker K. Quantitative Susceptibility Mapping: Contrast Mechanisms and Clinical Applications. *Tomography*. 2015;1(1):3–17.

39. Reichenbach JR, Venkatesan R, Schillinger DJ, Kido DK, Haacke EM. Small Vessels in the Human Brain: MR Venography with Deoxyhemoglobin as an Intrinsic Contrast Agent. *Radiology*. 1997;204:272–7.

40. Haacke EM, Mittal S, Wu Z, Neelavalli J, Cheng Y-CN. Susceptibility-weighted Imaging: Technical Aspects and Clinical Applications, Part 1. *Am J Neuroradiol*. 2009;30:19–30.

41. Wang Y, Liu T. Quantitative Susceptibility Mapping (QSM): Decoding MRI Data for a Tissue Magnetic Biomarker. *Magn Reson Med*. 2015;73:82–101.

42. Liu C. Susceptibility Tensor Imaging. *Magn Reson Med*. 2010;63:1471–7.

43. Ogawa S, Lee TM. Magnetic Resonance Imaging of Blood Vessels at High Fields: In Vivo and In Vitro Measurements and Image Simulation. *Magn Reson Med*. 1990;16:9–18.

44. Ogawa SJ, Tank DW, Menon R, Ellemann JM, Kim SG, Merkle H, et al. Intrinsic Signal Changes Accompanying Sensory Stimulation: Functional Brain Mapping with Magnetic Resonance Imaging. *Proc Natl Acad Sci USA*. 1992;89:5951–5.

45. Bandettini PA, Wong EC, Hinks RS, Tikofsky RS, Hyde JS. Time Course EPI of Human Brain Function During Task Activation. *Magn Reson Med*. 1992;25:390–7.

46. Boecker H, Kleinschmidt A, Requardt M, Haenicke W, Merboldt KD, Frahm J. Functional Cooperativity of Human Cortical Motor Areas During Self-paced Simple Finger Movements: A High-resolution MRI Study. *Brain*. 1994;117(6):1231–9.

47. Menon RS, Ogawa S, Hu X, Strupp JP, Anderson P, Ugurbil K. BOLD Based Functional MRI at Tesla Includes a Capillary Bed Contribution: Echo-planar Imaging Correlates with Previous Optical Imaging Using Intrinsic Signals. *Magn Reson Med*. 1995;33(3):453–9.

48. Biswal B, Yetkin FZ, Haughton VM, Hyde JS. Functional Connectivity in the Motor Cortex of Resting Human Brain Using Echo-planar MRI. *Magn Reson Med*. 1995;34(4):537–41.

49. Lv H, Wang Z, Tong E, Williams LM, Zaharchuk G, Zeineh M, et al. Resting-state Functional MRI: Everything that Nonexperts Have Always Wanted to Know. *Am J Neuroradiol*. 2018;39(8):1390–9.

50. Happer W, Miron E, Schaefer S, Schreifer D, van Wijngaarden WA, Zeng X. Polarization of the Nuclear Spins of Noble-gas Atoms by Spin Exchange with Optically Pumped Alkali-metal Atoms. *Phys Review A*. 1984;29(6):3092-110.

51. Driehuys B, Cates GD, Miron E, Sauer K, Walter DK, Happer W. High-volume Production of Laser Polarized 129Xe. *Appl Phys Lett*. 1996;69(12):1668-70.

52. Leawoods JC, Yablonskiy DA, Saam B, Gierada DS, Conradi MS. Hyperpolarized 3-He Gas Production and MR Imaging of the Lung. *Concepts Magn Reson*. 2001;13(5):277-93.

53. Albert MS, Cates GD, Drlehuys B, Happer W, Saam B, Springer CSJ, et al. Biological Magnetic Resonance Imaging Using Laser-polarized 129 Xe. *Nature*. 1994;370:199–201.

54. Middleton H, Black R, Saam B, Cates G, Cofer G, Guenther R, et al. MR Imaging with Hyperpolarized 3He Gas. *Magn Reson Med*. 1995;33:271.

55. Roos JE, McAdams HP, Kaushik SS, Driehuys B. Hyperpolarized Gas MRI: Technique and Applications. *Magn Reson Imaging Clin N Am*. 2015;23(2):217–29.

56. Li H, Zhao X, Wang Y, Lou X, Chen S, Deng H, et al. Damaged Lung Gas Exchange Function of Discharged COVID-19 Patients Detected by Hyperpolarized 129Xe MRI. *Sci Adv*. 2021;7:eabc8180.

57. Wagshul ME, Button TM, Li HF, Liang ZR, Springer CS, Zhong K, et al. In Vivo MR Imaging and Spectroscopy Using Hyperpolarized 129Xe. *Magn Reson Med*. 1996;36:183–91.

58. Carver TR, Slichter CP. Polarization of Nuclear Spins in Metals. *Phys Rev*. 1953;92:212.

59. Wang ZJ, Ohliger MA, Larson PEZ, Gordon JW, Bok RA, Slater J, et al. Hyperpolarized 13C MRI: State of the Art and Future Directions. *Radiology*. 2019;291:273–84.

17

Reading the Binary Data

Truly creative experiments in NMR and MRI involve the design of new pulse sequences that manipulate the spin system in a novel/unforeseen way. The design and implementation of a new pulse sequence and acquisition method depend on the individual instrument manufacturer and computer platform, which is beyond the scope of this book.

An equally true creativity comes from being able to analyze the acquired FID data and reconstruct the images in a totally new manner, when no commercial software is available to perform this particular analysis. To be able to carry out this type of creative analysis, one must understand the format of the raw data that are acquired by the spectrometer – the way the FID data are stored in the computer memory and the format of the storage. This chapter briefly discusses some of these aspects.

17.1 FORMATS OF DATA

First, we briefly review the format or convention in which we record/save the numbers. In our daily life the most common way is the decimal system, which uses the base of 10 and has 10 different numbers (0, 1, 2, 3, 4, 5, 6, 7, 8, 9). Any number bigger than 9 use two digits or numbers (e.g., 10, 11, 12, …). In computer memory the binary system is used, which uses the base of 2 and has only two different numbers (0, 1). These two numbers "0" and "1" can be considered to represent the "off" and "on" states of an electric circuit. A third relevant system also commonly used in computer science is the hexadecimal system, which uses a base of 16 and has 16 different digits (0, 1, 2, 3, 4, 5, 6, 7, 8, 9, A, B, C, D, E, F). The following table lists the conversion among these three systems.

Decimal (Base-10)	Binary (Base-2)	Hexadecimal (Base-16)
0	0	0
1	1	1
2	10	2
3	11	3
4	100	4
5	101	5
6	110	6

Essential Concepts in MRI: Physics, Instrumentation, Spectroscopy, and Imaging, First Edition. Yang Xia.
© 2022 John Wiley & Sons Ltd. Published 2022 by John Wiley & Sons Ltd.

Decimal (Base-10)	Binary (Base-2)	Hexadecimal (Base-16)
7	111	7
8	1000	8
9	1001	9
10	1010	A
11	1011	B
12	1100	C
13	1101	D
14	1110	E
15	1111	F
16	1 0000	10
17	1 0001	11
18	1 0010	12
.		
253	1111 1101	FD
254	1111 1110	FE
255	1111 1111	FF

Each number in the above table is a digit. Each hexadecimal digit represents four binary digits, which is one half of a byte and has 16 different combinations. One byte requires two hexadecimal digits or 8 binary digits, which can represent any value from 00 to FF in hexadecimal form, or 0000 0000 to 1111 1111 in binary form, or 0 to 255 in decimal form. (Note that you do not need to insert a space every four digits, but a space increases the readability of any long binary number.)

17.2   FORMATS OF DATA STORAGE

How do you crack open a boiled egg – from its big end or small end? The choice between these two possibilities was the cause of the long-running war between two fictional countries (Blefusu and Lilliput) in the novel *Gulliver's Travels* (Jonathan Swift, 1726). There were similar "wars" in computer science, related to how the binary data are stored in computer memory.

In computer memory, each cell (an individual memory bit for storage) is numbered, from 0 to *n*-1 where *n* is the total number of the storage cells in the memory (e.g., 2 MB or two megabytes). A multi-byte number takes more space and has to be stored in multiple cells. Hence one needs to know not only the value in a particular cell but also the sequence of this cell in relation with the other cells, in order to correctly read back the value of the cells. For example, a four-byte number could be stored in four consecutive cells (assume each cell can store each byte). In order to save a four-byte number written in the hexadecimal format (here we use hexadecimal numbers as the example, instead of binary numbers), "4D3C2B1A," into the computer memory, we need to not only save the values in four sequential cells (e.g., from the base address +0, +1, +2, +3) but also know how these four values are sequenced (i.e., whether the number is 1A2B3C4D or 4D3C2B1A). Note that in the original number "4D3C2B1A" as we have written down in our usual habit, "4D" is the most significant byte (MSB), while "1A" is the least significant byte (LSB). So for a four-byte number, we have three different ways to store it in the computer memory.

Big Endian means that the highest-order byte of the number (MSB) is stored in the memory at the lowest address and the low-order byte at the highest address. In this way, the big end appears first when you read the numbers from the memory. Our number would be stored as:

Base Address	+0	+1	+2	+3
Value	4D	3C	2B	1A

Examples of computer processors that use the "Big-Endian" byte order include some processors from Motorola and IBM.

Little Endian means that the low-order byte of the number (LSB) is stored in memory at the lowest address and the high-order byte at the highest address. In this way, the little end appears first when you read the numbers from the memory. Our number would be stored as:

Base Address	+0	+1	+2	+3
Value	1A	2B	3C	4D

Examples of computer processors that use the "Little-Endian" byte order include some products from Intel.

Middle Endian or Mixed Endian saves a four-byte number as two 16-bit words in the Little Endian, then themselves in the Big Endian, as:

Base Address	+0	+1	+2	+3
Value	3C	4D	1A	2B

This can be interpreted as storing the most significant "half" (16 bits) followed by the less significant half, which is in Big Endian, but with each half stored in Little-Endian format. This ordering is also known as PDP-Endianness.

In the Little-Endian form, if a number (2B1A) grows to be bigger (i.e., beyond what can be represented by a 16-bit number), you can simply add the extra information to the right (at a higher address), for example,

from

Add	0	1	2	3	4	5	6	7	8	9	A	B	C	D	E	F
#	00	00	1A	00	00	00	00	00	00	00	00	00	00	00	00	00

to

Add	0	1	2	3	4	5	6	7	8	9	A	B	C	D	E	F
#	00	00	1A	2B	00	00	00	00	00	00	00	00	00	00	00	00

to

Add	0	1	2	3	4	5	6	7	8	9	A	B	C	D	E	F
#	00	00	1A	2B	00	00	00	00	00	00	00	00	00	00	00	00

Therefore, the Little-Endian format allows us to extend the size of a number to the limits of memory without actually changing its value (i.e., "1A 2B," "1A 2B 00," or "1A 2B 00 00" are all the same number), as well as without moving the memory locations of the existing values.

In the Big-Endian format, in contrast, we would have to slide the first byte to the right, changing its address, and then extend the number toward the left. The above example would become (in the Big-Endian form),

Add	0	1	2	3	4	5	6	7	8	9	A	B	C	D	E	F
#	00	00	1A	00	00	00	00	00	00	00	00	00	00	00	00	00

to

Add	0	1	2	3	4	5	6	7	8	9	A	B	C	D	E	F
#	00	00	2B	1A	00	00	00	00	00	00	00	00	00	00	00	00

to

Add	0	1	2	3	4	5	6	7	8	9	A	B	C	D	E	F
#	00	00	00	2B	1A	00	00	00	00	00	00	00	00	00	00	00

This format keeps the digits in the correct order but forces a definite size into the number. In this format, "2B 1A," "2B 1A 00," and "2B 1A 00 00" refer to three different values.

Big-Endian numbers are easier to read when debugging a program (since the digit order is normal to a human eye), but one might think they are less intuitive because the most significant byte is at the *smaller* address (when the computer reads its memory). Little-Endian numbers enjoy some slight computational advantages since the variables in the memory do not have to be read and manipulated at their full widths. For example, a 32-bit variable in the memory such as 00 00 00 1A can be read at the same address as either 8 bits (1A), 16 bits (00 1A), or 32 bits (00 00 00 1A) as long as its value stays within bounds. Big Endian cannot do this because the relative location of the least significant byte(s) changes with the overall width of the variable. For example, 00 00 00 1A would become 00 when it is addressed as an 8-bit variable. Big-Endian numbers can therefore get corrupted if they are addressed as the wrong width.

If you think that this complication in the Endianness only occurs in digital computers, think again. In the United States and a few other countries, the dates are most commonly written as *Month, Day, Year* (e.g., "May 10th, 2006,", or "5/10/2006'), which is the Middle-Endian order. In most of Oceania and European countries, the dates are written as *Day, Month, Year* (e.g., "10th May 2006," "10/5/2006"), which is the Little-Endian order. In many other countries, including China and Japan, the ISO 8601 international standard in ordering of dates is prevalent, as *Year, Month, Day* (e.g., "2006 May 10th," or "2006/5/10"), which is the Big-Endian order.

17.3   READING UNKNOWN BINARY DATA

Once you are armed with the knowledge of the digital information storage in computers, you can start to crack-open the binary data from your imaging instruments. Many manufacturers provide you with the formats of their data; but some other manufacturers may claim proprietary privileges (which forces you to use their proprietary software). If you encounter some data with unknown format, here are the general steps to take, in order to read the data.

1. Gather as much information as possible from the manufacturer. Also search the internet – there is a good chance that someone else also had this desire earlier and has solved the format issue for you.
2. Obtain one or two sets of binary data, as completed and simple as possible. The completeness in NMR and MRI experiments means to obtain both the acquisition files (FID, pulse

sequence, acquisition parameters, machine model) and the processed files (spectrum or image, processing parameters). For any 2D image, make sure that you print the image in a hard copy and know the size of the image (e.g., an image of 256 pixels by 256 pixels). The simpleness means that your first trial should be data for a single-sliced 2D image, not the data from a multi-slice multi-echo (MSME) imaging experiment (or in spectroscopy, first look at a straightforward 1D FID, not a 2D FID).

3. On the monitor screen of an MRI instrument, open the image and move the cursor to the top left corner (which often has just noise) and write down the values of the first several pixels, one by one sequentially in the first row of the image.

4. Note that 1 kB of data has 1024 bytes. So 64 kB of data should be 65,536 bytes, not 64,000 bytes.

5. Do a quick calculation for the data: If the FID or image has a pixel array of 256 pixels by 256 pixels, you should have a total pixel number of 65,536 (= 256 × 256) in the entire data. If your data has a 32-bit or 4-byte depth per pixel (most common these days in modern instruments), the size of your FID file without any head or tail should be 256 kB or 262,144 bytes (= 256 × 256 x 4). If the data are saved in 16 bits (uncommon, in order to save computer memory), the size of your FID file without any head or tail should be 128 kB or 131,072 bytes (= 256 × 256 x 2). The sizes for a reconstructed 2D image should be the same when it is in any binary format (not any graphical format such as jpeg etc.). For 1D NMR spectroscopy data, the calculation is the same – an 8k 32-bit FID data should be 8 × 1024 x 4 = 32,768 bytes in size. You should be able to check the size of any digital file in the computer without opening it, from the desktop of your computer. Remember, the size has to be *exact*. Any size other than 262,144 bytes or 131,072 bytes means there are other elements in the file (e.g., a head, a tail, a certain format, a compression).

6. Note also that the raw FID data (images and spectrum data) in NMR and MRI experiments should contain both real and imaginary data, commonly stored in an alternating way like Re1, Im1, Re2, Im2, Re3, Im3, …. Hence, the file size for the FID data should be twice as big as the calculation in Step 5. The magnitude images and processed spectrum should not contain the real and imaginary components, hence have the same sizes as in step 5.

7. On your computer, download and set up two essential software programs: one is for image analysis such as *ImageJ*, which can be downloaded for free from the website of its developer (https://imagej.nih.gov/ij, accessed 2020.6.1); the second is a binary reader, such as a free hex editor *0xED* (https://www.suavetech.com/0xed, accessed 2020.6.1).

8. If the file has the exact size as in step 5, you are lucky, since you most likely only need to use the image analysis software. In this case, open *ImageJ*, go to File/Import/Raw, select the file you want to read, which leads you to a parameter panel (Figure 17.1a). From this you can define the size of your file (e.g., 256, 256), the Endianness (big, little), and the format of your bytes (Figure 17.1b). Although you would likely get the format all wrong in your first trial (you should know if it is wrong when you see a window of "noises" without any clearly defined object), the possible combinations are limited – so just keep trying for a different set of combinations when you open the file again. If you see a recognizable image on the computer window, compare the values of the first several pixels on your computer window with the values you have written down from the MRI scanner. If the values agree and the image looks good, you have succeeded in reading an unknown binary file.

9. If the file size does not fit your calculation in step 5, it is likely that the file has some extra elements or format. If the file size is actually smaller than your calculation, which is uncommon, this likely means either your calculation is wrong or the file has some compression. Mostly, the file size is larger than your calculation. In that case, you need to use the binary reader, which can open any file. After opening your file, you start to examine the patterns of the binary data. Two screen prints from the software *0xED* are shown in

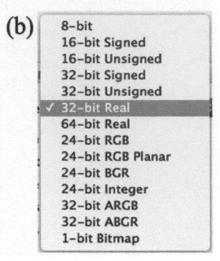

Figure 17.1 (a) The selections in the "File/Import/Raw" menu of *ImageJ*, which lets you specify several essential parameters. (b) From the selection of "Image type," one can specify the depth of the data.

Figure 17.2a (a binary file without any header) and Figure 17.2b (a binary file with a header). In each of the windows, the left part is the address of the file, which starts at 00; the middle part is the binary format of your file; the right part is the text format of your file; and the lines at the bottom show you the value of your selection if you use your cursor to select a few bytes in the binary file. Comparing these two files, you can recognize a clear difference between the two files: the data in the left window (Figure 17.2a) seems continuous without any clear pattern, while the data in the right window (Figure 17.2b) has a change in the pattern, starting from the final quarter of data in line 4, marked on the figure by an arrow. You should also scroll down the window to its end, to see if you recognize any similar patterns within the file (it would indicate some further format, e.g., a sub-header between images in a multi-imaging data) and at the end of the file (it would indicate a tail).

(10) Once you suspect a certain segment of the data is a header (or a tail), you can use the software to strip the header (or tail). After you are done with stripping all odd-looking patterns, save the worked file to a different name, and do another calculation to see if the size of the remaining file meets your expectation from step 5. If yes, try step 8 to open the file. It no, either there are still some remaining elements of format in the file or the file has some peculiar or proprietary structure.

Following these essential steps, you can successfully open and correctly read many unknown binary files and are ready to analyze the data in a totally new manner! It is great fun to do this kind of data opening, much like a true detective, in science!

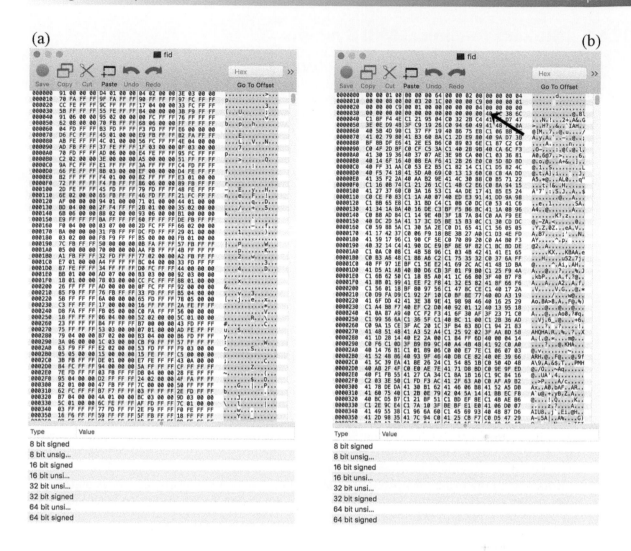

Figure 17.2 (a) Binary view of an FID data that does not have a header ("2018 μMRI-Bruker 4-C6R2-T2@0," the first part of a large data). (b) Binary view of an FID data that has a header ("2018 mMRI-Varian C5-Left-Coronal").

17.4 EXAMPLES OF SPECIFIC FORMATS

Over the course of the past 25 years, I helped my students to read and process various binary files generated on diverse instruments, from several different MRI and NMR systems to various optical imaging systems. Here is some information on the file formats in Bruker NMR and pre-clinical MRI instruments.

17.4.1   Bruker Pre-clinical MRI Systems

For the current versions of Bruker systems (AVANCE models), each experiment (spectroscopy or imaging) generates a directory. Inside the directory there is an FID file (named as *fid*), several small files that can be read by any text editor (which have the acquisition parameters), and a directory named as *pdata*, which stands for the processed data. Inside the *pdata* directory, one can find another set of directories, each directory corresponding to one round of post-acquisition processes. If you process the FID data only once, there should be only one sub-directory with the name *1*. Inside this *1* directory, there is a binary file with a name as *2dseq*, which is your processed image(s), as well as several small (text) files for processing parameters.

The modern Bruker data has the Little-Endian format; the *fid* data are saved as 32-bit signed. The best news for the Bruker pre-clinical data is that they are not structured and have no header or tail. (This is in contrast to the binary data from some other pre-clinical MRI scanners by Varian [a company that no longer exists], which has structured data with a header and some extra bytes inside.) If you have a choice, use the *2dseq* file, since it has been processed and hence you do not need to separate the real and imaginary pixels. If this *2dseq* file is not available, you have to work on the *fid* file, which should have both real and imaginary data, commonly existing alternatively in one single file, stored sequentially as Re1, Im1, Re2, Im2, Re3, Im3, ... The data shown in Figure 17.2a are from a *fid* file for a 2D imaging experiment. If the data are from an MSME experiment, there will be some type of data organization, depending upon how your pulse sequence is programmed to acquire the signal. For example, if the binary *fid* file comes from a 10-slice and 10-echo experiment using Bruker's MSME sequence, which means that there are 100 images in a single file, the first 100 lines in the *fid* file would be the first *k*-space FID line for the 100 images, and the next 100 lines are the second *k*-space FID line for the same 100 images, ... Since there is no header or structure in this massive file, to read it properly you need to figure out the way that the different loops in the pulse sequences have been set up – the data are saved as the pulse sequence is programmed in signal acquisition.

Note that on the Bruker AVANCE MRI system, the pixel location for any 2D image on the computer monitor starts with the address of 1, 1 (i.e., $x = 1$ and $y = 1$), while the pixel location of any 2D image in *ImageJ* starts with an address of 0, 0 (i.e., $x = 0$ and $y = 0$).

17.4.2   Bruker NMR Spectroscopy Systems

For the current versions of Bruker NMR systems (AVANCE models), each spectroscopy experiment also generates a directory, which has a near-identical structure as in the imaging systems (i.e., having a binary data named as *fid* and a number of small parameter files in text format). An important difference is in the *pdata* directory. For a 1D spectroscopy experiment, the *pdata* directory will have two files named as *1r* and *1i*, which are the real and imaginary spectra. If you open the *1r* file, it would be your final spectrum, properly phase adjusted. For a 2D spectroscopy experiment, the *pdata* directory will have a file named as *2rr*.

Similarly, it would be the easiest to work with the data files inside the *pdata* directory, since they have the real and imaginary parts already separated. You can use *ImageJ* to import the raw file named *1r*; enter the following parameters in importing: 32-bit signed, Little-Endian, Width = 1, Height = the number of data points (e.g., enter 32768 if the size of your *1r* file is 131 kB (= 131,072 bytes), since 131,072/4 = 32768). After importing the *1r* file into *ImageJ*, there is no need to look at it, since it is an elongated image with only one column. You just save it as a text image, which would be a structured file with a *tab* or *return* inserted between any two numbers. You can then use any graph software to open this text image and plot any segment of the spectrum.

If you must read the *fid* file from NMR spectroscopy, you can also use *ImageJ* to import the *fid* file, with the import parameters set at 32-bit signed, Width = 2, Height = # of data points per channel (1/2 of the total data points), and Little Endian. You would get an elongated image with only two columns; one column is for the real data and the other column for the imaginary data. You can use *ImageJ* to view the FID in each column, then save them as a text image, which can be read and processed by other generic software.

Appendices

Appendices 1–3 contain the background information on several topics that are relevant to the readers of this book. These appendices are not intended to be proper teaching materials on these topics; rather, they list some useful equations and conclusions and are intended as a quick reference. Appendix 4 has several sample syllabi that can be used for a one-semester course or a short introductory course. Appendix 5 contains a number of homework problems.

Appendix 1
Background in Mathematics

A1.1 ELEMENTARY MATHEMATICS

In physics, a coordinate system defines the space where an action is carried out and measured. The most common coordinate system is the three-dimensional (3D) Cartesian coordinate system, which has three orthogonal axes (commonly labeled as the *x, y, z* axes). One also uses three unit vectors *i*, *j*, *k*, to define the directions of these orthogonal axes. By eliminating the third axis *z*, one can have a two-dimensional (2D) Cartesian coordinate, commonly in the *x–y* plane. In addition to the Cartesian coordinate system, other commonly used systems are the cylindrical coordinate system (ρ, θ, z) and the spherical coordinate system (r, θ, φ).

A1.1.1 Scalars and Vectors

In physics, *scalars* are physical quantities that are fully described by a numerical value (often together with a unit of measurement), for example, 35 kg or 100 °C. Graphically, a scalar can be represented by a dot on a one-dimensional (1D) axis often together with a reference (e.g., 0 °C). In contrast, **vectors** are quantities that are described by both a numerical value (its magnitude) and a direction, which are commonly written with a bold font, such as *A* or *a*. For example, velocity is a vector, which has both the magnitude (e.g., 90 km/s) and direction (e.g., northwest). Graphically, a vector can be represented by an arrow in a 2D plane, where the arrow has both a length (the magnitude) and an angle in the 2D plane (the direction).

A generic vector in the 3D Cartesian coordinate system can be written as $\mathbf{A} = A_x\mathbf{i} + A_y\mathbf{j} + A_z\mathbf{k}$. When this vector *A* is presented in a different coordinate system, the vector *A* itself does not change, but the components of the vector *A* depend upon the coordinate system used. For example, the same vector *A* would have three different components in a cylindrical coordinate system, as $\mathbf{A} = A_\rho\mathbf{r} + A_\theta\boldsymbol{\theta} + A_z\mathbf{k}$.

A1.1.2 Dot and Cross Products in Vector Analysis

The dot product in the vector analysis refers to the multiplication of two vectors *A* and *B*, which yields a scalar,

$$\mathbf{A} \cdot \mathbf{B} = AB\cos\theta, \qquad (A1.1)$$

Essential Concepts in MRI: Physics, Instrumentation, Spectroscopy, and Imaging, First Edition. Yang Xia.
© 2022 John Wiley & Sons Ltd. Published 2022 by John Wiley & Sons Ltd.

where the angle θ is between the two vectors $\boldsymbol{A}$ and $\boldsymbol{B}$. The dot product is noted by a solid dot between the two vectors $\boldsymbol{A}$ and $\boldsymbol{B}$.

The standard basis vectors in the Cartesian coordinate system, $\boldsymbol{i}$, $\boldsymbol{j}$, and $\boldsymbol{k}$, satisfy the following equalities in a right-hand coordinate system:

$$\boldsymbol{i} \cdot \boldsymbol{i} = \boldsymbol{j} \cdot \boldsymbol{j} = \boldsymbol{k} \cdot \boldsymbol{k} = 1 \tag{A1.2a}$$

and

$$\boldsymbol{i} \cdot \boldsymbol{j} = \boldsymbol{j} \cdot \boldsymbol{k} = \boldsymbol{k} \cdot \boldsymbol{i} = 0, \tag{A1.2b}$$

so that the dot product of two vectors $\boldsymbol{A}$ and $\boldsymbol{B}$ can be written as

$$\boldsymbol{A} \cdot \boldsymbol{B} = A_x B_x + A_y B_y + A_z B_z. \tag{A1.3}$$

The cross product in the vector analysis refers to the multiplication of two vectors $\boldsymbol{A}$ and $\boldsymbol{B}$, which yields a vector quantity,

$$\boldsymbol{A} \times \boldsymbol{B} = \boldsymbol{A}\boldsymbol{B} \sin\theta \boldsymbol{n}, \tag{A1.4}$$

where the vector $\boldsymbol{n}$ is a unit vector along an axis that is orthogonal to the plane defined by the two vectors $\boldsymbol{A}$ and $\boldsymbol{B}$, as shown in Figure A1.1. The cross product is noted by a solid cross between the two vectors $\boldsymbol{A}$ and $\boldsymbol{B}$.

The standard basis vectors in the Cartesian coordinate system, $\boldsymbol{i}$, $\boldsymbol{j}$, and $\boldsymbol{k}$, satisfy the following equalities in a right-hand coordinate system:

$$\boldsymbol{i} \times \boldsymbol{j} = \boldsymbol{k}, \ \boldsymbol{j} \times \boldsymbol{k} = \boldsymbol{i}, \ \boldsymbol{k} \times \boldsymbol{i} = \boldsymbol{j}, \tag{A1.5a}$$

which imply, by the anticommutativity of the cross product, that

$$\boldsymbol{j} \times \boldsymbol{i} = -\boldsymbol{k}, \ \boldsymbol{k} \times \boldsymbol{j} = -\boldsymbol{i}, \ \boldsymbol{i} \times \boldsymbol{k} = -\boldsymbol{j}. \tag{A1.5b}$$

The anticommutativity of the cross product (and the obvious lack of linear independence) also implies that

$$\boldsymbol{i} \times \boldsymbol{i} = \boldsymbol{j} \times \boldsymbol{j} = \boldsymbol{k} \times \boldsymbol{k} = 0. \tag{A1.6}$$

The cross product of two vectors $\boldsymbol{A}$ and $\boldsymbol{B}$ can be written

$$\boldsymbol{A} \times \boldsymbol{B} = \begin{vmatrix} \boldsymbol{i} & \boldsymbol{j} & \boldsymbol{k} \\ A_x & A_y & A_z \\ B_x & B_y & B_z \end{vmatrix} = \boldsymbol{i}\left(A_y B_z - A_z B_y\right) + \boldsymbol{j}\left(A_z B_x - A_x B_z\right) + \boldsymbol{k}\left(A_x B_y - A_y B_x\right). \tag{A1.7}$$

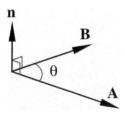

Figure A1.1 The directions in the vector cross product of two vectors $\boldsymbol{A}$ and $\boldsymbol{B}$, where $\boldsymbol{n}$ is a unit vector for the 2D surface defined by $\boldsymbol{A}$ and $\boldsymbol{B}$.

A1.1.3 Complex Numbers

A complex number z has a real part a and an imaginary part b, which can be visualized graphically in an Argand diagram (Figure A1.2) and written in the following different ways:

$$z = (a, b), \tag{A1.8a}$$

$$z = a + ib, \tag{A1.8b}$$

$$z = |z|(\cos\theta + i\sin\theta), \tag{A1.8c}$$

$$z = |z| \exp(i\theta), \tag{A1.8d}$$

where the complex i signals the imaginary part of the complex number z that has a 90° rotation in the Argand diagram. (In some texts, especially electronic and electrical engineering literature, a complex j instead of i is used to symbolize the imaginary part, since the symbol I is reserved for the electric current.) It is common to define $i = \sqrt{-1}$, and $i^2 = -1$. The exponential term $\exp(i\theta)$ in Eq. (A1.8d) is called a complex exponential, which has $\exp(i\theta) = \cos\theta + i\sin\theta$.

The modulus or absolute value of the complex number z is

$$|z| = r = \left(a^2 + b^2\right)^{1/2}. \tag{A1.9}$$

The format of Eq. (A1.8b) is convenient in the addition and subtraction of the complex numbers, since

$$z_1 \pm z_2 = (a_1 \pm a_2, b_1 \pm b_2) = (a_1 \pm a_2) + i(b_1 \pm b_2). \tag{A1.10}$$

The format of Eq. (A1.8d) is convenient in the multiplication and division of the complex numbers, since

$$z_1\, z_2 = |z_1|\, |z_2| \exp\left[i(\theta_1 + \theta_2)\right]. \tag{A1.11}$$

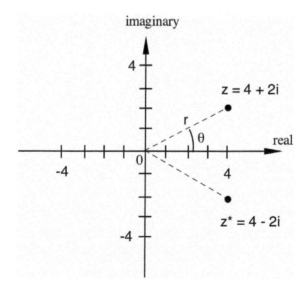

Figure A1.2 The Argand diagram of complex numbers z and z^*. z^* is the complex conjugate of z.

A complex conjugate of a complex number has the same real part but the opposite imaginary part, which is shown in Figure A1.2. A complex conjugate of z is indicated as z^*. The following descriptions apply to a complex conjugate of the complex number z, when z is defined in Eq. (A1.8):

$$(a + ib)^* = (a - ib), \tag{A1.12a}$$

$$z^* = |z| \exp(-i\theta), \tag{A1.12b}$$

$$z z^* = |z|^2 = a^2 + b^2. \tag{A1.12c}$$

If $z^* = z$, then z is a real number.

A1.1.4 Taylor Expansions

A Taylor expansion or Taylor series can replace a function by an infinite sum of terms, where each term has an increasingly larger exponent like x, x^2, x^3, and so on. A Taylor series is commonly used in physics as a way to simplify an equation. Although using an infinite sum of terms to replace an equation, no matter how complicated it is, does not sound like a good idea, the trick is to use only the first several terms in the infinite sum of terms, together with the knowledge of a certain small and finite error. A few useful expansions are

$$e^x = \exp(x) = 1 + \frac{x}{1!} + \frac{x^2}{2!} + \frac{x^3}{3!} + \frac{x^4}{4!} + \frac{x^5}{5!} + \ldots + \frac{x^n}{n!} + \ldots, \text{ for } -\infty < x < \infty, \tag{A1.13}$$

$$\ln(1 + x) = x - \frac{x^2}{2} + \frac{x^3}{3} - \frac{x^4}{4} + \frac{x^5}{5} - \ldots, \text{ for } -1 < x < 1, \tag{A1.14}$$

$$\sin x = x - \frac{x^3}{3!} + \frac{x^5}{5!} - \frac{x^7}{7!} + \frac{x^9}{9!} - \ldots, \tag{A1.15}$$

$$\cos x = 1 - \frac{x^2}{2!} + \frac{x^4}{4!} - \frac{x^6}{6!} + \frac{x^8}{8!} - \ldots, \tag{A1.16}$$

$$\frac{1}{1 \pm x} = 1 \mp x + x^2 \mp x^3 + x^4 \mp \ldots, \text{ for } -1 < x < 1, \tag{A1.17}$$

where the symbol $!$ is the factorial function (e.g., $4! = 4 \times 3 \times 2 \times 1 = 24$).

A1.1.5 Matrix and Operations

A matrix in mathematics is a rectangular array of numbers (or symbols, or expressions). The matrices in consideration are mostly square matrices, which have an equal number of rows (run horizontally) and columns (run vertically). For example, a 3 by 3 matrix can be defined as

$$A \equiv \begin{bmatrix} a_{11} & a_{12} & a_{13} \\ a_{21} & a_{22} & a_{23} \\ a_{31} & a_{32} & a_{33} \end{bmatrix}. \tag{A1.18}$$

In particular, three elements a_{11}, a_{22}, and a_{33} are called the diagonal elements. Some of the simple matrix operations are

$$\begin{bmatrix} a_{11} & a_{12} \\ a_{21} & a_{22} \end{bmatrix} \pm \begin{bmatrix} b_{11} & b_{12} \\ b_{21} & b_{22} \end{bmatrix} = \begin{bmatrix} a_{11} \pm b_{11} & a_{12} \pm b_{12} \\ a_{21} \pm b_{21} & a_{22} \pm b_{22} \end{bmatrix} \tag{A1.19}$$

and

$$\begin{bmatrix} a_{11} & a_{12} \\ a_{21} & a_{22} \end{bmatrix} \begin{bmatrix} b_{11} & b_{12} \\ b_{21} & b_{22} \end{bmatrix} = \begin{bmatrix} a_{11}b_{11} + a_{12}b_{21} & a_{11}b_{12} + a_{12}b_{22} \\ a_{21}b_{11} + a_{22}b_{21} & a_{21}b_{12} + a_{22}b_{22} \end{bmatrix}. \tag{A1.20}$$

In linear algebra, the trace of a matrix A is the sum of elements on the main diagonal (from the upper left to the lower right) of A, as

$$\text{Tr}A = \sum_i a_{ii} = a_{11} + a_{22} + a_{33}. \tag{A1.21}$$

A determinant is a scalar value that can be computed from the elements of a square matrix A, as

$$|A| = \begin{vmatrix} a & b \\ c & d \end{vmatrix} = ad - bc. \tag{A1.22}$$

A rotation matrix is a matrix that performs a rotation of another matrix in a space, for example, the Cartesian space. The following 2×2 matrix rotates another matrix in the 2D xy plane of a Cartesian coordinate system *clockwise* through an angle θ about the origin (Figure A1.3),

$$R_c = \begin{bmatrix} \cos\theta & \sin\theta \\ -\sin\theta & \cos\theta \end{bmatrix}. \tag{A1.23}$$

For a point (x, y) in the original coordinate system, the new coordinates (x', y') after the rotation are

$$x' = x\cos\theta + y\sin\theta, \tag{A1.24a}$$

$$y' = -x\sin\theta + y\cos\theta. \tag{A1.24b}$$

A 2×2 *counterclockwise* rotation matrix can be written as

$$R_{cc} = \begin{bmatrix} \cos\theta & -\sin\theta \\ \sin\theta & \cos\theta \end{bmatrix}. \tag{A1.25}$$

The rotation described in Eq. (A1.23) is consistent with the rotation convention specified in Figure 1.3. Note that the rotation references such as "*clockwise*" and "*counterclockwise*" have specific meanings, while the meaning of the descriptive terms such as "positive" and "negative" would need an additional reference. Whether θ in the rotation matrix is positive or negative will depend upon if one uses a left-handed or a right-handed rotation. When a right-handed notation is used as the rotation reference, a negative θ will result in a clockwise rotation, while a positive θ will result a counterclockwise rotation.

A1.2 FOURIER TRANSFORM

The Fourier transform is a mathematic operation that converts a waveform (a function, signal) in one domain/space into an alternate representation in another domain/space, with the use of sine and cosine functions. The most commonly used pair of FT domains are time t and frequency f,

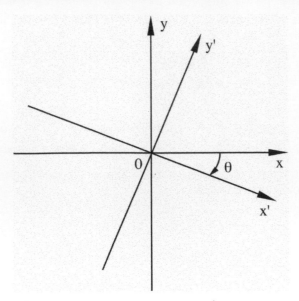

Figure A1.3 A 2D matrix rotation.

which are said to be conjugate to each other (which is used in the Fourier transform in NMR between the FID signal and the spectrum); position *r* and wave-vector *k* are also conjugate to each other, which are used in the Fourier transform in MRI.

A1.2.1 Definitions

Depending upon if the regular cyclic frequency f or the angular frequency ω is used in the transform equations ($\omega = 2\pi f$), the equations for Fourier transform can be written in different ways. When the cyclic frequency in the unit of hertz is used, the FT equations between two functions f(t) and F(f) are

$$F(f) = \int_{-\infty}^{+\infty} f(t)\exp(-i2\pi ft)dt, \qquad (A1.26a)$$

$$f(t) = \int_{-\infty}^{+\infty} F(f)\exp(i2\pi ft)df. \qquad (A1.26b)$$

Equation (A1.26a), which has the -i in the exponential, is commonly called the forward Fourier transform, and Eq. (A1.26b) is commonly called the inverse Fourier transform. Two functions f(t) and F(f) contain the same information.

The equations for Fourier transform in terms of the angular frequency ω in the unit of radians per second between two functions, f(t) and F(ω), can be written either asymmetrically or symmetrically, as

• **the asymmetric form:**

$$F(\omega) = \int_{-\infty}^{+\infty} f(t)\exp(-i\omega t)dt, \qquad (A1.27a)$$

$$f(t) = \frac{1}{2\pi} \int_{-\infty}^{+\infty} F(\omega)\exp(i\omega t)d\omega. \qquad (A1.27b)$$

• **the symmetric form:**

$$F(\omega) = \frac{1}{\sqrt{2\pi}} \int_{-\infty}^{+\infty} f(t)\exp(-i\omega t)dt, \qquad\qquad (A1.28a)$$

$$f(t) = \frac{1}{\sqrt{2\pi}} \int_{-\infty}^{+\infty} F(\omega)\exp(i\omega t)d\omega. \qquad\qquad (A1.28b)$$

Two short notations can also be used to indicate a Fourier relationship between the two related functions:

$$\mathcal{F}\{f(x)\} = F(k) \text{ and } \mathcal{F}^{-1}\{F(k)\} = f(x) \qquad\qquad (A1.29a)$$

or

$$F(k) \Leftrightarrow f(x). \qquad\qquad (A1.29b)$$

Of course, the dimension of the space can be more than one. Hence, in general, between two 3D spaces $\boldsymbol{r}$ and $\boldsymbol{k}$, we have

$$F(\boldsymbol{k}) = \int_{-\infty}^{+\infty} f(\boldsymbol{r})\exp(-i2\pi\boldsymbol{k}\cdot\boldsymbol{r})d\boldsymbol{r}, \qquad\qquad (A1.30a)$$

$$f(\boldsymbol{r}) = \int_{-\infty}^{+\infty} F(\boldsymbol{k})\exp(i2\pi\boldsymbol{k}\cdot\boldsymbol{r})d\boldsymbol{k}. \qquad\qquad (A1.30b)$$

Because of the use of sine and cosine functions, Fourier transform has some symmetry relationships that can be exploited. For example, if the original function $f(t)$ is real and even, then the FT function $F(k)$ should also be real and even. Most books about the Fourier transform have tables and figures that describe the symmetries in FT [1]. In fact, since $\exp(i\theta) = \cos\theta + i\sin\theta$, it is possible to use only the Fourier sine transform or the Fourier cosine transform, if the characteristics of the original function $f(t)$ are known:

$$F_c(\boldsymbol{k}) = 2\int_0^{+\infty} f_c(\boldsymbol{r})\cos(2\pi\boldsymbol{k}\cdot\boldsymbol{r})d\boldsymbol{r} \qquad\qquad (A1.31a)$$

and

$$f_c(\boldsymbol{r}) = 2\int_0^{+\infty} F_c(\boldsymbol{k})\cos(2\pi\boldsymbol{k}\cdot\boldsymbol{r})d\boldsymbol{k}, \qquad\qquad (A1.31b)$$

$$F_s(\boldsymbol{k}) = 2\int_0^{+\infty} f_s(\boldsymbol{r})\sin(2\pi\boldsymbol{k}\cdot\boldsymbol{r})d\boldsymbol{r} \qquad\qquad (A1.32a)$$

and

$$f_s(\boldsymbol{r}) = 2\int_0^{+\infty} F_s(\boldsymbol{k})\sin(2\pi\boldsymbol{k}\cdot\boldsymbol{r})d\boldsymbol{k}. \qquad\qquad (A1.32b)$$

A1.2.2 Convolution Theorem

The FT of a product of two functions is the convolution of the two transformed functions in conjugate space, which can be expressed as

$$f(\mathbf{r}) \otimes g(\mathbf{r}) \iff F(\mathbf{k})G(\mathbf{k}), \tag{A1.33}$$

where $\otimes$ is the symbol for a convolution; F and G are the FTs of f and g; and the convolution integral is given in 1D form by

$$f(x) \otimes g(x) = \int_{-\infty}^{+\infty} f(x')g(x-x')dx'. \tag{A1.34}$$

Convolution effects are always apparent when the signal is sampled in a conjugate space and a truncation is involved. The most common example is the truncation and zero-filling in NMR (see Chapter 6.8) and MRI (see Chapter 13.1). The convolution process can be thought of as a running average where $f(x)$ is called the weighting function and the output $f(x) \otimes g(x)$ is the weighted function $g(x)$. There are a number of consequences associated with the convolution theorem; for example, the Dirac delta function and the shift theorem. Some of these consequences are relevant to MRI applications; for example, the quantitative calculations of translational flow and self-diffusion in MRI, as shown in one example in Figure 15.4 [2].

A1.2.3 Digital Fourier Transform

All formulas of Fourier transform in the previous sections are analytical expressions, which are continuous with an infinite space dimension. In current practice, we are doing signal acquisition using computers and digital electronics, where the signal is digitized in a *finite, discrete* space. (The integral to + and − infinities can be avoided if the signal already decays to zero at finite values; but if it does not, one obtains the cutoff or truncation artifacts – see Figure 13.1.) Two functions, $f(nT)$ and $F(k/NT)$, given by

$$f(nT) \text{ where } n = 0, \cdots, N-1 \tag{A1.35a}$$

and

$$F\left(\frac{k}{NT}\right) \text{ where } k = -\frac{N}{2}, \cdots, \frac{N}{2} - 1, \tag{A1.35b}$$

are termed an FT pair if they are related with each other via the discrete Fourier transform, by the following two equations:

$$F\left(\frac{k}{NT}\right) = \sum_{n=0}^{N-1} f(nT)\exp(-i2\pi kn/N), \ k = 0, 1, 2, \cdots, N-1, \tag{A1.36a}$$

$$f(nT) = \frac{1}{N}\sum_{k=0}^{N-1} F\left(\frac{k}{NT}\right)\exp(i2\pi kn/N), \ n = 0, 1, 2, \cdots, N-1, \tag{A1.36b}$$

where N is the number of the discrete digits; T is the "time domain" sampling interval; $1/NT$ is the "frequency domain" sampling interval; n is the variable in the time domain; and k is the variable in the frequency domain. Figure A1.4 shows the discrete time and frequency domains [2,3].

(a) time domain

(b) frequency domain

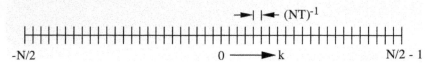

Figure A1.4   Discrete time and frequency domains in Fourier transform. Source: Adapted from Xia {2}.

References

1. Bracewell R. *The Fourier Transform and Its Applications*. New York: McGraw-Hill Book Company; 1965.
2. Xia Y. *Static and Dynamic Imaging Using Magnetic Field Gradients [MSc Thesis]*. Massey University, New Zealand; 1988.
3. Callaghan PT. *Principles of Nuclear Magnetic Resonance Microscopy*. Oxford: Oxford University Press; 1991.

315

Appendix 2

Background in Quantum Mechanics

In a quantum mechanical description of NMR, the nuclear spin systems are represented by the complex wave functions that are labeled with Greek letters Ψ, Φ, φ, ψ. These wave functions can be described by the Dirac notation, which uses a vectoral term called a ket and is written as $|\varphi>$. This vectoral ket has a conjugated vector called the bra and is written as $<\varphi|$. Since $<\varphi|$ is defined as the complex conjugate of $|\varphi>$, the scalar product of the two vectors $|\psi>$ and $|\varphi>$ is represented by the contraction of the bra $<\psi|$ by the ket $|\varphi>$ and can be written as $<\psi|\varphi>$.

The spin systems can be described by the Schrödinger equation ($\mathcal{H}\psi = E\psi$), where the operator $\mathcal{H}$ is a Hamiltonian that contains the spin angular momentum operators. The wave function itself cannot be observed, but the results of any observation can be predicted from it. The stable states of quantum mechanical systems are the eigenfunctions of the energy operator $\mathcal{H}$, hence to calculate NMR spectra, we must find the eigenvalues of $\mathcal{H}$.

A2.1 OPERATORS

An operator is a general term in mathematics. It acts on a number or function to generate another number or another function. For example, in a simple multiplication, $2 \times 3 = 6$, $\times$ is an operator that generates 6 by multiplying 3 by 2. A slightly more complicated example is a differential calculation $d(\sin(x))/dx = \cos(x)$, where d/dx is an operator, and $\sin(x)$ and $\cos(x)$ are both functions. In quantum mechanics, an operator $\mathbf{A}$ acts on a function to yield another function, such as

$$\mathbf{A}f(\mathbf{r}) = g(\mathbf{r}), \tag{A2.1}$$

where $f(\mathbf{r})$ and $g(\mathbf{r})$ are functions and $\mathbf{r}$ represents a mathematical space where the two functions exist.

Some properties of operators relevant to our topic are listed here.

1. Two operators $\mathbf{A}_1$ and $\mathbf{A}_2$ are said to "commute" if

$$(\mathbf{A}_1\mathbf{A}_2 - \mathbf{A}_2\mathbf{A}_1) = [\mathbf{A}_1, \mathbf{A}_2] = 0, \tag{A2.2}$$

where $[\mathbf{A}_1, \mathbf{A}_2]$ is called the *commutator*.

Essential Concepts in MRI: Physics, Instrumentation, Spectroscopy, and Imaging, First Edition. Yang Xia.
© 2022 John Wiley & Sons Ltd. Published 2022 by John Wiley & Sons Ltd.

In general, two operators cannot be assumed to be commute, as

$$(\mathbf{A_1 A_2} - \mathbf{A_2 A_1}) \neq 0. \tag{A2.3}$$

This equation implies that the order in which the operators operate on a function cannot be assumed to be switchable, unless they commute.

2. If an operator acts on a wave function and returns the product of a number and the wave function itself, as

$$\mathbf{A}|\psi> = a|\psi>, \tag{A2.4}$$

where $\mathbf{A}$ is an operator and a is a number, then $|\psi>$ is called an eigenfunction of $\mathbf{A}$ with an eigenvalue of a. The equation is called the eigenvalue equation. Note that the eigenvalues of an operator are intrinsic properties of the operator.

Example

$$\frac{d}{dx}\left(\exp(kx)\right) = k\exp(kx)$$

where d/dx is the operator, $\exp(kx)$ is the eigenfunction, k is the eigenvalue, and the equation is an eigenvalue equation. In this example, many k can satisfy this equation. In quantum mechanics, however, the number of eigenvalues is finite.

---

3. If two operators $\mathbf{A}$ and $\mathbf{B}$ do commute, we can choose states that are *simultaneous* eigenstates of both operators. The result of each operation is its eigenvalue and the order of the operation is immaterial. Hence we have

$$\mathbf{AB}|a,b> = \mathbf{A}b|a,b> = ab|a,b> \tag{A2.5a}$$

and

$$\mathbf{BA}|a,b> = \mathbf{B}a|a,b> = ba|a,b>, \tag{A2.5b}$$

where a and b are the eigenvalues.

4. The action of an operator $\mathbf{A}$ on a ket $|\psi>$ is, in general, to transform it into another ket $|\psi'>$, unless the situation is described previously by the eigenequation in Eq. (A2.4):

$$\mathbf{A}|\psi> = |\psi'> \tag{A2.6}$$

and

$$<\varphi|\mathbf{A} = <\varphi'|. \tag{A2.7}$$

5. The scalar product can be written as

$$<\varphi|\mathbf{A}|\psi> = \int \varphi^* \mathbf{A}\psi\, dt, \tag{A2.8}$$

where φ^* is the complex conjugate of φ.

6. If $A|\psi> = |\psi'>$ and $<\psi'| = <\psi|A^+$, then A and A^+ are said *Hermitian conjugate*. The consequences of being Hermitian conjugate are (a) the eigenvalues of Hermitian operators are real (i.e., $A|\psi> = a|\psi>$ and a is a real number), and (b) the eigenfunctions corresponding to different eigenvalues are orthogonal to each other (their scalar product is zero).

7. The trace of an operator is given by

$$\text{Tr}A = \sum_i <i|A|i>, \tag{A2.9}$$

where $<i|A|i>$ is the diagonal elements of the operator A.

Note that the trace of an operator is independent of the ket basis, although individual diagonal elements may depend on the ket basis. Hence we are free to choose the easiest basis for the calculation. An analogy similar to this statement is that a vector M can exist independent of the coordinate system it is presented (i.e., you can use either Cartesian or cylindrical coordinates to present the same vector), but the components of the vector will depend upon which coordinate system it is presented (i.e., $M_x, M_y, M_z \neq M_r, M_\phi, M_\theta$). The trace of the matrix, however, is independent of the coordinate system.

A2.2 EXPANSION OF A WAVE FUNCTION

A set of kets $|i>$ is an orthonormal basis for the kets if the kets have the following properties,

$$<i|j> = \delta_{ij} = \begin{cases} 1, \text{ if } i = j \\ 0, \text{ if } i \neq j \end{cases}. \tag{A2.10}$$

An analogy to this is the dot products among the unit vectors i, j, k, where only $i \cdot i, j \cdot j$, and $k \cdot k$ return a scalar 1 as the result.

In general, the spin state, $|\psi>$, of a nuclear system is described by a linear combination of its spin eigenstates:

$$|\psi> = \sum_m a_m |l\,m> \tag{A2.11}$$

where $|l\,m>$ is the set of orthonormal basis eigenkets; l is either integer or half-integer ($l = 0$, 1/2, 1, 3/2, ...) and termed the spin quantum number (or simply, the spin); m are the azimuthal quantum numbers with values of integer or half-integer between l and $-l$; and $a_m = <m|\psi>$ are the complex admixture amplitudes. An analogy to this equation is that a vector A can be expressed in a Cartesian coordinate system as $A = A_x i + A_y j + A_z k$.

For a set of eigenkets with two components, $|\psi>$ can be written as

$$|\psi> = a_1 |i> + a_2 |j>, \tag{A2.12}$$

where $|i>$ and $|j>$ are normalized eigenfunctions of A with differing eigenvalues.

The mean (or expectation) value, $<A>$, for the observation due to the operation of A is

$$<A> = <\psi|A|\psi> = a_1 a_1^* i + a_2 a_2^* j, \tag{A2.13}$$

where $<i|j>$ and $<j|i> = 0$. The coefficients $a_1 a_1^*$ and $a_2 a_2^*$ are the probability of observing i or j, respectively.

A2.3 SPIN OPERATOR I

The (dimensionless) spin angular momentum operator is given by

$$I = I_x \boldsymbol{i} + I_y \boldsymbol{j} + I_z \boldsymbol{k} \tag{A2.14}$$

with

$$I^2 = I_x^2 + I_y^2 + I_z^2. \tag{A2.15}$$

With dimensions, the above spin angular momentum operators are $\hbar I_x$, $\hbar I_y$, $\hbar I_z$, and $\hbar \boldsymbol{I}$.

Since $\boldsymbol{I}$ commutes with I_x, I_y, and I_z (i.e., the order of the operation does not matter), it is possible to define simultaneously both the magnitude of the spin angular momentum and its component in one direction. Therefore, we can choose a set of normalized functions $|I\ m\rangle$ that are eigenfunctions of both I^2 and I_z. These functions $|I\ m\rangle$ are the $2I + 1$ states with total angular momentum $\hbar\left[I\left(I+1\right)\right]^{1/2}$ and $\hbar m$ as its z component. The eigenequations for I_z and I^2 are

$$I_z |\psi\rangle = m|\psi\rangle \tag{A2.16}$$

and

$$I^2|\psi\rangle = I(I+1)|\psi\rangle. \tag{A2.17}$$

These three independent components of spin angular momentum act on the spin wave functions of the particles. The commutation relations for the spin operators are

$$[I_x, I_y] = iI_z, \tag{A2.18a}$$

$$[I_y, I_z] = iI_x, \tag{A2.18b}$$

$$[I_z, I_x] = iI_y, \tag{A2.18c}$$

where i is $\sqrt{-1}$. Note that the orders of the indexes in Eq. (A2.18) follow the orders of the *ijk* unit vectors in their cross products (Appendix 1.1). There are no states that are *simultaneous* eigenstates of I_x, I_y, and I_z (except states with zero-spin angular momentum). Therefore if a spin is an eigenstate of I_z with eigenvalue of m, the result of an observation of the z component of angular momentum can be predicted exactly, while the x and y components are uncertain.

A2.4 RAISING AND LOWERING OPERATORS I_+ AND I_-

A raising operator is used to convert the state $|I, m\rangle$ to $|I, m+1\rangle$, and a lowering operator is used to convert the state $|I, m\rangle$ to $|I, m-1\rangle$. The raising and lowering operators are defined by

$$I_+ = I_x + iI_y, \tag{A2.19a}$$

$$I_- = I_x - iI_y,$$ (A2.19b)

with the following commutation relations:

$$[I_+, I_z] = -I_+,$$ (A2.20a)

$$[I_-, I_z] = I_-,$$ (A2.20b)

$$[I_+, I_-] = 2I_z.$$ (A2.20c)

It can be shown that

$$I_z(I_\pm|l, m>) = (m \pm 1)(I_\pm|l, m>).$$ (A2.21a, b)

Example

$\because (I_+ I_z - I_z I_+)|l, m> = -I_+|l, m>$ [applying Eq. (A2.20a) to $|l, m>$]

$\because I_z|l, m> = m|l, m>$ [quoting Eq. (A2.16)]

$\therefore mI_+|l, m> - I_z I_+|l, m> = -I_+|l, m>$

$\therefore I_z(I_+|l, m>) = (m + 1)(I_+|l, m>)$ [Eq. (A2.21a)]

Hence $I_+|l, m>$ is an eigenfunction of I_z with eigenvalue of $(m+1)$. It can also be shown that the "new state" $(I_+|l, m>)$ is an eigenfunction of I^2 with the same eigenvalue as $|l, m>$, since I_+ commutes with I^2.

The maximum value of m, m_{max}, and the minimum value of m, m_{min}, can be derived using the raising and lowering operators as

$$m_{max} = -m_{min} = l$$ (A2.22)

A2.5 SPIN-1/2 OPERATOR (IN THE FORMALISM OF PAULI'S SPIN MATRICES)

Now we describe the spin-1/2 operation in Pauli's spin matrices formalism, by defining the angular momentum (or spin) quantum number $l = 1/2$, and spin states, as

$$|\alpha> = |1/2, 1/2> = |1/2> = \begin{bmatrix} 1 \\ 0 \end{bmatrix}$$ (A2.23a)

and

$$|\beta> = |1/2, -1/2> = |-1/2> = \begin{bmatrix} 0 \\ 1 \end{bmatrix},$$ (A2.23b)

where the corresponding bras are $<\alpha| = [1, 0]$ and $<\beta| = [0, 1]$.

Hence, we have

$$I_x \equiv \frac{1}{2}\begin{bmatrix} 0 & 1 \\ 1 & 0 \end{bmatrix}, \quad I_y \equiv \frac{1}{2i}\begin{bmatrix} 0 & 1 \\ -1 & 0 \end{bmatrix}, \quad I_z \equiv \frac{1}{2}\begin{bmatrix} 1 & 0 \\ 0 & -1 \end{bmatrix}, \tag{A2.24a–c}$$

$$I_+ \equiv \begin{bmatrix} 0 & 1 \\ 0 & 0 \end{bmatrix}, \quad I_- \equiv \begin{bmatrix} 0 & 0 \\ 1 & 0 \end{bmatrix}, \quad I^2 \equiv \frac{3}{4}\begin{bmatrix} 1 & 0 \\ 0 & 1 \end{bmatrix}. \tag{A2.25a–c}$$

For a given operator **O**, the four elements of any 2 × 2 matrix are

$$\begin{bmatrix} <\beta|O|\beta> & <\beta|O|\alpha> \\ <\alpha|O|\beta> & <\alpha|O|\alpha> \end{bmatrix}. \tag{A2.26}$$

If we want to observe the angular momentum component along, say, the z axis, for a single nucleus in a basis state $|m>$, we apply it with an operator:

$$I_z |m> = m|m>, \tag{A2.27}$$

where m is the result of the observation.

Example

For a spin-1/2 system in the "spin-up" state, we can use the eigenequation to write down the observation of the z component of the angular momentum as

$$I_z |1/2> = 1/2|1/2>.$$

Or, we could quote Eq. (A2.13) and write down

$$<1/2|I_z|1/2> = \begin{bmatrix} 1 & 0 \end{bmatrix} \frac{1}{2}\begin{bmatrix} 1 & 0 \\ 0 & -1 \end{bmatrix}\begin{bmatrix} 1 \\ 0 \end{bmatrix}$$

$$= \begin{bmatrix} 1 & 0 \end{bmatrix} \frac{1}{2}\begin{bmatrix} 1 \\ 0 \end{bmatrix}$$

$$= \frac{1}{2}.$$

More generally, given $|\psi> = \sum_m a_m |m>$, we have

$$<\psi|I_z|\psi> = \sum_{(m,m')} a_m a_{m'}^* <m'|I_z|m>$$

$$= \sum_{(m,m')} a_m a_{m'}^* m <m'|m> \tag{A2.28}$$

$$= \sum_m |a_m|^2 m,$$

where the last step is because basis vectors are orthonormal, hence there are no $m \neq m'$ terms. For a single nucleus, $|a_m|^2$ is the probability of returning m; for a nuclear ensemble, $|a_m|^2$ is the normalized probabilities of returning m(s).

A2.6 DENSITY MATRIX OPERATOR ρ

In classical physics, when the information of the system is not complete, statistical physics can be applied to study the system. Similarly, in quantum mechanics, when the information about the states is incomplete, a density matrix formalism can be used in the description. In NMR, it is convenient to employ the density matrix operator ρ [1] to treat *ensemble averages*. By definition, we introduce the operator as

$$\rho = \sum_{\psi} P_{\psi} |\psi><\psi|. \tag{A2.29}$$

The elements of the density matrix may be defined by any convenient but complete set of basis states by quoting Eq. (A2.13), as

$$\rho_{mn} = <m|\rho|n>.$$
$$= \overline{a_m a_n^*} \tag{A2.30}$$

In particular, the diagonal elements of the density matrix are given by

$$\rho_{mm} = <m|\rho|m>.$$
$$= \overline{|a_m|^2} \tag{A2.31}$$

Hence ρ, in the chosen representation for non-interacting spin-1/2 particles, is given by

$$\rho = \left| \begin{array}{cc} \overline{|a_{1/2}|^2} & \overline{a_{1/2}^* a_{-1/2}} \\ \overline{a_{1/2} a_{-1/2}^*} & \overline{|a_{-1/2}|^2} \end{array} \right|, \tag{A2.32}$$

where the diagonal elements represent the population (or probability) in the corresponding eigenstates, and the off-diagonal elements are related to the coherence between the states.

The usefulness of ρ lies in the fact that the expectation value of any operator **A** may be written as

$$\overline{<\psi|\boldsymbol{A}|\psi>} = \text{Tr}(\boldsymbol{A}\rho) \tag{A2.33}$$

where $\text{Tr}(\boldsymbol{A}\rho)$ represents the trace of (or diagonal sum over) the matrix product $(\boldsymbol{A}\rho)$. We will soon see what this means (or how powerful the density matrix is).

Some Properties of ρ:

1. While the Schrödinger equation is used to describe the evolution of the wave function $|\psi>$, the evolution of ρ with time can be deduced from the Liouville equation [1],

$$\frac{\partial \rho}{\partial t} = \frac{1}{i\hbar}[\mathcal{H}, \rho]. \tag{A2.34}$$

2. If $\mathcal{H}$ is constant, the solution of the Liouville equation is

$$\rho(t) = U(t)\rho(0)U^{-1}(t)$$

$$= \exp\left(-(i/\hbar)\mathcal{H}t\right)\rho(0)\exp\left((i/\hbar)\mathcal{H}t\right). \tag{A2.35}$$

323

3. The sum of diagonal elements is just the sum of probability of the states, and hence

$$\mathrm{Tr}(\rho) = 1. \tag{A2.36}$$

Note that the point of employing quantum statistical mechanics is that whatever the uncertainty in the spin system (due to probability and too many objects, $N \sim 10^{23}$!) on the value of the variable for the individual objects, its bulk value for the N objects is known to a very great accuracy.

Example

The usefulness of the density matrix concept can be seen, for example, if one wants to calculate the expectation value of I_z for a spin-1/2 system [Eq. (3.12)]. Using the Pauli's matrices formalism, one can use Eq. (A2.28), as

$$< \Psi | I_z | \Psi > = P_{1/2} \left\{ a_{1/2}[1,0] + a_{-1/2}[0,1] \right\} \left\{ \tfrac{1}{2} \begin{bmatrix} 1 & 0 \\ 0 & -1 \end{bmatrix} \right\} \left\{ a_{1/2} \begin{bmatrix} 1 \\ 0 \end{bmatrix} + a_{-1/2} \begin{bmatrix} 0 \\ 1 \end{bmatrix} \right\}$$

$$+ P_{-1/2} \left\{ a_{1/2}[1,0] + a_{-1/2}[0,1] \right\} \left\{ \tfrac{1}{2} \begin{bmatrix} 1 & 0 \\ 0 & -1 \end{bmatrix} \right\} \left\{ a_{1/2} \begin{bmatrix} 1 \\ 0 \end{bmatrix} + a_{-1/2} \begin{bmatrix} 0 \\ 1 \end{bmatrix} \right\}$$

$$= \dots \text{(very lengthy matrix operations)}$$

$$= \tfrac{1}{2} \left[\overline{|a_{1/2}|^2} - \overline{|a_{-1/2}|^2} \right].$$

The same calculation can be carried out much more easily, by the density matrix operation, as

$$< \Psi | I_z | \Psi > = \mathrm{Tr}(I_z \rho)$$

$$= \mathrm{Tr} \left\{ \tfrac{1}{2} \begin{bmatrix} 1 & 0 \\ 0 & -1 \end{bmatrix} \begin{bmatrix} \overline{|a_{1/2}|^2} & \overline{a_{1/2}^* a_{-1/2}} \\ \overline{a_{1/2} a_{-1/2}^*} & \overline{|a_{-1/2}|^2} \end{bmatrix} \right\}$$

$$= \mathrm{Tr} \left\{ \tfrac{1}{2} \begin{bmatrix} \overline{|a_{1/2}|^2} & \overline{a_{1/2} a_{-1/2}^*} \\ -\overline{a_{1/2}^* a_{-1/2}} & -\overline{|a_{-1/2}|^2} \end{bmatrix} \right\}$$

$$= \tfrac{1}{2} [\overline{|a_{1/2}|^2} - \overline{|a_{-1/2}|^2}],$$

which is much easier!

Reference

1. Blum K. *Density Matrix Theory and Applications*. New York: Plenum; 1981.

Appendix 3

Background in Electronics

This chapter briefly presents the essential concepts in electronics [1] that are needed for some chapters of this book.

A3.1 OHM'S LAW FOR DC AND AC CIRCUITS

For direct current (DC) circuits, where the electric charge (which forms the electric current) only flows in one direction, the most fundamental equation is given by Ohm's law, which states that the ratio between the voltage across the conductor and the current through a conductor equals to the resistance of a conductor,

$$R = \frac{V}{I},$$
(A3.1)

where R is the resistance of the conductor in ohms (Ω), V is the voltage measured across the conductor in volts (V), and I is the current through the conductor in amperes (A). The resistance is the property of a material that opposes the flow of electric current through it. For a copper wire with a circular cross section in a DC circuit, the current density is uniformly distributed over the cross section of the wire. R is therefore a function of the length, the cross-sectional area of the conductor, and the resistivity of the conducting material.

An alternating current (AC) is a current that is constantly changing both magnitude and phase between the negative and positive maxima a certain number of times per second. (The regular 120-volt house electricity in the North America has a nominal 60-Hz frequency, which means it changes its phase 60 times per second.) For AC circuits, Eq. (A3.1) could stay the same, as long as the resistance R is replaced by the impedance Z of the circuit, and V and I are the root-mean-square (rms) values of the voltage and current,

$$Z = \frac{V_{rms}}{I_{rms}},$$
(A3.2)

Essential Concepts in MRI: Physics, Instrumentation, Spectroscopy, and Imaging, First Edition. Yang Xia.
© 2022 John Wiley & Sons Ltd. Published 2022 by John Wiley & Sons Ltd.

where Z has the same units as the resistance (Ω). For a pure sine wave [e.g., $y = A \sin(\omega t)$], the rms value of the wave is simply $A/\sqrt{2}$. The impedance Z in the AC circuits is a complex term, which contains the usual resistance R as the real part and the reactance X as the imaginary part (Figure A3.1a), and can be defined as

$$Z = R + jX, \tag{A3.3}$$

where $j = \sqrt{-1}$. (Note that in the previous sections in mathematics [e.g., Eq. (A1.8)] and quantum mechanics [e.g., Eq. (A2.34)], $i = \sqrt{-1}$ is used. In electronics, it is common to use $j = \sqrt{-1}$ to avoid confusion with the electric current I and i.)

The reactance X can have two components: the inductance X_L, which is the property of a coil to resist any change in the flow of electric current through it, and the capacitance X_C, which is the amount of charge that can be stored in a capacitor. In a regular resistor, both the current and the voltage change simultaneously, which are said to be in phase. In a coil, the voltage is ¼ wavelength (i.e., 90°) ahead of the current; while in a capacitor, the voltage is ¼ wavelength behind the current. These phase differences are shown in Figure A3.1b. Noting $j = \sqrt{-1}$, we have

$$X_L = 2\pi f L j = j\omega L, \tag{A3.4a}$$

$$X_C = 1/(2\pi f C j) = -j/\omega C, \tag{A3.4b}$$

where L is the inductance of a coil in henries (H) and C is the capacitance of a capacitor in farads (F). These two equations illustrate that as the frequency increases, the impedance of an inductor increases, and the impedance of a capacitor decreases. Together with R, we have

$$Z^2 = R^2 + \left(X_L - X_C\right)^2. \tag{A3.5}$$

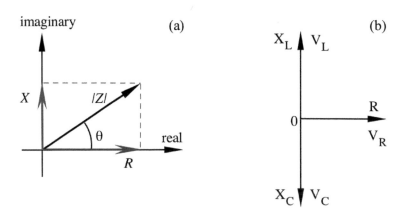

Figure A3.1 (a) The real and imaginary parts of an impedance Z. (b) The phase differences between the voltages (V_R, V_C, V_L) in resistor (R), capacitor (C), and inductor (L).

326

A3.2 ELECTRONICS AT RADIO FREQUENCY

When the frequency *f* in Eqs. (A3.4) and (A3.5) reaches megahertz or higher, "unexpected" things happen; for example, the effects of stray capacitance and stray inductance (wiring inductance) begin to dominate the circuit behavior. At these frequencies, electronics enters a subfield, *rf electronics*, where the circuit design departs rapidly from those used at low frequencies. For the purpose of this book, a few concepts in rf electronics [2, 3] are briefly mentioned.

A3.2.1 The Frequency Dependency of Wavelength

One reason for the *new* characteristics of rf circuits at these *high* frequencies is the fact that the wavelength is becoming comparable to the length of a conductor, as

$$\lambda = \frac{c}{f\sqrt{\varepsilon}},$$

(A3.6)

where λ is the wavelength in meters, c is the speed of light (3×10^8 m/s), f is the frequency in hertz, and ε is the dielectric constant of the medium. $1/\sqrt{\varepsilon}$ is approximately 1 for air and 0.66 for a coaxial cable. Using this equation, the wavelength is about 3.3×10^6 meters in a copper cable at 60 Hz (the regular AC frequency in household power) but becomes a mere 0.66 m for a 300-MHz signal in the coaxial cable (a common NMR and MRI frequency). Since the wavelength is the distance over which the wave's shape repeats (Figure 15.3), a finite wavelength at rf frequencies limits our ability to use a single wire to make a large rf coil.

A3.2.2 The Skin Effect

One additional reason that rf circuitry is so complex is that the electric current is no longer uniformly distributed over the cross section of a wire; instead, the current will have a higher density in the outer parts of a conductor than in the inner parts, due to the counter-emf generated by the AC current in a conductor electromotive force (emf). This effect increases as frequency increases. One could visualize this non-uniform current density as if the current flows only in a thin sheet near and below the surface of a solid conductor, an effect that is typically called the "skin effect." This effect is determined by the depth (δ) in the conducting wire at which the current density drops to 1/e = 0.37, given by

$$\delta = \sqrt{\frac{2\rho}{\omega\mu}},$$

(A3.7)

where ρ is the resistivity of the material, and μ is the magnetic permeability. For copper, the skin depth is about 8.47 mm at $f = 60$ Hz, 210 μm at 100 kHz, 6.6 μm at 100 MHz, and 3.76 μm at 300 MHz. This means that at radio frequencies, a thin-wall copper tube conducts electricity just as well as a solid copper wire.

A3.2.3 Concepts of Transmission Line in rf Electronics

A transmission line, which is typically a coaxial cable in the current context, is the means to deliver rf energy from one place to another. The characteristic features of a transmission line depend upon where and how the line is terminated. According to the theory of transmission lines [3], the line is considered matched if the line is connected to or terminated with, at its output end, a pure resistive load of a value equal to the characteristic impedance of the line (e.g., a

50Ω cable). To current traveling along this matched line, the line would appear to be infinitely long. Consequently, the energy will be delivered from the source to the load, where it is completely absorbed. By contrast, a mismatched transmission line will produce a reflected wave. The delay time is determined by the electrical length of the line, while the polarity of the reflected wave depends upon the nature of the termination. For a line terminated in a short circuit, a wave of opposite phase is created, while an open-circuit line produces a non-inverted reflection of the amplitude equal to the applied signal.

For a lossless line, one can write down the input impendence of the cable as (Figure A3.2)

$$Z_{in} = Z_0 \frac{Z_L \cos(\beta l) + Z_0 j \sin(\beta l)}{Z_0 \cos(\beta l) + Z_L j \sin(\beta l)}, \tag{A3.8}$$

where Z_0 is the characteristic impedance of the coaxial cable (typically 50Ω or 75Ω), $j = \sqrt{-1}$, Z_L (Z_{load}) is the equivalent impedance of the resonant circuit, β is the wavenumber ($2\pi/\lambda$), and l is the length of the cable. An example is given in the following table for a quarter-wave cable (the value of $\beta l = \pi/2$).

Cable length (l)	Z_{in}	Value of Z_{in}	Termination
$\lambda/4$	$Z_{in} = Z_0^2/Z_L$	0	Open (i.e., $Z_L = \infty$, not connected)
		∞	Short (i.e., $Z_L = 0$, grounded)

These results for the quarter-wave cable in the transmission line are important to the understanding of rf resonant circuitry in Chapter 5.2.

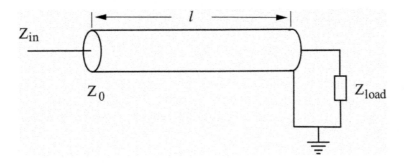

Figure A3.2 A schematic of a transmission line in rf electronics. Z_0 is the characteristic impedance of the coaxial cable, Z_{load} (Z_L) is the equivalent impedance of the resonant circuit, and l is the length of the cable.

References

1. Horowitz P, Hill W. *The Art of Electronics*. 2nd ed. New York: Cambridge University Press; 1989.
2. Terman FZ. *Radio Engineer's Handbook*. New York: McGraw-Hill; 1943.
3. ARRL. *The ARRL Handbook for Radio Amateurs*. 70th ed. Newington, CT: American Radio Relay League, Inc.; 1993.

Appendix 4

Sample Syllabi for a One-semester Course

Our university has a semester system, where each semester is 14 weeks long. A typical 4-credit course meets twice per week, with a total of 4 hours of teaching time each week. The following syllabi are suggestions for the adaption of this book for a one-semester course, based on my teaching of these materials since 1995. The course could be tailored to several different levels:

- Schedule 1: A graduate-level course in MRI and MRI physics, which uses all components of this book, and assuming the concepts in Appendices 1–3 are known.
- Schedule 2: A senior undergraduate-level-course in MRI and MRI physics, which covers properly the concepts in Appendices 1–3 but goes briefly on quantum mechanics (Chapter 3) and spectroscopy (Chapters 8–10).
- Schedule 3: An MRI/imaging course for MRI technologists and medical physicists, which covers properly the concepts in some appendices, goes briefly on quantum mechanics (Chapter 3) and spectroscopy (Chapters 8–10), and goes deep into the instrumentation, experimental methods, and pulse sequences.
- Schedule 4: An MRI course for medical students, which is often a much shorter survey course. This course can be taught without mentioning any equation except $\omega_0 = \gamma B_0$. Try to use as many figures as possible, and include the features and conclusions of each major concept.

You might notice that most of these syllabi include some lab sessions, which was done on purpose. Although the course I have taught at our university has been labeled as a classroom course, I have always included a lab component in my teaching – to bring the students to the machine, at least twice, once for NMR spectroscopy and once for MRI. I have always asked my students to bring to the lab sessions whatever has been fascinating them in their home, garden, or supermarket. We have measured all kinds of liquids using NMR spectroscopy (e.g., expensive perfume, soy sauce, various kinds of energy drinks, etc.), and we have imaged all kinds of objects using MRI (e.g., various bugs, vegetables, fruits, etc.). Some of these spectra and images are shown in this book as the examples. Even though many students will not study NMR and MRI in their next academic level, getting them into the lab and getting their hands dirty at the spectrometer and imager is always an eye-opening opportunity for them.

Essential Concepts in MRI: Physics, Instrumentation, Spectroscopy, and Imaging, First Edition. Yang Xia.
© 2022 John Wiley & Sons Ltd. Published 2022 by John Wiley & Sons Ltd.

Four Schedules for Syllabi

Week	Schedule 1 Graduate-Level NMR and MRI	Schedule 2 Undergraduate Physics Course	Schedule 3 MRI & Imaging Technology	Schedule 4 MRI & Imaging Medical Students
1a	Ch 1, Appendices	Ch 1, Appendices	Ch 1, Appendices	Ch 1, Appendices
1b	Ch 2	Appendices	Appendices	Ch 2, Ch 4
2a	Ch 2	Ch 2	Appendices	Ch 5, Ch 6
2b	Ch 3	Ch 2	Ch 2	Ch 7
3a	Ch 3	Ch 3	Ch 2	Ch 8, Ch 9
3b	Ch 4	Ch 4	Ch 3, Ch 4	Ch 11, Ch 12
4a	Ch 5	Ch 5	Ch 5	Ch 12
4b	Ch 6	Ch 6	Ch 5	Ch 13, Ch 14
5a	Ch 7	Ch 7	Ch 6	Ch 15
5b	Test (Ch 1 – Ch 7)	Test (Ch 1 – Ch 7)	Ch 6	Ch 15
6a	Ch 8	Ch 8	Ch 7	Ch 16
6b	Ch 8	Ch 8	Test (Ch 1 – Ch 7)	Assessment
7a	Ch 9	Ch 9, Ch 10	Ch 8	
7b	Ch 10	Lab: Spectroscopy	Ch 9, Ch 10	
8a	Lab: Spectroscopy	Lab: Spectroscopy	Lab: Spectroscopy	
8b	Lab: Spectroscopy	Ch 11	Lab: Spectroscopy	
9a	Ch 11	Ch 12	Ch 11	
9b	Ch 12	Ch 13	Ch 12	
10a	Ch 13	Ch 13	Ch 13	
10b	Ch 13	Ch 14	Ch 13	
11a	Ch 14	Ch 14	Ch 14	
11b	Ch 15	Ch 15	Ch 15	
12a	Ch 15	Ch 15	Ch 15	
12b	Ch 16	Ch 16	Ch 16	
13a	Ch 17	Ch 17	Ch 17	
13b	Lab: Imaging	Lab: Imaging	Lab: Imaging	
14a	Lab: Imaging	Lab: Imaging	Lab: Imaging	
14b	Final Exam	Final Exam	Final Exam	

Appendix 5
Homework Problems

As the materials in this book have been taught extensively, I have accumulated a number of homework problems. Since the principles of NMR and MRI are deep in both classical and quantum physics, ample high-level math skills are required if the equations are being derived in exercise (all equations can be derived from the first principles). However, this book can be taught at different levels and tailored to different audiences, in which the instructor needs to be careful for the levels of the math and physics in the homework problems. Here is a collection of them that suits several levels of students.

SAMPLE PROBLEMS FOR PART 1 ESSENTIAL CONCEPTS IN NMR

Many equations in Chapter 2 can be derived with the knowledge of classical mechanics, calculus, and differential equations. In addition, one can ask some descriptive questions or even multiple-choice questions, which still need a good understanding of the theory.

1. Since the properties of a single nucleus belong to a discrete set of possibilities (i.e., quantum states), a precise understanding of NMR therefore lies in the realm of quantum mechanics. How can we justify the use of the classical vector approach in most practical NMR and MRI experiments?
2. For a spin-half particle such as a proton in the hydrogen atom, there are two possible orientations of the spin state, spin up or spin down. For NMR and MRI experiments using samples containing protons, why is the macroscopic magnetization in thermal equilibrium oriented along the magnetic field direction?
3. Many magnets in modern NMR and MRI instrument have the "fringe fields" that extend a large distance into the nearby space. If you are standing on the "5-Gauss line" of a 7 Tesla magnet, what is the resonant frequency of water molecules in your body in hertz?
4. Derive Eq. (2.29a) and Eq. (2.29b) from Eq. (2.19).
5. (a) Solve the Bloch equation after a 180° rf pulse, starting from Eq. (2.19).
   (b) Draw a schematic for the motion of **M** after a 180° pulse.
6. Given $g(x) = \exp(-(x - x_0)^2/a^2)$, proof that its FWHM is $2a\sqrt{\ln(2)}$.

Many equations in Chapter 3 can be derived with the working knowledge of quantum mechanics.

Essential Concepts in MRI: Physics, Instrumentation, Spectroscopy, and Imaging, First Edition. Yang Xia.
© 2022 John Wiley & Sons Ltd. Published 2022 by John Wiley & Sons Ltd.

7. Show $I^2 \equiv I_z^2 + \dfrac{1}{2}(I_+I_- + I_-I_+)$.

8. Show $I^2 \equiv I_z^2 + I_z + I_-I_+$.

9. Given $[I_-, I_z] = I_-$, proof that I_- is a lowering operator (show all intermediate steps).

10. Given $I_x \equiv \dfrac{1}{2}\begin{bmatrix} 0 & 1 \\ 1 & 0 \end{bmatrix}$, show the expectation value of I_x.

11. Given a B_0 field of (a) 1 Tesla and (b) 7 Tesla, calculate ratios of the energy difference between the two states for protons, respectively. [Hint: use the room temperature.]

Many different problems can be designed for the nuclear interactions discussed in Chapter 4, for example, to derive the equations, to calculate some constants (e.g., the dipolar coupling constant R for protons), to estimate the sizes of the various interactions, and to summarize each individual interaction in words or in equations.

12. For a spin-½ system where the two normalized populations at the thermal equilibrium are given by Eq. (3.14), proof that the population difference $n = N_0\lambda/2$, where $\lambda = \dfrac{\gamma\hbar B_0}{k_B T}$. [Hint: start from Eq. (3.26b), where $n = N_+ - N_-$, and N_0 is the total population.]

13. Describe the main/critical features of dipolar interaction, chemical shift interaction, scalar interaction, and nuclear quadrupole interaction (1-2 pages total).

The working knowledge of Fourier transform is valuable to the students, in and out of this class. One can easily find a commercial software that can do FFT.

14. Find a commercial software (or write your own program) that can do 1D FFT, generate the following functions, and obtain their FT pairs (use 256 points as the total points in both domains, use and plot both real and imaginary parts):
    (a) A centered square pulse with the duration of 32, 16, and 8 points, respectively.
    (b) A centered sinc function $(\sin(\theta)/\theta)$ with a central peak width of 8 points and side lobe of 3 and 0, respectively.
    (c) A fully decayed and centered cosine wave that can be described by $f(x) = A\cos(\theta)\exp(-t/b)$.
    (d) As in (c), but truncate $f(x)$ at 50% of its max amplitude.
    (e) Describe briefly how the FFT routine works (its structures, not the programming language itself).

15. Find or write a program that can do 2D FFT, to FT the following 2D data (i.e., 2D image) into the reciprocal domain (use 256 points as the total points in both domains, plot both real and imaginary parts in 2D):
    (a) A flattop circular cylinder at the center of the 2D space, with a diameter of 32, 16, and 8 points, respectively.
    (b) A 2D sinc function $(\sin(\theta)/\theta)$ with a central peak width of 8 points and side lobe of 3 and 0, respectively.
    (c) A fully decaying 2D cosine wave that can be described by $f(x) = A\cos(\theta)\exp(-t/b)$.
    (d) As in (c), but truncate $f(x)$ at 50%.
    (e) Perform the inverse FT using the results of (a)-(d). Will you get back the same functions? Discuss the results and offer suggestions.

16. Can you write/assemble an FFT routine, in any software or language? You should consult books such as *Numerical Recipes* (Press et al., Cambridge University Press, 1989) to find out the procedures involved. Describe how your routine works (its structures, not the programming language itself), write a short program to send several *time-domain* data into the program and read out the *frequency-domain* data from your program, and plot your data in both domains. Discuss the features of your results.

SAMPLE PROBLEMS FOR PART 2 ESSENTIAL CONCEPTS IN NMR INSTRUMENTATION

Problems for Chapter 5 and Chapter 6 likely require the knowledge of electricity and magnetism, electronic circuitry, and computer essentials. Problems for Chapter 7 are numerous, from the description of these and additional pulse sequences, to the on-paper operation using a sequence.

17. Show that for any given experimental time T, the best SNR can be obtained at TR ~ $1.26 \times T_1$. [T >> T_1; SNR increases as the square root of the number of accumulations.]

18. The inversion recovery pulse sequence (IR) was used to verify the T_1 of a non-viscous liquid sample, which has a known T_1 of 2 s. The experiment used 10 different τ (the time between 180° and 90°) and a standard CYCLOPS cycle and was repeated at a rate of $5 \times T_1$. [You can ignore the effect of T_2 in this exercise.]

    (a) Choose the timing of the data points so that the magnetization decreases by an approximately equal amount between any two neighboring points, with the first point having the shortest delay and the last point taken when the signal reaches 80% of its equilibrium magnetization.

    (b) Draw the pulse sequence and provide all the timings of the pulse sequence (no need to scale the timings).

    (c) Make a plot so that the relaxation time can be calculated as the slope of a straight line, and show that you can indeed calculate the $T_1 = 2$ s from the slope of this simulated experiment.

    (d) Draw schematically the motion of magnetization for this pulse sequence when the separation between the 180° and 90° pulses is about 1.4 s.

    (e) What is the total experiment time in minutes?

19. Draw schematically a saddle coil that is used to generate a transverse field. Mark the directions of the electric current in the four vertical segments, and show schematically that these four segments can indeed produce a transverse field at the central *x-y* plane in the coil.

(Note, you can repeat problem 19 for a different type of coil; if your students have the knowledge of electricity and magnetism, you can ask them to calculate the magnetic field at the center or over a central plane.)

SAMPLE PROBLEMS FOR PART 3 ESSENTIAL CONCEPTS IN NMR SPECTROSCOPY

If you have access to a spectrometer in a teaching or research lab, you can study many common chemicals in the laboratories, which have the first-order peak patterns. In addition, you can acquire spectrum from many household liquids from supermarkets and pharmacies. Over the years, we did many spectroscopy experiments using all kinds of liquids, for example, various drinks, cooking condiments, perfumes and cosmetic liquids, wines, and common medicines. You would be surprised what you could find in them (mostly water!).

20. For an $A_3M_4X_2$ spin system and assuming $J_{AM} > J_{AX} > J_{MX} > 0$, derive the first-order pattern of lines, with the intensities relative to the weakest line in the whole spectrum. (Draw to scale the relative intensities for all spins.)

21. Describe the different influences of paramagnetic and diamagnetic materials to chemical shift.

22. Describe an experiment for 2D homonuclear J-resolved spectroscopy for an AX system.
    (a) Draw a pulse sequence.
    (b) Describe, schematically with brief explanation if necessary, how the magnetization vector from the X nuclei behaves during the experiment.

(Note, you can easily change the combination of n in problem 20 to a different AMX system, and use either $J_{AM} > J_{AX} > J_{MX} > 0$ or $J_{AM} > J_{AX} = J_{MX} > 0$.)

SAMPLE PROBLEMS FOR PART 4 ESSENTIAL CONCEPTS IN MRI

*At this point, your students would like to know how an imaging pulse sequence works. Starting from something simple, you can dissect many imaging pulse sequences by understanding the usefulness of each pulse or component, drawing the **k**-space maps, determining the relationships among various pulses, and, finally, following the phase cycling steps. Similarly, do your best to take your students to an MRI instrument, and run some specimens if possible!*

23. How do we select a slice electronically in MRI?

24. Describe what mathematic formulas are needed to generate a spiral trajectory in **k** space (e.g., the patterns in Chapter 11).

25. Understand the EPI pulse sequence. Draw a detailed schematic diagram for the **k**-space trajectories. Make sure that you clearly mark the relationships among all gradients (i.e., whether some gradient pulses should have an equal area etc). You can use a total of 16 echoes, with each echo having 16 data points.

26. Understand the FLASH pulse sequence. Draw a detailed schematic diagram for the **k**-space trajectories. Make sure that you clearly mark the relationships among all gradients (i.e., whether some gradient pulses should have an equal area etc). You can use a total of 16 echoes, with each echo having 16 data points.

27. Using a specimen of 9-mm diameter, one finds that the 2D image of the specimen has 230 pixels across its diameter in the 256×256 digital array.
    (a) Calculate the transverse pixel resolution of the experiment.
    (b) Calculate the field-of-view of the experiment.
    (c) What is the slice thickness if the bandwidth of the sinc pulse is 8515 Hz and the slice gradient is 0.2 T/m?

28. Describe 2D Fourier imaging pulse sequence.
    (a) Draw schematically the pulse sequence. Label the three gradients as the slice, phase, and read; mark the timing gaps for the repetition time (TR) and the echo time (TE) in the sequence.
    (b) If a patient whose max dimension is 60 cm has a tumor less than 1 mm in size, can an MRI scan using a standard body coil and a 512 digital array detect this tumor? Why?

SAMPLE PROBLEMS FOR PART 5
QUANTITATIVE AND CREATIVE MRI

This is the fun part of this book – the exercises are numerous. Each contrast/method/protocol in Chapter 15 and Chapter 16 can become a set of exercises to work on. You could start with some on-paper experiments for which you know the results. Then you should find the original papers on a particular imaging method/protocol, which generally describe the method in detail, and ask the students to re-discover the method/protocol.

29. The proton resonance frequency of oil and water is 3 ppm apart. Design an MRI pulse sequence that contains the following functions.
    (a) Suppression of oil component to the signal.
    (b) Slice selection in the z direction;
    (c) 2D imaging using the Cartesian-raster sampling in the k-space.
    (You need to provide the necessary descriptions/legends/labels for all pulses, timings, channels in the drawing. Also mark the echo time and repetition time.)

30. In a proton PGSE experiment, we have $\delta = 2$ ms, $\Delta = 10$ ms.
    (a) If the self-diffusion of the sample is 2.3×10^{-9} m^2/s, how strong must the gradient be to get a 50% signal attenuation?
    (b) If D is 2.3×10^{-10} m^2/s, how strong must the gradient be to get the same attenuation?

31. A spin-echo sequence is repeated five times to measure T_2 of water (with a known $T_2 = 400$ ms).
    (a) Choose the time delays so that five data points have an equal amplitude difference between any neighboring points [choose the shortest delay = 0 and the longest delay that gives a signal at 20% of its equilibrium value].
    (b) Present the pulse sequence with timings.
    (c) Make a rough plot so that the relaxation time can be calculated from the slope of a straight line, and estimate the T_2 using the slope of the line.

32. A sample has two proton components with the following spin characteristics:

	Component A	Component B
Spin-lattice relaxation (s)	3.0	1.0
Spin-spin relaxation (ms)	200.0	50.0
Spin density	1.0	1.0

    (a) Design schematically a 2D imaging sequence that has two pulse segments: a leading inversion-recovery sequence and a subsequent 2D imaging sequence.
    (b) What is the τ value (the timing between the leading 180° and the 90°) in the sequence if we want the final image to contain the signal only from component B?
    (c) What is the percentage of component B that remains at the end of τ?
    (d) What is the final percentage of component B that can contribute towards the FID, if TE of the imaging segment is 25 ms?

33. Many biological tissues, such as articular cartilage and tendon, are load-bearing. What are the consequences of compression (of articular cartilage) and stretching (of tendon) towards the MRI properties such as T_1 and T_2?

34. Assign one or two students to a topic of your/their interest; ask them to work on it and to present it to the class. I have used this approach several times – students learn a lot more than from the problem you assign to them. Just make sure to give the students some time on this – it takes time for them to be able to explain things to their classmates and you.

35. If your students like to be MRI detectives, give them a set of raw data and ask them to show you what it is, based on the knowledge of Chapter 17.

Index

Essential Concepts in MRI: Physics, Instrumentation, Spectroscopy, and Imaging, First Edition. Yang Xia.
© 2022 John Wiley & Sons Ltd. Published 2022 by John Wiley & Sons Ltd.

347